Soft Matter in Plants
From Biophysics to Biomimetics

Soft Matter Series

Series editors:
Hans-Jürgen Butt, *Max Planck Institute for Polymer Research, Germany*
Ian W. Hamley, *University of Reading, UK*
Howard A. Stone, *Princeton University, USA*
Karen J. Edler, *University of Bath, UK*
Amy Shen, *Okinawa Institute of Science and Technology Graduate University, Japan*

Titles in this series:
1: Functional Molecular Gels
2: Hydrogels in Cell-based Therapies
3: Particle-stabilized Emulsions and Colloids: Formation and Applications
4: Fluid–Structure Interactions in Low-Reynolds-Number Flows
5: Non-wettable Surfaces: Theory, Preparation and Applications
6: Wormlike Micelles: Advances in Systems, Characterisation and Applications
7: Electrospinning: From Basic Research to Commercialization
8: Polymer-modified Liquid Crystals
9: Polymer Colloids: Formation, Characterization and Applications
10: Bijels: Bicontinuous Particle-stabilized Emulsions
11: Peptide-based Biomaterials
12: Droplet Microfluidics
13: Soft Matter for Biomedical Applications
14: Drying of Complex Fluid Drops: Fundamentals and Applications
15: Soft Matter in Plants: From Biophysics to Biomimetics

How to obtain future titles on publication:
A standing order plan is available for this series. A standing order will bring delivery of each new volume immediately on publication.

For further information please contact:
Book Sales Department, Royal Society of Chemistry, Thomas Graham House, Science Park, Milton Road, Cambridge, CB4 0WF, UK
Telephone: +44 (0)1223 420066, Fax: +44 (0)1223 420247
Email: booksales@rsc.org
Visit our website at books.rsc.org

Soft Matter in Plants
From Biophysics to Biomimetics

Edited by

Kaare H. Jensen
Technical University of Denmark, Denmark
Email: khjensen@fysik.dtu.dk

and

Yoël Forterre
CNRS Aix-Marseille University, France
Email: yoel.forterre@univ-amu.fr

ROYAL SOCIETY
OF **CHEMISTRY**

Soft Matter Series No. 15

Print ISBN: 978-1-78801-724-4
PDF ISBN: 978-1-83916-116-2
EPUB ISBN: 978-1-83916-117-9
Print ISSN: 2048-7681
Electronic ISSN: 2048-769X

A catalogue record for this book is available from the British Library

The Royal Society of Chemistry is a charity, registered in England and Wales, Number 207890, and a company incorporated in England by Royal Charter (Registered No. RC000524), registered office: Burlington House, Piccadilly, London W1J 0BA, UK, Telephone: +44 (0) 20 7437 8656.

For further information see our website at www.rsc.org

For general enquiries, please contact books@rsc.org

For EU product safety enquiries, please email books@rsc.org or contact Royal Society of Chemistry Worldwide (Germany) GmbH, Römischer Hof, Unter den Linden 10, 10117 Berlin.

Preface

Vascular plants have inhabited Earth's surface for about 400 million years. They lead the evolutionary race by all accounts,[1] and their importance to the terrestrial ecosystem, geology, and human civilization requires no emphasis. Yet, plants remain mysterious when observed from the vantage point of a physicist or an engineer. How do organisms made from just a few soft polymers, minerals, and water perform such striking physical and chemical feats? The remarkable relationship between structure, deformation, flow, and function in plants is the topic of this book.

A myriad of reasons has been put forward to justify the scientific study of mechanics and fluid flow in plants. In most instances, they fall within the four categories proposed by T. J. Pedley:[2] (1) physiology and ecology: understanding how plants work and interact with other living organisms and the environment; (2) stress: understanding how strong fluctuations impact plants; (3) pathology: discerning the origins and development of diseases; (4) bioengineering: exploring synthetic changes to the plant genome. It is evident that scientists working in these diverse research fields will be motivated by different research questions and applications of their work. However, the essential fluid and solid mechanical principles are clearly the same for all. Exploring the physics of both equilibrium and extreme cases as they relate to the mechanical integrity and transport capacity of plants is therefore worthwhile.

From a societal perspective, it is appears that plants will become more (or at least not less) significant to humans in the coming decades and centuries. To this end, substantial resources are being invested into research programmes with the goal of improving plant productivity, water and fertiliser efficiency, disease resistance, and so on. The critical question, however, is how best to achieve this goal. On the one hand, our

---

Soft Matter Series No. 15
Soft Matter in Plants: From Biophysics to Biomimetics
Edited by Kaare H. Jensen and Yoël Forterre

Published by the Royal Society of Chemistry, www.rsc.org

understanding of molecular processes in plants is steadily advancing, and many transformation tools have become more widely available. Also, certain key activities, such as photosynthesis or starch biosynthesis, are being elucidated with ever greater accuracy. On the other hand, the strong focus in research and funding on molecular biology has yet to achieve a Moon-shot jump in productivity. The reason for this is, of course, as of yet unknown. One cannot help wonder, however, if pursuing a complementary approach to understand the mechanics of plants at mesoscopic and microscopic scales is worthwhile. Such an approach would bridge the intellectual gap between the action of genes and proteins and the biophysical or biochemical processes with which they are ultimately associated. It would also help us better understand plants' uniquely distributed organismal architecture. However, this portfolio of research questions are, we believe, too often overlooked.

This book is written in an effort to place soft matter physics, which naturally focuses on mesoscopic and microscopic scales, in the context of modern plant science. As such, our objective is twofold: first, we aim to introduce physicists to a myriad of fascinating and important phenomena in plants. Second, we seek to highlight the benefits of using reasonably simple models to describe tangled biological systems. Physics applied to biology has often involved bringing new instrumentation to tackle biological questions. However, we do not believe this should be the sole focus. Physics represents a way of thinking that can shed new light on biological questions:[3] the desire to find the simplest possible description, with a few essential ingredients, of complex phenomena is a unique perspective that has important advantages (and yes, limitations as well). Equally important is the fact that biology can inspire physicists to address new questions in physics, *e.g.*, related to transport networks or efficient design strategies.

Having established the need for a concise introduction to *soft matter in plants*, we were first to recognize our own inability to provide such an overview. We are therefore extremely grateful to our editor Michelle Carey who permitted us to commission chapters from some of the leading researchers working in the field. Apart from two introductory chapters written by us, their work comprises the main body of this book, and should be recognised as such.

The book begins with a general introductory chapter giving the basic physical concepts needed for describing important processes in plant biophysics and biomechanics, such as water transport, growth or plant movements (Chapter 1). It is then followed by six independent chapters covering a wide range of applications and scales, from cell and tissue physics to engineering applications. Chapter 2 focuses on fluid–structure interaction phenomena in plants, in connection with signaling, vascular transport or intracellular flows. Chapter 3 is devoted to the mechanics of growth and morphogenesis, a central topic in modern biology. Starting from the seminal work of Lockhart on single plant cell growth, it provides the basic theoretical tools and concepts for modeling growing plant tissues. Chapter 4 provides an overview of the rich physics of water stress and cavitation, as found in the vascular system of plants during sap ascent, with particular emphasis on cavitation in plant-relevant confined and deformable environments. The book then moves underground to explore the fascinating world of plant roots (Chapter 5). This chapter details how soil affects the physics of root growth and the feedback of roots on the mechanical properties of the soil. Root penetration in the soil can be seen as a particular case of invasive growth – a topic of broad importance in biology and related to reproduction, nutrition and disease transmission. Chapter 6 discusses the experimental strategies developed to investigate the biomechanics of invasive growth at the cellular level in plants and fungi. The book closes with a discussion on biomimetic applications of plant movements, focusing on passive movements induced by humidity change in botanical and artificial systems (Chapter 7).

Kaare H. Jensen and Yoel Forterre

References

1. Y. M. Bar-On, R. Phillips and R. Milo, The biomass distribution on earth, *Proc. Natl. Acad. Sci.*, 2018, **115**(25), 6506–6511.
2. T. J. Pedley, Blood flow in arteries and veins, in *Perspectives in fluid dynamics: a collective introduction to current research*, ed. G. K. Batchelor, H. K. Moffatt, and M. G. Worster, Cambridge University Press, 2000, pp. 105–158.
3. National Academies of Sciences, Engineering, and Medicine, *Physics of Life*, The National Academies Press, Washington, DC, 2022, DOI: 10.17226/26403.

Contents

Soft Matter Series No. 15
Soft Matter in Plants: From Biophysics to Biomimetics
Edited by Kaare H. Jensen and Yoël Forterre

Published by the Royal Society of Chemistry, www.rsc.org

CHAPTER 1

Basic Soft Matter for Plants

YOËL FORTERRE

Aix-Marseille University, CNRS, IUSTI, Marseille 13013, France
Email: yoel.forterre@univ-amu.fr

This chapter provides an overview of fluid and solid mechanical concepts applied to plants. The objective is to present the main physical mechanisms and laws needed to describe some important physiological processes in plants, such as water and solute transport, growth, rapid movements and the feedback between mechanical signals and biology. Emphasis is given on presenting the various laws in their simplest mathematical form (avoiding tensorial formulation as much as possible), discussing the scaling laws and orders of magnitude relevant to plants. Many of the physical and biological concepts introduced in the chapter will be studied in more detail in subsequent chapters.

1.1 Fluids

Plants manipulate and move water to perform almost all their physiological functions. Yet, there is no microscopic "water pump" in plants that would "actively" transport water molecules across cells, as there is for proteins or ions. Therefore, whether minute transmembrane flows or large bulk transport through trees, all water movement in plants occurs passively, according to the well-known principles of thermodynamics and fluid mechanics. In this section, we review those mechanisms and present the main physical laws of water transport in plants. Readers interested in a more comprehensive presentation of both the biological and physical contexts

Soft Matter Series No. 15
Soft Matter in Plants: From Biophysics to Biomimetics
Edited by Kaare H. Jensen and Yoël Forterre

Published by the Royal Society of Chemistry, www.rsc.org

may wish to consult recent reviews on the topic, such as Stroock *et al.*[153] or Jensen *et al.*[75]

1.1.1 Water Potential and Turgor Pressure

1.1.1.1 Some Definitions

In plant science, it is common to characterize the thermodynamic status of water using *water potential* Ψ, defined as the chemical potential of water relative to a reference state, per unit volume:[35,85,110,112]

$$\Psi = \frac{\mu_w - \mu_w^0}{\nu_w}. \tag{1.1}$$

Here μ_w is the chemical potential of water, *i.e.* the Gibbs free energy per unit mole, μ_w^0 is the chemical potential of water in the reference state, chosen to be pure liquid at atmospheric pressure and at ambient temperature, and $\nu_w \approx 18\ \text{cm}^3\,\text{mol}^{-1}$ is the molar volume of liquid water. Using this definition, the water potential has the dimension of a pressure (unit Pa or $\text{J}\,\text{m}^{-3}$) and is null for pure water at 1 atmosphere. The water potential depends on the thermodynamic state of water, *i.e.* its solid, liquid or gaseous state, the presence of other molecular constituents interacting with the water and the possible presence of an external field such as gravity acting on the water. We provide below the main expressions of water potential useful for plants.

- For a diluted aqueous solution under gravity, as found in the vessels of the vascular system or in the large vacuoles of cells (Box 1.1), the water potential may be decomposed as:

$$\Psi_{\text{liq}} = \Psi_{\text{p}} + \Psi_{\text{os}} + \Psi_{\text{g}} = P - c\mathcal{R}T + \rho g z. \tag{1.2}$$

The first term $\Psi_p = P$ is the enthalpic or pressure contribution to the water potential†, where P is defined as the total pressure of the solution minus the atmospheric pressure. When referring to the pressure inside plant cells, this pressure difference is usually called the *turgor pressure*. The second term $\Psi_{os} = -c\mathcal{R}T$ is the osmotic contribution corresponding to the free enthalpy of mixing of the solute in water, as given by the van't Hoff's law. Here c is the solute concentration (unit $\text{mol}\,\text{m}^{-3}$), $\mathcal{R} \simeq 8.31\ \text{JK}^{-1}\,\text{mol}^{-1}$ is the ideal gas constant and T is the absolute temperature (unit Kelvin), such that $\mathcal{R}T \simeq 2500\ \text{J}\,\text{mol}^{-1}$ at ambient temperature. The opposite of the osmotic term $\Pi = -\Psi_{os} = c\mathcal{R}T$, thereafter called the osmotic pressure, is often used instead of Ψ_{os}. Finally the last term $\Psi_g = \rho g z$ is the contribution of the external gravity field, where $\rho \simeq 10^3\ \text{kg}\,\text{m}^{-3}$ is the water density, $g \simeq 9.8\ \text{m}\,\text{s}^{-2}$ is the intensity of gravity and z is the altitude above ground.

† The compressibility of water is ignored in this expression; see Chapter 4.

Box 1.1 Biological background.

Figure 1.1a shows a typical plant cell as described in a biology textbook.[155] It consists of a water pocket – the vacuole – which can represent up to 95% of its volume, and the cytoplasm, which contains the cytoskeleton and the organelles necessary for the functioning of the cell, such as the nucleus, the chloroplasts and the mitochondria. Plant cells have various sizes depending on the organism, the type of cell and the stage of development, ranging from a few microns for the stem cells of buds to several centimeters for the giant cells of some aquatic algae§§§. Like any cell, the plant cell is surrounded by a semi-permeable plasma membrane, whose pore size allows water molecules to pass freely but prevents the passage of large solute molecules.

The main structural difference between animal cells and plant cells is the existence, in the latter, of a stiff wall surrounding the plasma membrane¶¶¶. This wall, whose thickness is of the order of 0.1 to 1 μm, allows the plant cell to sustain a very high internal hydrostatic pressure of the order of several bars, called turgor pressure (Section 1.1.1.2). The cell wall of plants is made of cellulose microfibrils embedded in a matrix of polysaccharides and proteins (Figure 1.1b). The cellulose microfibrils – the stiffer element of the wall – are produced from a protein complex that crosses the cell membrane and runs like a cargo along the cortical microtubules located inside the cell (Figure 1.1b). As a result, the orientation of the microfibrils deposited in the wall corresponds to the orientation of the cortical microtubules. This may impart mechanical anisotropy to the cell wall if the microtubule arrangement is anisotropic (Section 1.2.2.1).

Water in plants is generally taken up at the root level in the soil and transported throughout the plant body to the organs and leaves, where most of the absorbed water evaporates (Section 1.1.4). In compact, non-vascular tissue, such as in roots for radial transport or in the shoot growth zone, water follows two main pathways (Figure 1.1c). The first pathway, called the apoplast pathway, corresponds to transport in the cell wall only, the apoplast being defined as the plant volume contained outside the plasma membrane of cells. This pathway avoids the high hydraulic resistance of the plasma membrane but has a very small volume: the volume of the apoplast in non-vascular tissues is only 1–10% of the total tissue volume. In addition, the apoplast pathway does not allow for osmotic control of flow, because the cell wall is not very selective for solutes. The second pathway is the cell-to-cell pathway and includes the symplast and the transmembrane pathways. The symplast is defined as the volume contained within the plasma membrane of cells, and is thus the dual of

§§§Green algae and land plants belong to the Plantae kingdom, or "green plants", characterized by the existence of chloroplasts containing green chlorophyll.

¶¶¶Other examples of walled organisms are fungi, bacteria and archaea.

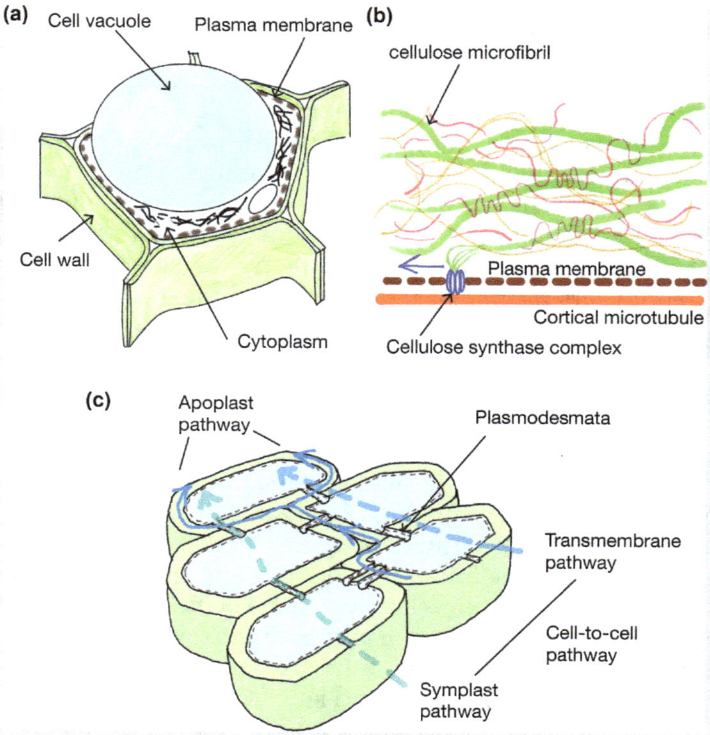

Figure 1.1 Basics of plant anatomy. (a) The plant cell. (b) The primary cell wall and the synthesis of the cellulose microfibrils, showing the cellulose microfibrils (green), hemicellulose (pink), pectin (beige), the plasma membrane (brown), the cellulose synthase complex (blue) and the cortical microtubule (orange). (c) Water pathways in non-vascular tissues. Drawings in (a) and (c) are inspired from the book.[155]

the apoplast. It forms a continuous medium because the cytoplasms of the cells are connected to each other by plasmodesmata, small nanoscale pores whose permeability to water and solutes can be regulated[68] (see Chapter 2). The cell-to-cell pathway is under direct osmotic control by the cells.

As well as these pathways, most land plants have evolved specialized vascular tissues to carry water and other fluids over long distances. The xylem, on the one hand, is the pipe network that transports the sap, made of almost pure water, from the roots to the organs and leaves. It consists of long interconnected conduits made of dead cells typically 10 to 500 μm in width. Here transport is driven by a pressure gradient induced by evaporation and capillary cohesion in the leaves, which pulls water under negative pressures (Section 1.1.1.2 and Chapter 4). The phloem vascular network, on the other hand, is adjacent to the xylem and distributes the photosynthesis product created in the leaves (mainly sugar) to the rest of the tissue for growth and storage. It consists of long living cells about 10 μm wide and 1 mm long separated by perforated sieve plates, forming a continuous symplastic pathway. Transport through the phloem is driven by an osmotic mechanism as explained in Section 1.1.3.2.

- The water potential of a humid vapor, as found inside leaves where gas exchange with the atmosphere takes place, is given by:

$$\Psi_{\text{vap}} = \frac{\mathcal{R}T}{v_{\text{w}}}\ln[\text{RH}] = \frac{\mathcal{R}T}{v_{\text{w}}}\ln\left(\frac{P_{\text{vap}}}{P_{\text{sat}}(T)}\right), \qquad (1.3)$$

  where $[RH] = P_{\text{vap}}/P_{\text{sat}}(T)$ is the relative humidity of the atmosphere, P_{vap} is the partial pressure of water in the vapor, $P_{\text{sat}}(T)$ is the saturation pressure of water in air[‡] under atmospheric pressure and ambient temperature T ($P_{\text{sat}} = 2.3 \times 10^3$ Pa at 20 °C) and $\mathcal{R}T/v_w \approx 135$ MPa at 300 K[§].

- Finally, water is also found inside the cell wall of plants, where it binds with the cellulose network and other macromolecules such as hemicelluloses, pectins, and xyloglucans. The cell wall can be highly hydrated, as in the primary wall of growing cells, where about 75–80% of the wall volume is water,[155] or less hydrated, as in the lignified secondary wall of wood. The water potential of a gel-like medium akin to the cell wall is called the matrix (or matric) potential and denoted Ψ_{m}. The matrix potential of the cell wall depends on its water content, pH, temperature and the chemical affinity between the water molecules and the polymer constituents of the wall, mainly.[112] We provide in Box 1.4 an expression of the matrix potential for an ideal elastomeric hydrogel, derived from the theory of polymer solutions.

1.1.1.2 Consequences of Water Balance

At equilibrium, the chemical potential of water, and thus the water potential, must be uniform in all regions that can freely exchange water molecules. This basic thermodynamic statement has key consequences for the mechanical status of water in plants, which we briefly discuss below.

---

‡The fact that the saturation pressure $P_{\text{sat}}(T)$ of water vapor *in air* appears in eqn (1.3), instead of the saturation pressure of a *pure* water vapor, comes from the choice of the reference state for the definition of the water potential (pure liquid under atmospheric pressure). In practice, these two definitions of the saturation pressure differ by less than 0.1%, so they are usually not distinguished (see the discussion in Chapter 4).

§The water potential of the vapor should include an additional gravity term $\Psi_{\text{g}} = \rho g z$, having the same expression as for a liquid phase. However, this contribution is usually negligible compared to the humidity term (eqn (1.3)).

First consider a plant cell containing solutes of concentration c immersed in an external bath of pure water at ambient atmosphere ($\Psi_{\text{bath}} = 0$) (Figure 1.2a). We assume that the cell is surrounded by an ideal semi-permeable membrane that allows water to pass freely through it but keeps solutes inside. Using eqn (1.2), equality of the water potential between the cell and the bath ($\Psi_{\text{cell}} = \Psi_{\text{bath}}$) implies that there exists a positive pressure difference, or turgor pressure, between the inside of the cell and the outside, given by $P = \Pi = c\mathcal{R}T$. At equilibrium, the turgor pressure is thus equal to the osmotic pressure. For a solute concentration $c = 0.2\ \text{mol L}^{-1} = 200\ \text{mol m}^{-3}$, the turgor pressure is $P = 0.5$ MPa. This is twice the pressure of a car tyre and larger than the typical tension $T_{\text{muscle}} \sim 0.2$ MPa produced by actin-myosin molecular motors and muscle-like fibers.[134] Such a large value of turgor

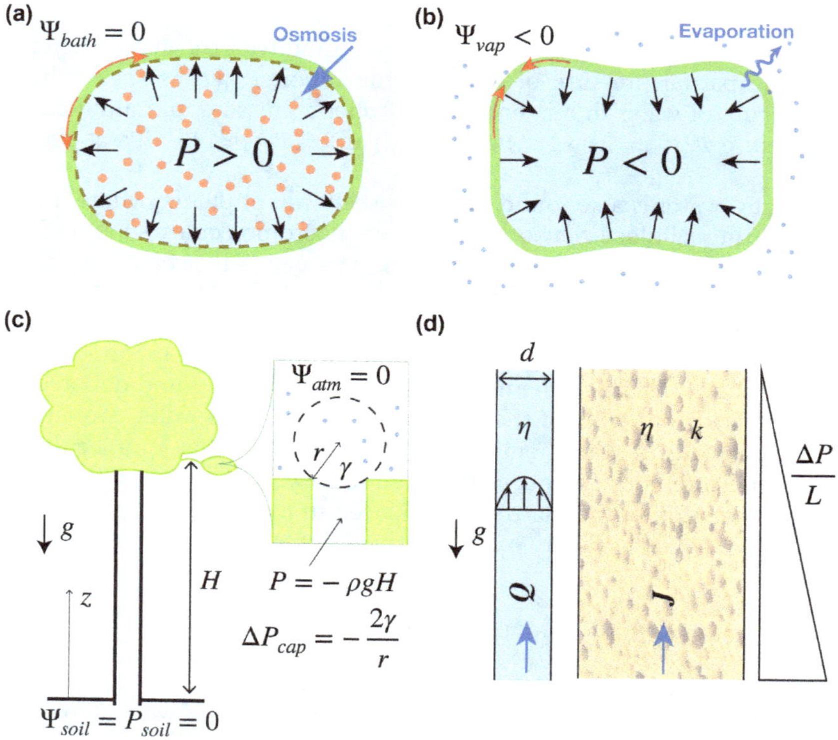

Figure 1.2 (a–c) Water potential balance and turgor pressure in plants. (a) Positive turgor pressure driven by osmosis for an isolated plant cell immersed in pure water (red arrows: tension in the cell wall). (b) Negative turgor pressure driven by evaporation for an isolated cell in equilibrium with a humid atmosphere (red arrows: compression in the cell wall). (c) The water column in the xylem of a tree is under tension (negative pressure). Mechanically, the column is held owing to a tiny capillary meniscus located at the cell wall/air interface (close-up sketch). (d) Pressure-driven flows: Poiseuille flow in a pipe (left) and Darcy flow in a porous medium of permeability k (right).

pressure is found routinely in hydrated plant cells, and can even reach up to 4 MPa – 40 bars! – in the stomata[55] or in the shoot apical meristem of *Arabidopsis*.[93] These high positive values of the turgor pressure can be achieved in plants because the cell is surrounded by a stiff cell wall (see Box 1.1). As we shall see in this chapter and throughout the book, turgor pressure plays a key role in plant biomechanics and physiology. It determines the rigidity of non-woody plant tissues and drives irreversible cell-wall deformation during growth. Change of turgor pressure driven by osmotic or evaporation gradients is also responsible for many reversible movements in plants.[39,71]

The second application we consider is an evaporating cell in contact with a humid atmosphere at 1 bar (Figure 1.2b). If thermodynamic equilibrium is reached, equality of the water potential between the cell and the vapor ($\Psi_{cell} = \Psi_{atm}$) implies from eqn (1.2) and (1.3) a turgor pressure inside the cell given by $P = c\mathcal{R}T + (\mathcal{R}T/v_w)\ln[\text{RH}]$. For a 90% humidity, *i.e.* $[\text{RH}] = 0.9$, and a solution concentration $c = 0.2$ mol L^{-1} as before, the turgor pressure is $P = -14.5$ MPa or -145 bars! In this situation, the turgor pressure is therefore highly negative, which means that the liquid water inside the cell is in tension and pulls on the cell wall. Large negative water pressures due to evaporation are indeed observed in plant cells with very rigid cell walls, such as in the woody cells of the xylem in trees. However, liquid water in tension is in a metastable state. If a gas germ of sufficient size nucleates or is already present inside the cell, cavitation may occur: the microscopic gas germ suddenly expands into a macroscopic bubble. The mechanism of bubble cavitation and its consequence for embolism formation in trees is discussed in Chapter 4. Negative water pressure, and more generally water tension in the cell wall of dead tissues, is also responsible for many passive movements in response to humidity change,[19,76] such as the opening and closing of pine cones,[36] the bending of wheat awns[45] or the curling of seed pods.[4,51] These hygroscopic movements are discussed in Chapter 7.

In the aforementioned examples, the change of water pressure is induced by osmotic or humidity gradients. We conclude by considering the more common situation of a pure water column in static equilibrium under gravity, as found in the vessels of a tall tree in the absence of flow (Figure 1.2c). At equilibrium, the water potential of the sap $P + \rho gz$ given by eqn (1.2) must be uniform along the column. Assuming that the water potential is zero, which corresponds to a well-watered soil, the water pressure at the top of the tree is negative and given by $P = -\rho gH \simeq -0.5$ MPa for a 50 meter-tall tree of height H. It is interesting to compute the size of the capillary meniscus needed to support this negative pressure difference mechanically. The Young–Laplace law states that the pressure difference ΔP_{cap} between water and air at a curved spherical interface is given by:

$$\Delta P_{cap} = -\frac{2\gamma}{r}, \tag{1.4}$$

where $\gamma \sim 0.07\ \mathrm{N\,m^{-1}}$ is the surface tension of the water/air interface and r is the radius of curvature of the meniscus, counted positive if the meniscus is curved toward the air. For $\Delta P_{\mathrm{cap}} \sim -\rho g H \sim -0.5$ MPa, the radius of curvature of the meniscus is about 0.3 μm. This value is at least one order of magnitude smaller than the size of the smallest conductive vessels of the xylem found in leaves. Therefore, air/liquid capillary suction in the xylem cannot be the mechanism of sap ascent in trees. By contrast, the wall of plant cells is a hydrogel with nanometer-size pores that are small enough to support a very large mismatch of capillary pressure across the water/air interface (Figure 1.2c, close-up). Therefore, the negative water pressure in the vascular system of plants is mechanically balanced at the cell wall level, thanks to the reduction of the matrix water potential: $\Psi_{\mathrm{m}} = -\rho g H$. This equilibrium of water under negative pressure holds only if there exists a continuous path of liquid from the ground to the leaves in contact with the atmosphere, where water eventually evaporates. The coupling between water in tension inside the xylem, capillary suction by the cell wall and evaporation at the cell wall/air interface is the basis of the cohesion-theory of sap ascent.[38] The reader interested in xylem flows can consult Tyree and Zimmermann's book[160] or recent reviews such as ref. 75 and 153.

In the previous examples we have considered situations of thermodynamic equilibrium, corresponding to a uniform water potential and no water flux. Water transport requires the presence of water potential gradients, which can arise from pressure gradients, solute concentration gradients, or humidity gradients. We now discuss the laws relating these gradients to water fluxes.

1.1.2 Pressure-driven Flows

1.1.2.1 Viscous Flows

Flows inside plants occur in tiny pipes. The largest conducting vessels in trees are barely larger than 0.1 mm in diameter (with a range of diameters from 10 μm to 500 μm[160]), while water pores inside the cell wall or the cell membrane are nanometric in size. Hence, although the pressure difference driving water transport in plants may be very large (~MPa range), the viscous dissipation is huge, leading to small flow velocities. For example, the typical velocity U of the sap inside the xylem of trees during the day is a few meters per hour (~1 $\mathrm{mm\,s^{-1}}$) for a vessel diameter $d \sim 100$ μm.[160] The Reynolds number of this flow, which compares inertial to viscous effects, is:

$$\mathrm{Re} = \frac{\rho U d}{\eta}, \tag{1.5}$$

where $\rho \simeq 10^3\ \mathrm{kg\,m^{-3}}$ is the sap density and $\eta \simeq 10^{-3}$ Pa s is the sap viscosity, giving a Reynolds number $\mathrm{Re} \sim 0.1$ smaller than 1. Similarly, change of velocity usually occurs on timescales T from minutes to a day, which are much

Box 1.2 Methods for measuring the turgor pressure at cellular resolution.

Turgor pressure plays a central role in plant physiology and biomechanics. The development of accurate methods to measure turgor has thus always been a goal for plant scientists. Most methods give access, not directly to the turgor pressure P, but to the water potential Ψ or the osmotic pressure Π. If both quantities are known, then P is simply deduced from $P = \Psi + \Pi$. The main methods for measuring Ψ and Π are described, for example, in Taiz and Zeiger's book.[155] These include (i) plasmolysis, by bathing the cells in osmotic solutions of varying concentrations and determining the onset for plasmolysis, *i.e.* the bath concentration for which cell turgor drops to zero; (ii) the use of pressure chambers, or pressure bombs, which pressurize tissues until water is expelled, when the applied pressure is equal to the initial tissue water potential; (iii) psychrometers or cryoscopic osmometers, which deduce the osmotic pressure from its effect on the shift in some thermodynamic properties. Besides being indirect, these methods only give access to the global water status of the tissue, assuming the equilibrium of the water potentials. By contrast, the two techniques described below can be used to determine the turgor pressure at cellular resolution, without assuming thermodynamic equilibrium.

The first method, known as the pressure probe, consists in introducing a microcapillary directly inside the cell vacuole to record the pressure (Figure 1.3a). The first detailed measurements were made in 1967 by Green and Stanton on giant algal cells, using a capillary filled with air and closed at one end.[62] The size of these cells (diameter of about 2 mm for a length of several centimeters) was large enough to neglect the loss of cell sap upon insertion of the capillary and to obtain the turgor pressure directly from the change of air volume in the capillary, using the ideal gas law $PV = nRT$. For smaller cells, however, this technique cannot be used, because the compressibility of the measuring system is too large compared to the compressibility of the cell. To overcome this difficulty, the cellular pressure probe was developed in the 1970s by Steudle and Zimmermann[150] (a detailed presentation of the method can be found in ref. 149). The tip of the capillary is reduced to a few microns and the capillary is filled with a liquid – silicone oil – and connected to a closed oil-filled chamber containing a pressure transducer and a piston. When the tip of the capillary is inserted into the cell, cell sap compresses the oil inside the capillary, forming a sap/oil meniscus near the tip that is visible under a microscope. At equilibrium, when the meniscus is maintained at a fixed position with the piston, the pressure measured in the chamber is equal to the turgor pressure inside the cell (the capillary pressure jump across the meniscus is usually negligible).

The pressure probe not only provides access to the turgor pressure, but also enables the measurement of the elastic and hydraulic properties of the cell. For this purpose, a rapid increase of the cell volume ΔV is applied by means of the piston, after which the pressure relaxation is measured by keeping the meniscus position fixed (Figure 1.3a). The initial relationship between ΔV and ΔP gives access to the elastic bulk modulus of the cell: $B_{\text{cell}} = V(\Delta P/\Delta V)$, providing that the cell volume V is known (eqn (1.33) in Section 2.2). The relaxation timescale of the pressure, on the other hand, gives access to the cellular relaxation time $\tau_{\text{cell}} = V/[L_p S(B_{\text{cell}} + \pi)]$ discussed in Section 2.3 (eqn (1.39)), enabling measurements of the plasma membrane hydraulic permeability L_p. The cellular pressure probe has been used to determine water relations in various plant cells and organs, including negative pressure in the xylem.[165] Recently, a somewhat simplified version has been developed, the pico gauge,[82] in which most of the volume of the capillary is filled with a resin that solidifies under UV light, the oil volume being reduced to the very tip of the capillary. As in Green's pioneering experiments, the pressure is deduced from the compression of the oil volume upon insertion into the cell, obtained from image analysis.

The cell pressure probe is unique in giving a direct measure of the cell turgor. However, this method is invasive and time-consuming and becomes increasingly difficult as the cell size decreases typically below 20 μm. This has motivated the development of alternative indentation-based methods in recent years.[56,97,164] Lintilhac *et al.* in 2000 were the first to use indentation to access the turgor of onion epidermal cells, using beads in the range 50–500 μm in diameter (ball tonometry method[91]). The turgor was simply derived from the indentation contact area A measured optically and the indentation force F assuming $F = PA$. However, for smaller cells, probes or indentation depths, the determination of the contact area may be difficult. The influence of the wall tension and cell wall elasticity, ignored in their analysis, can also be important.

Recently, the use of atomic force microscopy (AFM) techniques with smaller probes (submicron in size) has made it possible to determine, from a single indentation curve, both the elastic properties of the wall and the cell turgor, providing that the cell topography is known[11] (Figure 1.3b). The Young's modulus of the wall is first determined from the small indentation-depth portion of the curve, for indentation δ much smaller than the cell wall thickness h, using the classical contact Hertz's law: $F \propto E_{\text{wall}} \sqrt{R_{\text{probe}}} \delta^{3/2}$. Interpreting the larger depth portion of the curve requires a model for the indentation of an inflated shell.[161] Assuming the radius of the probe to be much smaller than the cell size and indentation not too large ($\delta \sim h$), a linear relationship is predicted between force and indentation: $F = k\delta$. The apparent stiffness k is a function of the cell topography (obtained with the AFM), the cell wall elasticity (determined previously) and the turgor pressure, from which P is obtained. When these assumptions are not fulfilled, disentangling turgor pressure and wall elasticity from the indentation curve is not obvious and mostly requires the use of finite element method (FEM) simulations.[93,135]

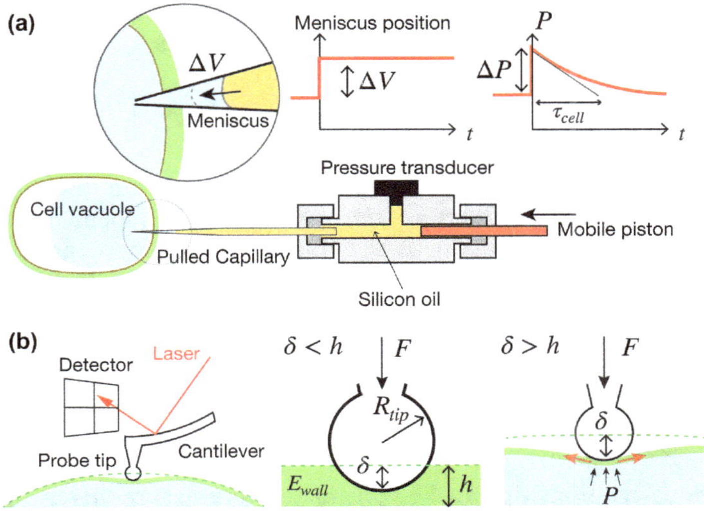

Figure 1.3 Measurements of the turgor pressure at cellular resolution. (a) Cellular pressure probe. (b) Nanoindentation (figure redrawn and adapted from ref. 93). Left: principle of the AFM technique. Center: small indentation depths δ give access to the cell wall elasticity. Right: large indentation depths are sensitive to both the local cell geometry, cell wall elasticity and turgor pressure.

longer than the timescale of momentum diffusion across the diameter of the pipe given by $\tau \sim \rho d^2/\eta \sim 0.01$ s, such that the Stokes number, defined as:

$$\mathrm{St} = \frac{\tau}{T} = \frac{\rho d^2}{\eta T} \tag{1.6}$$

is much smaller than 1.

Therefore, inertial effects can usually be ignored when dealing with water transport in plants. In this situation, the Navier–Stokes equation of fluid mechanics reduces to the Stokes equation, balancing the pressure gradient and the viscous stress:

$$-\nabla P + \eta \Delta \mathbf{U} + \rho \mathbf{g} = 0 \tag{1.7}$$

whereas mass conservation imposes that:

$$\nabla . \mathbf{U} = 0 \tag{1.8}$$

where **U** is the local velocity field of the flow and **g** the gravity vector.

1.1.2.2 Hagen–Poiseuille's Law

Integrating eqn (1.7) and (1.8) in the case of an infinitely long, vertical pipe of circular cross-section of diameter d, like the conducting vessels of the xylem (Figure 1.2d), gives Hagen–Poiseuille's law, relating the flow rate Q (unit $m^3\,s^{-1}$) and the pressure gradients along the pipe:

$$Q = -\frac{\pi d^4}{128\eta}\nabla(P + \rho g z). \tag{1.9}$$

One recognizes a linear relationship between the flow rate (the "flux") and the gradient of water potential (the "force")¶. The coefficient of proportionality defines the conductivity K of the pipe (conductance per unit length), which varies as the vessel diameter to the power 4. Hence, when the vessel diameter decreases in an ideal branch by N identical pipes in parallel, an increasing number of vessels $N \propto d^{-4}$ is needed to achieve the same flow rate under a given pressure gradient.

1.1.2.3 Darcy's Law

The previous relationship can be generalized to pressure-driven flows in a continuum porous material (Darcy's law):

$$\boldsymbol{J} = \frac{\kappa}{\eta}\nabla(P + \rho g z), \tag{1.10}$$

where $\boldsymbol{J}$ is the flow rate per unit area (unit $m\,s^{-1}$) and κ (unit m^2) is the hydraulic permeability of the porous material (Figure 1.2d). The hydraulic permeability scales with the pore size d as $\kappa \propto d^2$ and may be anisotropic, like in the xylem where flow in the longitudinal direction along the vessels is much easier than in the transverse direction. Assuming an ideal tree branch made of N identical pipes of diameter d in parallel, it is possible to explicitly find the relationship between κ and d. The flow rate per unit area is $J = NQ/S$, where Q is the flow rate across a single pipe and S is the area of the cross-section of the branch. Using Hagen–Poiseuille's law (eqn (1.9)) and identifying with Darcy's law gives an effective longitudinal permeability for the branch $\kappa_{xylem} = (d^2/32)\phi_{xylem}$, where $\phi_{xylem} = N\pi\ (d^2/4)/S$ is the surface fraction of xylem. Typical order of magnitudes for $d \sim 10$–100 μm and $\phi_{xylem} \sim 0.2$ give permeability values in the range $\kappa_{xylem} \sim 10^{-13}$–$10^{-11}$ m^2, similar to the permeability of fine granular soils (silt). By contrast, the hydraulic permeability of the cell wall is much lower, $\kappa_{wall} \simeq 10^{-17}$ m^2, which corresponds to water pores in the cellulosic matrix of nanometric size $d \sim 3$–4 nm.[20,49] For such an hydrogel made of microscopic pores, Darcy's law can be generalized as $J = -(\kappa/\eta)\nabla\Psi_{wall}$, with $\Psi_{wall} = \Psi_m + \rho g z$ (see Box 1.4).

¶Note that the gradient of solute does not appear in this expression. Indeed, the diffusion of solute alone, without pressure gradient, is not associated with a global volume flux in the absence of semi-permeable membranes. This is because the volume flux of solute in this case is exactly compensated by an equal and opposite volume flux of water molecules, such that $Q = 0$ (this corresponds to a reflection coefficient $\sigma_S = 0$; see Section 1.1.3). This may not be the case if the solute interacts with an external field in very narrow channels, as in diffusiophoresis and electro-osmosis.

1.1.3 Osmotic Flows and Solute Transport

Pressure-driven flows are very efficient in carrying water over long distances and are the main mode of water transport in the apoplast of plants (xylem, cell wall). Another important mode of transport in plants is osmosis, which occurs when solute concentration gradients exist across the semi-permeable membranes of living cells.

1.1.3.1 Water Transport Across Cell Membranes

The transport of water and solute across a semi-permeable membrane like the plasma membrane of cells can be described using the framework of irreversible thermodynamics and the linear response between "forces" and "flux" (Onsager relations).[35,47,77,148] In the presence of a pressure difference ΔP and concentration difference Δc across the membrane, a volume flux (essentially water) J (unit m s^{-1}) and a molar flux of solute j_S (unit mol m^{-2} s^{-1}) cross the membrane (Figure 1.4a), given by[‖]:

$$J = L_p \Delta P - L_p \sigma_S \mathcal{R} T \Delta c, \tag{1.11}$$

$$j_S = P_S \Delta c + (1 - \sigma_S)\bar{c}J + j_S^*. \tag{1.12}$$

On the right-hand side of the water transport eqn (1.11), the first term represents the pressure-driven flow, where L_p is the hydraulic conductivity of the membrane (unit m s^{-1} Pa^{-1}), the second term is the osmotic flow driven by the solute concentration difference, and σ_S is the reflection coefficient of the solute ($0 \leq \sigma_S \leq 1$)**. Physically, osmosis arises because some of the momentum associated with solute thermal agitation is taken up by the cell membrane and not by the water molecules, creating a net suction of water across the membrane to the more solute-concentrated compartment (see Box 1.3 and ref. 84).

On the right-hand side of the solute transport eqn (1.12), the first term corresponds to the passive diffusion of solute along the concentration gradient (Fick's law type), where P_S (unit m s^{-1}) is the membrane permeability to solute. In plant membranes, both P_S and σ_S are solute dependent and can be regulated through the plasmodesmata permeability, in a complex feedback with turgor pressure.[68] The second term corresponds to the coupling between water transport and solute transport across the membrane, where $\bar{c}$ is the mean concentration of solute across the membrane. Finally, the third term j_S^* corresponds to the molar flux of solute actively pumped against the solute gradient using specialized channels and external chemical energy. Values ranging from $j_S^* = 10^{-7}$ to 10^{-6} mol m^{-2} s^{-1} are found for the

‖These relationships hold for non-electrolytic solutions. In the case of ions, the electric potential difference across the membrane must be added in the driving forces.

**Eqn (1.11) can be rewritten with the water potential explicitly shown as: $J = L_p \Delta\Psi + (1 - \sigma_S)\mathcal{R}T\Delta c$. The second term corresponds to the volume dragged by the solutes when they cross a non-impermeable membrane ($\sigma_S < 1$). In the extreme case where $\sigma_S = 0$ (free diffusion), we recover the fact that only a pressure gradient can induce a net flow.

potassium ion flux responsible for stomatal movement[141] and in the motor cells of *Mimosa pudica*.[152] The highest solute conductance in plant membranes is probably found in MscS-like (small conductance mechanosensitive ion channel) stretch-activated channels.[65]

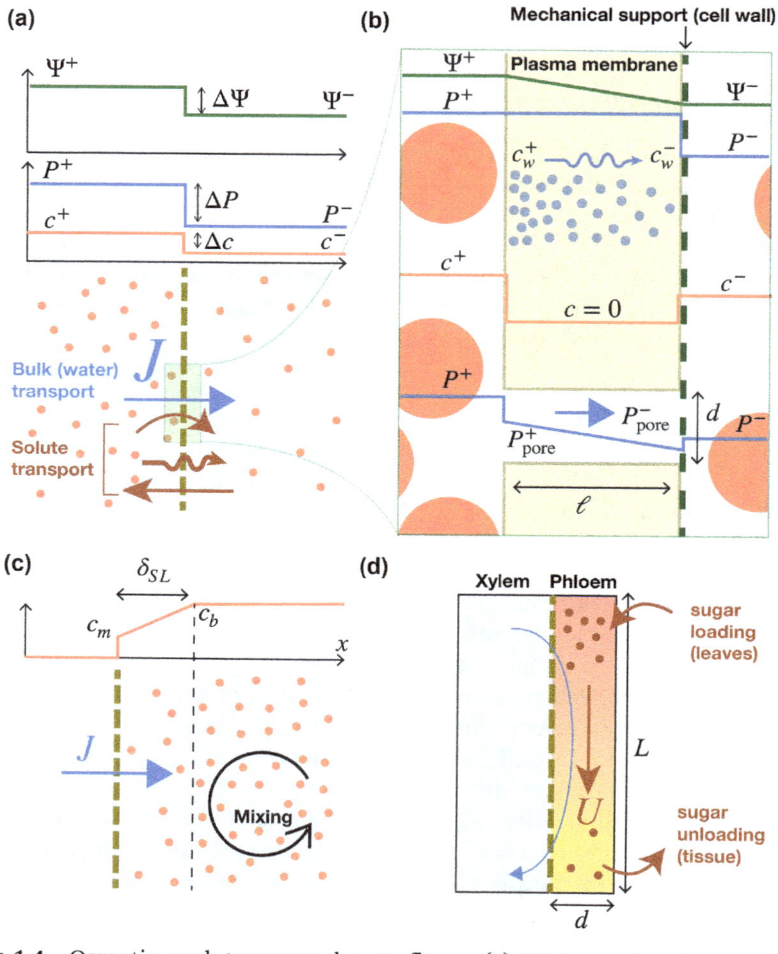

Figure 1.4 Osmotic and transmembrane flows. (a) Water and solute transport across a semi-permeable membrane driven by a gradient of water potential. Solute transport (red arrows) can be split into active transport (straight arrow) and passive transport if the membrane is leaky (undulating arrow: diffusion along gradients of concentration, curved arrow: solute carried by water flow). (b) Close-up showing two modes of water transport for an ideal semi-permeable membrane. Top: diffusive transport driven by a water concentration gradient inside a low-water-solubility membrane, such as the lipid bilayer of the plasma membrane. Bottom: pressure-driven flow inside water pores, such as aquaporins. (c) Stirring layer effects at the vicinity of a semi-permeable membrane. (d) Osmotically driven transport of sugar and Münch's mechanism.

In many situations (salts, sugar, metabolites) the reflection coefficient of the solute is close to unity and the membrane may be assumed perfectly impermeable to the solutes ($\sigma_S = 1$, $P_S = 0$). For such an ideal semipermeable membrane, the water flux across the membrane is just given by the water potential difference across the cell and the hydraulic conductivity of the membrane L_p:

$$J = L_p(\Delta P - \mathcal{R}T\Delta c) = L_p\Delta\Psi. \tag{1.13}$$

Typical values of the membrane conductivity in plant cells are $10^{-13} < L_p < 10^{-11}$ m s^{-1} Pa^{-1} depending on the physiological state of the cell.[147] These values actually reflect two distinct modes of water transport at the microscopic level (see Box 1.3 and Figure 1.4b). On the one hand, water can cross the cell membrane by molecular diffusion inside the lipid bilayer, which acts as a low-solubility medium for water. On the other hand, water can flow in bulk through selective water channels of high conductivity in the cell membrane, called aquaporins,[96] whose shape is optimized for water flow.[59] The opening and closing of these channels is under tight physiological control.

We conclude this brief survey by noting that the fluxes in eqn (1.11) and (1.12) are set by the value of the concentration jump determined exactly *at* the membrane. However, any osmotic flow – or solute diffusion if the membrane is leaky – tends to pile up solutes on one side of the membrane and sweep them away on the other side. The concentration difference across the membrane is then smaller than the concentration difference imposed in the bulk, which in turn reduces the osmotic flow. To minimize this effect, some mixing is necessary to homogenize the concentration field and bring the bulk concentration c_b to a distance δ_{SL} as close as possible to the membrane, called the "unstirred layer" thickness[35,120] (Figure 1.4c). In plant cells, such mixing can be achieved thanks to intracellular flows called "cytoplasmic streaming" that are actively driven by molecular motors (see Chapter 2). If δ_{SL} is known, it is possible to estimate the real solute concentration at the membrane $c_m = c(x=0)$ by balancing, in the unstrirred layer, the diffusive flux and the advection flux: $Ddc/dx \sim Jc$ with the boundary condition $c(x=\delta_{SL}) = c_b$, where D (unit m^2 s^{-1}) is the diffusion coefficient for the solutes and $J = L_p\mathcal{R}Tc_m$ is the osmotic flux (assuming an ideal semi-permeable membrane and no pressure difference across the membrane). We deduce a transcendental equation for c_m:

$$c_m = c_b \exp -\frac{J(c_m)\delta_{SL}}{D}, \tag{1.14}$$

which can be used to estimate the reduction of osmotic flux causes by unstirring layer effects. In practice, for a typical osmotic flux $J \sim 1$ μm s^{-1} (corresponding to an osmotic pressure $\Pi \sim 1$ MPa with $L_p \sim 10^{-12}$ m s^{-1} Pa^{-1}), an unstirred layer $\delta_{SL} \sim 10$ μm (a value similar to the cell width) and a coefficient of diffusion $D \sim 10^{-10}$ m^2 s^{-1}, we find that $c_m \sim 0.9\ c_b$, meaning that unstrirring layer effects on osmotic flows are likely to be small for plant cells. This may, however, not be the case for solute transport if the membrane is leaky.[35]

1.1.3.2 Osmotically Driven Transport of Sugar and Münch's Mechanism

Osmotic flows are the basis for the building of the positive turgor pressure in plant cells discussed in Section 1.1.1. By adjusting their solute content, plant cells pump in and out water to maintain cell hydration in dry conditions or change their volume. One of the most important cell movements driven by osmosis in plants is the closing and opening of the stomata, the cellular valves that control the exchange of gas between the plant and the atmosphere at the surface of leaves.[55]

Another important example of osmotically driven flow in plants is the long-distance transport of sugar (Figure 1.4d). Sugar is produced locally in leaves from photosynthesis. To ensure its transport, sugar is first loaded inside living pipe-like cells called the phloem, which are adjacent to the water-filled xylem. This high concentration of sugar in the phloem decreases its water potential and creates an osmotic flow coming from the xylem, where the water potential is larger. This inward flow creates, in turn, a positive turgor pressure at the top of the phloem and thus a pressure-driven flow that transports sugar along the phloem toward the places where it is unloaded and used for growth or storage. This mechanism was first proposed by Münch in 1930.[107] The velocity of the flow transporting sugar is controlled by the Münch number:

$$\mathrm{Mu} = \frac{64\eta L_{\mathrm{p}} L^2}{d^3}, \tag{1.15}$$

where L_{p} is the membrane permeability of the phloem tube, L the length of the phloem and d its diameter. For $\mathrm{Mu} \gg 1$, the viscous resistance of the phloem tube dominates. The mean flow velocity is then given by the Hagen–Poiseuille's law (see eqn (1.9)): $U = Q/(\pi d^2/4) \sim d^2/(32\eta)c\mathcal{R}T/L$, where c is the sugar concentration. By contrast for $\mathrm{Mu} \ll 1$ the dissipation from the cell membrane dominates and the velocity U is imposed by the water flux across the membrane $J = L_{\mathrm{p}} c\mathcal{R}T$ (see eqn (1.13)) and volume conservation $U\pi d^2/4 \sim \pi L d J$, giving $U \sim (4L/d)L_{\mathrm{p}} c\mathcal{R}T$. The Münch number is just the ratio of these two limiting case velocities. A typical value of the Münch number for the phloem is $\mathrm{Mu} \sim 10^{-4} \ll 1$ (taking $L = 1$ cm, the typical length of a leaf stem, $d \sim 20$ μm, $\eta \sim 2\times10^{-3}$ Pa s, $L_{\mathrm{p}} \sim 10^{-13}$ m s^{-1} Pa^{-1} and $c\mathcal{R}T \sim 1$ MPa[75]), which gives a phloem's velocity $U \sim 0.2$ mm s^{-1} consistent with observation.

The transport of sugar along the phloem is governed by an advection-diffusion equation:

$$\frac{\partial c}{\partial t} + U\nabla c = D\Delta c, \tag{1.16}$$

where D [unit $m^2\,s^{-1}$] is the diffusion coefficient of sugar molecules in water (here no sink or source of sugar are considered along the path). The ratio of the advective to the diffusive term is quantified by the Péclet number:

$$\mathrm{Pe} = \frac{UL}{D}. \tag{1.17}$$

Taking for the diffusion coefficient of sucrose in water $D = 5\times10^{-10}\ m^2\,s^{-1}$, $L = 1$ cm (leaf stem) and $U = 0.2\ mm\,s^{-1}$ gives $\mathrm{Pe} = 4\times10^3 \gg 1$. Therefore, advection by the osmotic flow clearly dominates diffusion in the long-distance sugar transport. In-depth discussion of this fascinating mode of transport can be found in ref. 75.

1.1.4 Evaporation and Vapor Diffusion

We have seen that the water potential within plant tissues is generally negative, with values ranging from 0 MPa for roots immersed in a well-watered soil to few -MPa at the tops of trees or for plants living in arid or saline environments. However, the water potential of the atmosphere is much lower, with $\Psi_{vap} = -70$ MPa for a typical 60% humidity (see eqn (1.3)). Water loss by evaporation is thus a critical issue faced by all terrestrial plants, which they partly solve by covering most of their surfaces with a waxy hydrophobic layer called cuticle (Figure 1.5a). However, some gas exchange with the atmosphere must be maintained in order to capture the carbon dioxine and other gases necessary for photosynthesis and plant metabolism. This exchange occurs on the surface of leaves through specialized pores called stomata. Still, a trade-off is required as many water molecules are lost through evaporation when the stomata open (typically 100 kg of water are lost for 1 kg of glucose synthesized in plants, see ref. 112), which is achieved by fine regulation of stomatal function. A general review on leaf hydraulics, from water movements through the leaf xylem to transport in the air space between the leaf cells can be found in ref. 137. The complex interplay between biological signaling and stomata dynamics is discussed in ref. 88. Here we focus on the basic physics of vapor diffusion in relation to plant evaporation.

1.1.4.1 Fick's Law

Evaporation at the liquid/air interface within leaves and diffusion of water through the stomata represent the final mode of water transport in plants as

Box 1.3 Osmosis and water transport across cell membranes.

We have seen that water transport across an ideal semi-permeable membrane is characterized by a single parameter, the hydraulic permeability of the membrane L_p, which relates the bulk flux J to the difference in water potential $\Delta\Psi$, regardless of whether it comes from a hydrostatic ΔP or osmotic $\mathcal{R}T\Delta c$ pressure gradient (eqn (1.13)). Microscopically, however, L_p can describe two very different modes of water transport, which we briefly discuss here (more details can be found in the excellent book by Finkelstein[47]).

The first situation concerns a membrane of low solubility for water, which typically corresponds to the lipid bilayers of the plasma membrane (Figure 1.4b). In this "oil membrane" model, the concentration of water molecules in the membrane is low, and water crosses the membrane from one side to another mainly through diffusion, following a water concentration gradient inside the membrane. Fick's law for the molar flux of water inside the membrane reads $j_{water} = D_w^{oil}(c_w^+ - c_w^-)/\ell$, where D_w^{oil} is the diffusion coefficient of water molecules in the oily membrane, ℓ is the membrane thickness and $(c_w^+ - c_w^-)$ is the difference in water molar concentration at both end of the membrane. This molar flux is associated with a volume flux of water:

$$J = \nu_w j_{water} = D_w^{oil}\nu_w \frac{c_w^+ - c_w^-}{\ell}. \tag{1.18}$$

It is possible to relate this concentration difference inside the membrane to the external water potential difference $\Delta\Psi$, and thus to relate L_p and D_w^{oil}. Indeed, the continuity of the water potential at the interface of both membranes implies that (see Figure 1.4b for notation):

$$P^+ - c^+\mathcal{R}T = -\frac{\Delta\mu^0}{\nu_w} + \frac{\nu_{oil}}{\nu_w}P^+ + \frac{\mathcal{R}T}{\nu_w}\ln(c_w^+\nu_{oil}), \quad \text{(left)} \tag{1.19}$$

$$-\frac{\Delta\mu^0}{\nu_w} + \frac{\nu_{oil}}{\nu_w}P^+ + \frac{\mathcal{R}T}{\nu_w}\ln(c_w^-\nu_{oil}) = P^- - c^-\mathcal{R}T. \quad \text{(right)} \tag{1.20}$$

Here, we have used the expression of the chemical potential of water inside the oily membrane in the dilute limit, $\mu_w^{oil} = \mu_w^{0,oil} + \nu_{oil}P + \mathcal{R}T\ln c_w\nu_{oil}$, where ν_{oil} is the molar volume of the membrane and $\Delta\mu^0 = \mu_w^0 - \mu_w^{0,oil}$ is the difference between the chemical potential of pure water in the reference state (atmospheric pressure at ambient temperature) and the chemical potential of water in the membrane in the reference state and for a water activity equal to one.[47] Importantly, since there is a difference of hydrostatic pressure across the membrane, an external "grid" (*e.g.* cell wall) must exist on the right-hand side of the membrane in order to balance the pressure difference. As a result, the hydrostatic pressure inside the whole membrane is uniform and equal to the left-hand side pressure P^+. Assuming that $\nu_w \Delta\Psi/\mathcal{R}T \ll 1$, one may linearize the exponential and obtain:

$$c_w^+ - c_w^- = \frac{\nu_w c_0}{\mathcal{R}T}(P^+ - c^+\mathcal{R}T - P^- + c^-\mathcal{R}T) = \frac{\nu_w c_0}{\mathcal{R}T}\Delta\Psi, \tag{1.21}$$

where $c_0 = (1/\nu_{oil})\exp[(\Delta\mu + P^+(\nu_w - \nu_{oil}))/\mathcal{R}T]$ is the molar concentration of water in the membrane for a membrane in equilibrium with pure water in the reference state. Introducing eqn (1.21) in (1.18) and identifying with the transport eqn (1.13) gives the following expression for the hydraulic permeability in the case of lipid membranes:

$$L_p^{\text{lipid bilayer}} = \frac{D_w^{oil}\nu_w^2 c_0}{\mathcal{R}T\ell}. \tag{1.22}$$

Taking $c_0 = 40$ mol m^{-3}, $D_w^{oil} = 10^{-10}$ m^2 s^{-1} and $\ell = 3$ nm (see ref. 47) gives $L_p \sim 10^{-13}$ m s^{-1} Pa^{-1}, a correct order of magnitude for the permeability of lipid bilayer membranes.

The other situation corresponds to the existence of water pores inside the membrane, such as aquaporin channels, which let water move freely but exclude large solutes or ions. In this case, water transport does not occur through diffusion but more like a pressure-driven bulk flow in a pipe, although the molecular size of the pore may complicate the description.[47,59] The gradient of hydrostatic pressure driving the flow inside the pore can be estimated as before, by assuming the continuity of the water potential on both sides of the pore. Since solutes are excluded from the pore, the sharp drop of solute concentration on both sides of the pore induces a sharp drop of hydrostatic pressure, such that $P_{pore}^+ = P^+ - c^+\mathcal{R}T$ and $P_{pore}^- = P^- - c^-\mathcal{R}T$ (Figure 1.4b). The gradient of hydrostatic pressure in the water pore is thus $(P_{pore}^+ - P_{pore}^-)\ell = \Delta\Psi/\ell$. We recover that, although the flow in the pore is driven by a purely hydrostatic pressure gradient, its value is given by the jump of water potential outside the pore, as expected from the transport eqn (1.13). Assuming Hagen–Poiseuille's law to hold (eqn (1.9)) and a surface density of n_s pores per unit surface, the total water flux per unit membrane area is $J = n_s(\pi d^4/128\eta)(\Delta\Psi/\ell)$. Identifying this expression with the transport eqn (1.13) gives the following expression for the hydraulic permeability in the case of transports through water pores:

$$L_p^{\text{water pores}} = \frac{n_s \pi d^4}{128\eta\ell}. \tag{1.23}$$

Taking $n_s = 10^6$ pores m^{-2}, $d = 0.2$ nm, $\ell = 3$ nm and $\eta = 10^{-3}$ Pas gives again $L_p \sim 10^{-13}$ m s^{-1} Pa^{-1}.

it moves from the soil to the atmosphere. The diffusive flux of water molecule in a humid atmosphere j_{water} (number of moles crossing a unit area per unit time, or mol m^{-2} s^{-1}) is given by Fick's law:

$$j_{water} = -D_w \nabla c_w, \tag{1.24}$$

where $D_w = 2.4 \times 10^{-5}$ m^2 s^{-1} is the diffusion coefficient of the water molecule in air at 20 °C and c_w is the molar concentration of water in air (unit mol m^{-3}). Fick's law can be equivalently expressed in term of the vapor pressure gradient, using the gas state equation, $P_{vap} = c_w \mathcal{R}T$, or in term of the water potential of the vapor using eqn (1.3):

$$j_{water} = -\frac{D_w}{\mathcal{R}T} \nabla P_{vap} = -\left[\frac{D_w v_w P_{sat}(T)}{(\mathcal{R}T)^2} \exp\left(\frac{v_w}{\mathcal{R}T}\Psi\right)\right] \nabla \Psi. \tag{1.25}$$

However, when expressed in term of the water potential, the coefficient of proportionality between the water molecular flux and the gradient of water potential is not constant but varies exponentially as: $\exp\left(\frac{v_w}{\mathcal{R}T}\Psi\right)$. Hence, for a given gradient of water potential, a dry atmosphere (very negative Ψ) has a much lower "conductivity" than a humid one ($\Psi \simeq 0$).

1.1.4.2 Evaporation from Stomata

As a first example of the application of Fick's law, we consider evaporation from a surface of pure water subjected to a side wind of velocity U_{wind} in an atmosphere of relative humidity $[RH]_{atm} = P_{vap}/P_{sat}(T)$ (Fig. 1.5b). A diffusive boundary layer of size $\delta \sim \sqrt{D_w L / U_{wind}}$ developed above the surface, where L is the typical surface size. The molecular flux of water is thus given by:

$$j_{water}^{BL} \sim D_w \frac{c_{sat} - c_{atm}}{\delta} \sim \frac{D_w P_{sat}(T)}{\mathcal{R}T}(1 - [RH]_{atm})\sqrt{\frac{U_{wind}}{D_w L}}. \tag{1.26}$$

Taking a wind velocity $U_{wind} = 10$ m s^{-1} and a lateral size $L = 10$ cm (leaf size) gives a diffusive boundary layer of thickness $\delta \sim 0.5$ mm. For a very dry atmosphere ($[RH]_{atm} \sim 0$), this gives a molecular flux $j_{water} \simeq 10^{-2}$ mol m^{-3} s^{-1} and an evaporative mass flux $J_m = M_w j_{water} \simeq 2 \times 10^{-4}$ kg m^{-3} s^{-1}, where $M_w = 18$ g mol^{-1} is the molar mass of water. This value is surprisingly close to the maximal evaporating rate reported per surface leaf area in plants.[75] This is all the more remarkable that stomatal pores covers only 1–10% of the leaf surface, the rest being almost impermeable to evaporation.

To understand this feature, we consider an ideal leaf made of a flat impermeable sheet perforated with holes (stomata) of pore size a (Figure 1.5c). The sheet is in contact with pure water at the bottom and placed in an atmosphere of water vapor concentration c_{atm} above. For an isolated stomata, the size of the diffusive boundary layer is fixed by the pore size a††. Assuming that stomata do not interact with each other, the total evaporation rate of the leaf per unit area is thus:

$$j_{\mathrm{water}}^{\mathrm{stomata}} \sim n_{\mathrm{s}} a^2 D_{\mathrm{w}} \frac{c_{\mathrm{sat}} - c_{\mathrm{atm}}}{a} \sim n_{\mathrm{s}} a D_{\mathrm{w}} (c_{\mathrm{sat}} - c_{\mathrm{atm}}). \tag{1.27}$$

where n_{s} is the number of stomata per unit leaf area. We find that the evaporative flux from the leaf is equal to the evaporative flux from a surface of pure water ($j_{\mathrm{water}}^{\mathrm{stomata}} \sim j_{\mathrm{water}}^{\mathrm{BL}}$) when $n_{\mathrm{s}} \sim 1/(a\delta)$ or $\phi_{\mathrm{s}} \sim a/\delta$, where $\phi_{\mathrm{s}} \sim n_{\mathrm{s}} a^2$ is the surface fraction covered by the stomata. Taking $\delta \sim 1$ mm and $a \sim 10$ µm, we find that this condition is fulfilled with stomata covering only 1% of the leaf area ($\phi_{\mathrm{s}} = 0.01$). Indeed, due to their small size, stomata impose a diffusion length scale much smaller than the air boundary layer, resulting in high local evaporative gradients that compensate for their small area.

Eqn (1.27) predicts the paradoxical result that, for a sufficiently high concentration of stomata, the evaporative flux from the leaf could be greater than that from of a pure water surface. This is because we have assumed that the external vapor concentration fixing the stomatal vapor flux is c_{atm}. This assumption is valid only if the distance between stomata, $s \sim 1/\sqrt{n_{\mathrm{s}}} \sim a/\sqrt{\phi_{\mathrm{s}}}$, is much larger than the air boundary layer δ. In practice, however, $s \ll \delta$. In this situation, the total resistance to evaporation from the leaf, defined as $R_{\mathrm{Leaf\ Evap}} = (c_{\mathrm{sat}} - c_{\mathrm{atm}})/j$, is the sum of the stomatal resistance $R_{\mathrm{stomata}} \sim 1/(n_{\mathrm{s}} a D_{\mathrm{w}}) \sim a/(\phi_{\mathrm{s}} D_{\mathrm{w}})$ and the resistance of the air boundary layer $R_{\mathrm{BL}} = \delta/D_{\mathrm{w}}$ (Figure 1.5d). We find that both resistances are equal when $\phi_{\mathrm{s}} \sim a/\delta$, recovering the previous condition. A detailed discussion of the role of stomata interaction on leaf vapor diffusion can be found in ref. 89.

1.1.4.3 *Diffusive* versus *Bulk Transport: Plants as Water Valve*

As a second application of Fick's law, it is instructive to compare, for the *same* difference in water potential, the mass of water transported in bulk by a pressure gradient with that transported in vapor form by diffusion in a humidity gradient (Figure 1.5e). According to Poiseuille's law (eqn (1.9)), the mass bulk flow through a pipe of diameter d and length L is $Q_{\mathrm{m}}^{\mathrm{bulk}} = \rho(\pi d^4/128\eta) \times (\Delta\Psi/L)$, where ρ is the density of the liquid water, η is the water viscosity and $\Delta\Psi = \Delta P$ is the pressure difference across the pipe. On the other hand, transport of water vapor in the same pipe by diffusion from a vapor at saturation to an external atmosphere of humidity $[\mathrm{RH}]_{\mathrm{atm}}$ gives, using Fick's law, a diffusive mass flux: $Q_{\mathrm{m}}^{\mathrm{diffusion}} = M_{\mathrm{w}}(\pi d^2/4)(D_{\mathrm{w}}/\mathcal{R}T)P_{\mathrm{sat}}(T)[(1 - [\mathrm{RH}]_{\mathrm{atm}})/\Delta\Psi] \times (\Delta\Psi/L)$. Therefore, for the same gradient of water potential, the ratio of the bulk to diffusive transport is:

$$\frac{Q_{\mathrm{m}}^{\mathrm{bulk}}}{Q_{\mathrm{m}}^{\mathrm{diffusion}}} = \frac{\rho \mathcal{R} T d^2 \Delta\Psi}{32\eta M_{\mathrm{w}} D_{\mathrm{w}} P_{\mathrm{sat}}(T)(1 - [\mathrm{RH}]_{\mathrm{atm}})}. \tag{1.28}$$

††This is valid as long as $a \ll \delta_{\mathrm{v}}$ where $\delta_{\mathrm{v}} \sim \sqrt{\nu L / U_{\mathrm{wind}}}$ is the viscous boundary layer and $\nu \sim 10^{-5}$ m^{-2} s^{-1} is the air kinematic viscosity.

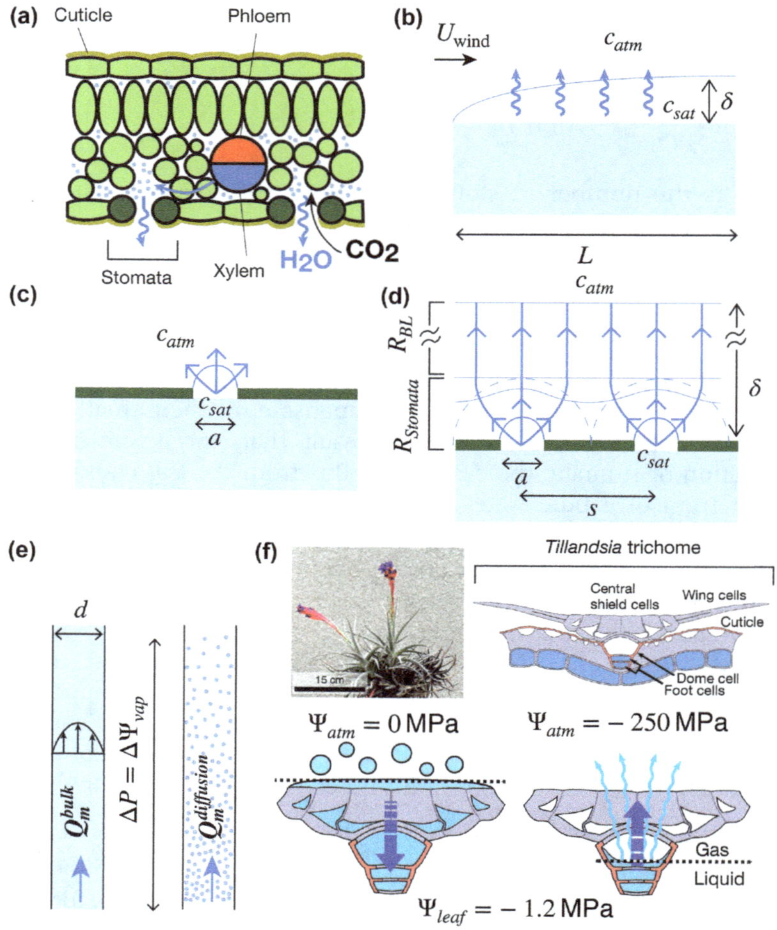

Figure 1.5 Evaporation in plants. (a) Cross-section of a leaf showing the water transport path from the xylem (liquid state) to the stomata (gas state). (b) Diffusive air boundary layer above an evaporative bath of water subjected to a cross-wind. (c) Evaporation from an isolated stomata modeled as a hole of size a in an impermeable plate. (d) Pattern of vapor diffusion above a "leaf" made of several "stomata". Thin solid lines give show the iso-concentration of vapor while thick solid lines show the direction of the vapor flux. (e) Bulk flow in the liquid state *versus* diffusive transport in the gas state for a given water potential difference $\Delta P = \Delta\Psi_{vap}$. (f) Asymmetric water transport ("water valve") at the trichomes of the plant *Tillandsia aeranthos* (photo) living in the Atacama desert in Chile. Reproduced from ref. 128, https://doi.org/10.1038/s41467-019-14236-5, under the terms of the CC BY 4.0 license https://creativecommons.org/licenses/by/4.0/.

Taking $d \sim 1$ μm and a water potential difference for an atmosphere of 50% humidity gives: $Q_m^{bulk}/Q_m^{diffusion} \sim 10^7$! This shows that, even in very small conduits, bulk transport driven by a pressure gradient in the liquid state is much more efficient than vapor diffusion driven by a humidity gradient.

This simple principle is harnessed by the plant *Tillandsia*, which lives in the Atacama desert in Chile, to design an asymmetrical "water valve" on the surface of its leaves[128] (Figure 1.5f). The Atacama desert is probably the driest place on Earth, with a humidity as low as 15%, which corresponds to a water potential $\Psi_{atm} = -250$ MPa. However, a few days a year, fog from the coast brings a very humid atmosphere full of tiny droplets of liquid water ($\Psi_{atm} \sim 0$ MPa). These droplets condense on the very hydrophilic surface of *Tillandsia* leaves and are absorbed by the leaf cells through osmosis, due to their high concentration of solute which creates a negative water potential Ψ_{leaf} of about -1 MPa. Remarkably, although the difference in water potential causing water absorption during wet periods ($\Delta\Psi_{abs} = |\Psi_{atm} - \Psi_{leaf}| = +1$ MPa) is a hundred times *smaller* than the difference in water potential causing water evaporation during dry periods ($\Delta\Psi_{evap} = +249$ MPa), the absorption flow rate is found to be a hundred times *greater.*[128] This corresponds to a difference in effective conductance $Q/\Delta\Psi$ between the humid and dry states a few thousands.

This strong asymmetry is explained by the existence of specialized structures at the surface of *Tillandsia*'s leaves, called trichomes, where water exchanges takes place.[128] These trichomes consist of a row of empty dead cells covered by a very thick cell wall, which separate the leaf cells from the atmosphere (Figure 1.5f). During wet periods, the wall and dead cell cavities are completely soaked with liquid water due to their hydrophilic nature; absorption then occurs in the liquid state, through an osmotically driven bulk flow limited by the leaf cell membrane permeability. By contrast, during the dry periods, the liquid/vapor interface shifts within the trichome. Transport to the outside then occurs by diffusion in the vapor form which, as we have seen, is very inefficient.

1.2 Solids

So far in the chapter, we have considered plants as rigid and static containers, in which water and solutes are transported according to the laws of thermodynamics and fluid mechanics. However, in order to absorb water, grow or generate movement, the cells and tissues of plants must deform. This chapter discusses the relationships between stress and deformation in the plant solid body, and how this deformation is coupled with water transport. Other examples of fluid–structure interaction in plants are discussed in Chapter 2.

1.2.1 The Wall Stress and the Force Balance

1.2.1.1 Force Balance at the Cellular Level

The solid phase of plants is essentially made by the wall material that surrounds all plant cells (Box 1.1 and Figure 1.1). Plant cells have two types of wall.[155] Young growing cells and the mature cells of some tissues (leaf) are surrounded by a thin primary cell wall made of stiff cross-linked cellulose microfibrils embedded in a highly hydrated matrix of polysaccharides (hemicelluloses and pectins mainly),[32,33,72] whose thickness h is typically between 0.1 μm and a few μm. When growth stops, a thicker and more rigid secondary cell wall of lignin and other woody components may be added, giving additional strength.

Under physiological conditions, the wall of living cells is stretched by the internal, osmotically induced, turgor pressure (see Section 1.1). From a mechanics perspective, a plant cell thus behaves like a pressurized thin shell (Figure 1.6a). The tensile stress σ_{wall} (force per unit area) stretching the wall can be estimated by modeling the cell as a spherical shell of radius R and uniform wall thickness $h \ll R$ under a pressure difference P between the inside and outside. The balance of force on the half-shell imposes that the external pressure force $\pi R^2 P$ is equal and opposite to the tensile force in the wall integrated along the wall perimeter $2\pi R h \sigma_{\text{wall}}$ (Figure 1.6a), giving:

$$\sigma_{\text{wall}} \sim \frac{R}{2h} P. \tag{1.29}$$

This Young–Laplace-like relationship shows that the tensile stress in the wall may be much larger than the turgor pressure due to the geometrical amplification factor R/h. Taking a typical plant cell geometry $R = 10$ μm, $h = 1$ μm and turgor pressure $P = 0.5$ MPa gives $\sigma_{\text{wall}} = 2.5$ MPa, which is much larger than the maximal tensile stress generated by animal muscle ($T_{\text{muscle}} \sim 0.2$ MPa[134]).

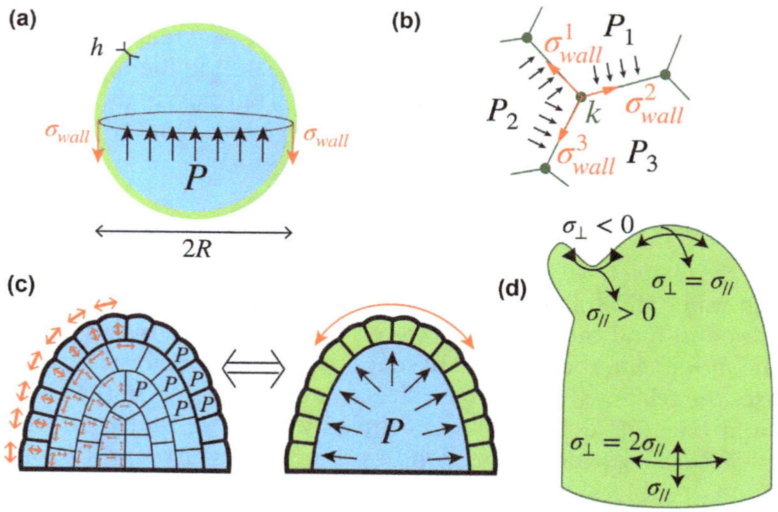

Figure 1.6 Force and stress balance in plant tissues. (a) Force balance in a cell viewed as a pressurized shell in order to compute the wall stress σ_{wall}. (b) Vertex model of a plant tissue (redrawn and adapted from ref. 26). (c) Tissue-tension approximation: the epidermis, which is much stiffer than the inner cells, carries most of the turgor-driven wall stress (red arrows). The left-hand drawing is inspired by ref. 117. (d) Surface stress distribution in the tissue-tension approximation, depending on the local geometry of the organ.

We will see that this stress level is enough to significantly deform the plant cell wall and change its cell volume.

1.2.1.2 Force Balance in Plant Tissues

The balance of forces written here for an isolated cell can be generalized to a tissue composed of several cells. In plants, each cell wall is glued to its neighbor by an adhesive layer called the middle lamella, which prevents the cells from sliding against each other or becoming detached (the situation of invasive growth, discussed in detail in Chapter 7, is an exception). However, unlike the case of an isolated cell, the balance of forces is generally not sufficient to determine the stress distribution in all the walls of the tissue. Even when the cell geometry and the turgor pressure are known, additional knowledge of the constitutive relation of the cell wall is usually required (see Sections 1.2.2 and 1.2.4). In most situations, the complete computation of the tissue stress can only be done by means of heavy and expensive numerical simulations, using for example Finite Element Methods.[15] Modeling plant solid mechanics at the tissue and organ level thus often relies on simplified models.

The first class of model, broadly called cell-based models, considers the discrete nature of cells with different degrees of approximation. Of these methods the vertex model is one of the most popular.[26,63,93] In its two-dimensional version, the tissue is modeled as a collection of polygons representing cells, in which linear segments representing the wall are in tension and connected to vertices representing the wall junction (Figure 1.6b). The global force balance implies that, at each vertex k, the (half) sum of the pressure difference force at each face adjacent to k balances the sum of the wall tension applied to k. A detailed presentation of vertex-like approaches is given in Chapter 3.

The second class of model uses continuum mechanics to define an average stress tensor $\boldsymbol{\sigma}$ which integrates both the wall stress and the turgor pressure at the tissue level (see Chapter 3 for a detailed discussion of the averaged stress tensor in plant tissue). In the absence of body forces, the internal force balance is then given by:

$$\nabla \cdot \boldsymbol{\sigma} = 0. \tag{1.30}$$

1.2.1.3 The Tissue–Tension Assumption: Plant Tissues as Pressurized Shells

An approximation often used to describe stresses in plant tissues is based on the fact that epidermal cells are generally much stiffer than the cells of the underlying tissue, due to their small size and thick outer wall. The epidermal layer is then assumed to support most of the turgor pressure generated by the internal cells and the entire tissue is modeled as a pressurized shell[14,67,86] (Figure 1.6c). Using this tissue–tension assumption and neglecting the bending stiffness of the epidermis, the force balance in the direction perpendicular to the epidermis reduces to a generalized Young–Laplace equation:[157]

$$h\left(\frac{\sigma_{\parallel}}{R_{\parallel}}+\frac{\sigma_{\perp}}{R_{\perp}}\right)=P. \tag{1.31}$$

Here $(\sigma_{\parallel}, \sigma_{\perp})$ are the local principal in-plane stresses in the epidermis; $(R_{\parallel}, R_{\perp})$ are the local principal radius of curvature of the organ's surface, and P represents the turgor pressure of the cells in the bulk below the epidermis. This equation must be complemented by the force balance in the plane of the shell given by $(\partial\sigma_{\parallel}/\partial_{s\parallel}=0)$ and $\partial\sigma_{\perp}/\partial_{s\perp}=0$, where $(s_{\parallel}, s_{\perp})$ is the arc-length along the principal directions. The pressurized-shell model has been widely used in the context of plant morphogenesis to determine the tissue stress distribution in organs of various shapes, such as the shoot apical meristem or in leaves[63,69,138] (Figure 1.6d). For a spherical cap of radius R like the tip of a growing shoot $(R_{\parallel}=R_{\perp}=R)$, the tissue stress is uniform and we recover the relationship (1.29): $\sigma_{\perp}=\sigma_{\parallel}=PR/2h$. For a cylinder of radius R like a stem or a root $(R_{\parallel}=\infty, R_{\perp}=R)$, eqn (1.31) predicts that the perpendicular, hoop stress is twice the longitudinal stress: $\sigma_{\perp}=2\sigma_{\parallel}=PR/h$. Finally, for a saddle crease geometry like at the cusp of a bud $(R_{\perp}=-r<0$ and $R_{\parallel}=+R\gg r)$, one of the stresses is then compressive: $\sigma_{\perp}\approx -Pr/h$ while the other is in tension: $\sigma_{\parallel}\approx +PR/h$ (Figure 1.6d). Hence, although the internal driving force – the turgor pressure – is isotropic, stress anisotropy is more the rule than the exception in plants.

The tension of the epidermal tissue predicted by the pressurized shell model is an example of residual stress, *i.e.* a non-zero internal stress in a body at equilibrium and in the absence of external loading. Such residual stresses are common in plants due to wall adhesion and the absence of sliding between cells. In fact, any material inhomogeneity induced by gradients of structure, elasticity, swelling rate or growth in a plant tissue is susceptible to generate residual stresses, which may or may not be resolved by tissue buckling or stress relaxation. The feedback between the tissue shape, residual stress, and the biological response (growth law, mechanotransduction) is central to plant morphogenesis and development.[14]

1.2.2 Elasticity

1.2.2.1 Hooke's Law and Cell Wall Elasticity

As the plant wall is a living material, its mechanical description must be approached with caution, especially for growing cells. However, as long as the deformations are small enough (typically less than 5–10% in living cells), and the timescales of observation not too large, the cell wall may be approximated as an elastic solid. The simplest elastic constitutive law is Hooke's law, in which deformations are reversible and proportional to the applied stresses. For uniaxial deformation, Hooke's law reduces to:

$$\sigma=E\varepsilon_{\mathrm{el}}, \tag{1.32}$$

where $\varepsilon_{el} = \Delta L/L$ is the elastic deformation or elastic strain and E (unit Pa) is the Young's modulus of the material (Figure 1.7a). The Young's modulus is a measure of the stiffness or compliance/softness of the material: the larger E is the stiffer the material, while the lower E is the more compliant or soft it is.

As a first application of Hooke's law, we can calculate the reversible volume change ΔV induced by a change in turgor pressure ΔP in a plant cell modeled as a thin spherical shell of thickness $h \ll R$ (Figure 1.7b). The force balance implies $\sigma_{wall} = (R/2h)P$ (eqn (1.29)) and Hooke's law can be written approximately as follows[‡‡]: $\sigma_{wall} \approx E_{wall}\varepsilon_{el} = E_{wall}(R - R_0)/R_0$, where E_{wall} is the Young's modulus of the wall and R_0 the radius of the cell in its rest state when the turgor pressure $P = 0$. Identifying these two expressions and assuming small deformation $(R - R_0) \ll R_0$ gives: $\Delta P \approx (2h/R)E_{wall}(\Delta R/R) \approx (2h/3R)E_{wall}(\Delta V/V)$. The coefficient of proportionality of this pressure/volume relationship is called the cellular bulk modulus B_{cell} (unit Pa):

$$B_{cell} = V\frac{\Delta P}{\Delta V} \approx \frac{2h}{3R}E_{wall}. \tag{1.33}$$

The cellular bulk modulus determines the water storage capacity of the cells and, as we shall see, controls the timescale of cell swelling. For xylem cells under negative pressure, it also influences the dynamics of cavitation, as discussed in Chapter 4. Importantly, the cellular bulk modulus is a measure of the Young's modulus E_{wall} of the cell wall, up to a geometric factor $\sim h/R$. Measurements of B_{cell} in a wide variety of cells has been made using the cell pressure probe described in Box 1.2.[30,80,151] A typical value for a growing plant cell is $B_{cell} \sim 5$ MPa. Using eqn (1.33) for a plant cell of thickness $h \sim 1$ μm and radius $R \sim 10$ μm gives $E_{wall} \sim 75$ MPa. This order of magnitude is consistent with direct measurements of cell wall Young's modulus using tensile assays, which gives $E_{wall} \sim 50$–150 MPa for onion epidermal walls[167] (Figure 1.7c). The Young's modulus of the wall is even larger in giant algal cells ($E_{wall} \sim 0.5$–2 GPa[125]) and can exceed 25 GPa for wood fibers.[57]

Of course, the real mechanical behavior of the cell wall is more complex than the ideal Hooke's law considered here. First, the wall of living cells typically exhibits a nonlinear strain-hardening behavior, *i.e.* an increase of Young's modulus with strain, together with a dissipative behavior (viscoelasticity, plasticity),[70,78,151,167] as found in other polymeric and gels materials (Figure 1.7c). Second, the mechanical response of the cell wall is generally anisotropic, with Young's modulus being much larger in the direction parallel to the mean orientation of the stiff cellulose microfibrils than in their perpendicular direction. Therefore, even for an isotropic loading such as that imposed by the internal turgor pressure, the deformation of a cell is generally

---

[‡‡]For this geometry and assuming an isotropic and homogeneous material, Hooke's law in the plane of the shell also involves Poisson's ratio ν of the material ($-1 \leq \nu \leq 0.5$). The exact expression of the cellular bulk modulus in this case is $B_{cell} = [2h/(3(1-\nu)R)]E_{wall}$; see ref. 87.

non-isotropic. This anisotropy, combined with the microscopic heterogeneity of the wall, makes the characterization of the wall by a single Young's modulus too simplistic. For example, wall stiffness deduced from indentation methods using very small probes (sub-micrometer in size) are generally much lower (of the order of few MPa[98,166]) than those obtained from tensile micro-machines or inferred from the value of the cellular bulk modulus. This is because indentation at nanoscales primarily probes the transverse elastic properties of the wall and the polysaccharide matrix between the cellulose nanofibrils, rather than in-plane global tensile properties.[166]

1.2.2.2 *Elasticity of a Single Cell*

The cellular bulk modulus (eqn (1.33)) characterizes the elastic response of a plant cell to volume change, *i.e.* when the cell wall is forced to stretch further during the deformation. It is interesting to study a situation of cell deformation without change of stretch, such as a single turgid cell squeezed between two plates by a displacement δ. When the internal turgor pressure P is high and the cell wall thickness is thin ($h \ll \delta$), the problem is similar to that of a squeezed elastic balloon (Figure 1.7d). The force then mainly comes from the internal pressure acting against the plates: $F \sim P\pi a^2$, where a is the radius of the contact area between the cell and the plates. Assuming that δ is small compared to the radius of the cell R gives the geometrical relation $a^2 \sim 2R\delta$ and thus a linear relationship between the force and the displacement: $F \sim 2\pi RP\delta$. From this result, one can define an effective external stress acting on the cell $\sigma = F/(\pi R^2)$, an effective elastic strain: $\varepsilon_{el} = \delta/R$ and thus an apparent Young's modulus of the cell $E_{cell} = \sigma/\varepsilon_{el}$ given by:

$$E_{cell} \sim 2P. \tag{1.34}$$

Interestingly, this apparent Young's modulus depends on the turgor pressure but not on the cell wall elasticity, a property that can be used to infer the turgor pressure in experiments on single cells[41,91] (see Box 1.2). When the turgor pressure drops to zero, the apparent cell Young's modulus predicted by eqn (1.34) vanishes. In reality, there is always a small elastic resistance coming from the bending of the cell wall, which we have ignored previously. The force needed to bend a plate of size R and thickness h over a displacement δ is $F \sim E_{wall}(h^3/R^2)\delta$,[87] yielding an apparent bending cell modulus $E_{bending} \sim [F/(R^2)]/(\delta/R)$ given by:

$$E_{bending} \sim \left(\frac{h}{R}\right)^3 E_{wall}. \tag{1.35}$$

1.2.2.3 *Elasticity of Plant Tissues*

The previous analysis at the cellular level is helpful to understand the relationship between the macroscopic Young's modulus of a multicellular turgid

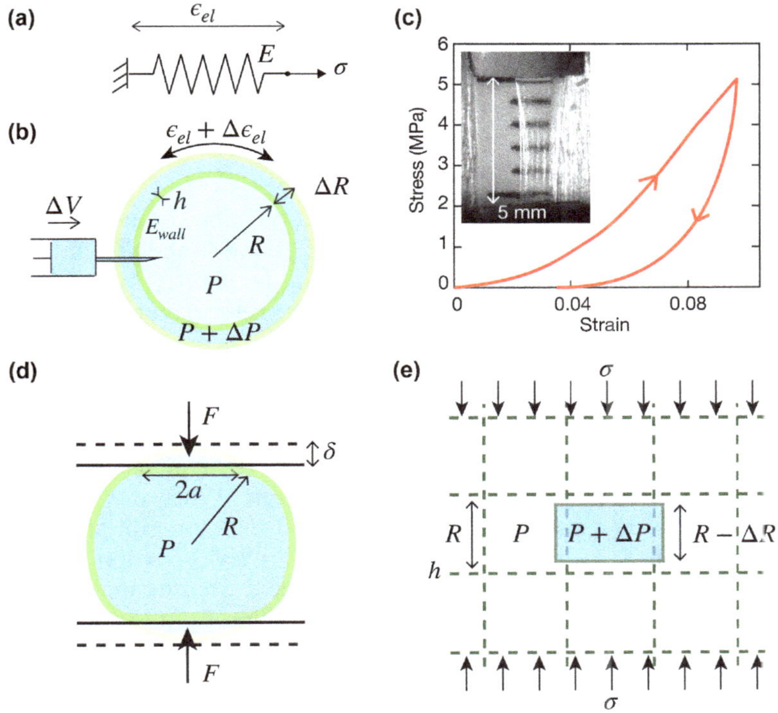

Figure 1.7 Elasticity. (a) Mechanical model of Hooke's law. (b) Pressure–volume relationship for a single cell in the elastic regime, characterized by the cellular bulk modulus $B_{\text{cell}} = V(\Delta P/\Delta V) \approx (h/R)E_{\text{wall}}$. (c) Stress–strain relationship for onion epidermal walls during a loading and unloading cycle in a tension assay. Adapted from ref. 167 with permission from AAAS, Copyright © 2021 The Authors. Photo: Onion epidermal strip stretched by ~10% (image credit: Daniel Cosgrove). (d) Elasticity of a single plant cell squeezed between two rigid plates, yielding an apparent cell Young's modulus $E_{\text{cell}} \sim P$. (e) Elasticity of an ideal plant tissue. Under compression, the cell wall is stretched and the turgor pressure changes, yielding an effective tissue Young's modulus $E_{\text{tissue}} \sim B_{\text{cell}}$.

plant tissue and the microscopic properties of its constituents (Young's modulus of the cell wall, cell geometry). This link is, however, not straightforward. Even for an idealized tissue composed of identical cells of internal turgor P, wall thickness h and size R, dimensional analysis states that the tissue Young's modulus should depend on both the turgor pressure and the cell properties through two independent dimensionless numbers, namely P/E_{wall} and h/R. It is possible to simplify the prediction and derive scaling relations when the turgor pressure is higher than the bending stiffness ($P \gg E_{\text{bending}}$) so that the cell walls are all pre-stretched and in tension. As the liquid is incompressible, it is generally not possible to deform such a tissue without further stretching the walls, unlike the case of an isolated cell

(Figure 1.7e). The tissue's Young's modulus E_{tissue} is then dominated by the stretching mode of deformation, as in the case of the cellular bulk modulus, which gives:

$$E_{\mathrm{tissue}} \sim B_{\mathrm{cell}} \sim \left(\frac{h}{R}\right) E_{\mathrm{wall}} \sim \phi E_{\mathrm{wall}}, \tag{1.36}$$

where $\phi \sim h/R$ is the solid (wall) volume fraction of the tissue.

This scaling law is compatible with values of Young's modulus of turgid tissues $E_{\mathrm{tissue}} \sim 10$ MPa and is valid as long as $E_{\mathrm{bending}} \ll P \ll B_{\mathrm{cell}}$. When the turgor pressure drops to zero, for *e.g.* by plasmolyzing the cells in an osmotic solution, the tissue modulus sharply drops because the cell walls are no longer stretched and can bend easily. In real plant tissue, the situation is complicated by the fact that the cell wall's modulus E_{wall} itself is not constant but strongly varies with the deformation (Figure 1.7c), yielding complex turgor pressure dependences. Note that in all this discussion, we have assumed that the tissue remains always close to full hydration, even when the turgor drops to zero. When the tissue further loses water below the onset of plasmolysis, as during drying in air, the tissue Young's modulus E_{tissue} may rise again because the solid fraction ϕ increases, until it reaches the wall Young's modulus E_{wall} (see the discussion in Chapter 5 in the context of roots). More details on this question from the standpoint of cellular materials can be found in ref. 57, 111 and 163.

1.2.3 Poroelasticity: From Cell to Tissue

So far, we have discussed the elastic behavior of a plant cell and tissue without considering the flow of water within it. However, we have seen that the cell wall and plasma membrane are permeable to water. Therefore, any deformation resulting in a local change in turgor pressure, such as the bending of a stem (Chapter 2) or the squeezing of a fruit, must induce a movement of water within the medium. The same is true for a plant cell or tissue immersed in a solution of different osmolarity. To restore the water potential balance, the tissue must absorb or expel water and therefore must swell or shrink elastically. This coupling between flow and deformation in elastic porous media like plant tissues is described by poroelasticity or poromechanics theory.[34] As we shall see, poroelasticy sets the timescale of response of plant tissues to sudden change of water potential, and thus provides a bound for all water-driven movements.[39,50,146]

1.2.3.1 The Cellular Relaxation Time

We first consider poroelasticity at the cellular level. Consider an isolated plant cell of initial volume V_i immersed in a bath at thermodynamic equilibrium ($\Psi_{\mathrm{bath}} = \Psi_{\mathrm{cell},0} = P_0 - c_0\mathcal{R}T$), where P_0 is the initial turgor pressure of the cell and c_0 is its initial solute concentration (Figure 1.8a). The water potential balance is suddenly perturbed at time $t = 0$, for example by

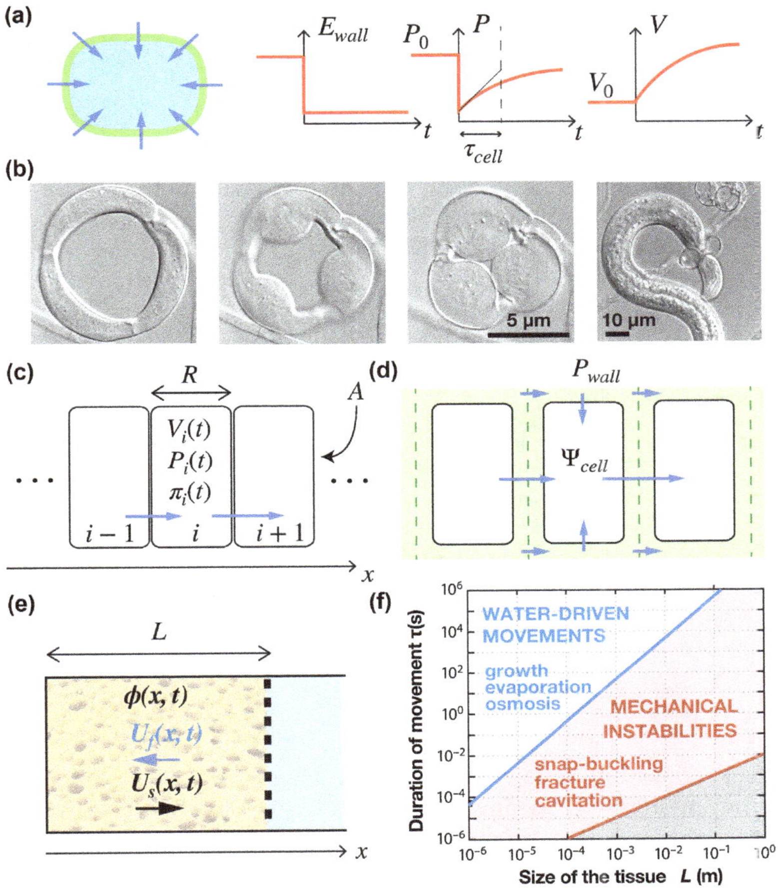

Figure 1.8 Poroelasticity. (a) Cellular relaxation timescale τ_{cell} induced by a sudden change of water potential (here a sudden drop of the cell wall Young's modulus). (b) Rapid swelling of the ring cells of the trap of the carnivorous fungus *Arthrobotrys brochopaga*. Reproduced from ref. 114 with permission from TIB, Copyright 1995. (c) Philip's model of water transport in a one-dimensional tissue (transmembrane transport only). (d) Molz and Ikenberry's model[100] taking into account both the apoplast and transmembrane pathway. (e) Two-fluid continuum model for a poroelastic medium. (f) Physical classification of plant movements showing the poroelastic time $\tau_p \propto L^2$ (blue line) and the inertial time $\tau_{\text{inertia}} \sim L\sqrt{\rho / E}$ (red curve), below which no motion is possible.[146]

suddenly changing the internal turgor pressure (a similar result holds for a perturbation of solute concentration or external water potential). What is the swelling/shrinking dynamics of the cell in response to this perturbation?

To answer this question, we use mass conservation and state that the change in volume V of the cell per unit time is equal to the flow rate of water through its surface S. Using eqn (1.13) of the water flux J across an ideal semi-permeable membrane, this gives:

$$\frac{\mathrm{d}V}{\mathrm{d}t} = -J \times S = -L_{\mathrm{p}}(\Psi_{\mathrm{cell}} - \Psi_{\mathrm{bath}})S = -L_{\mathrm{p}}(P - c\mathcal{R}T - \Psi_{\mathrm{bath}})S, \tag{1.37}$$

where L_{p} is the hydraulic conductivity of the membrane. For small perturbations around the initial state, we can write: $V = V_0 + \delta V(t)$, $S = S_0 + \delta S(t)$, $P = P_0 + \delta P(t)$ and $c = c_0 + \delta c(t)$. Since inertia is negligible, the mechanical equilibrium is always satisfied and we can use the pressure/volume relationship $\delta P = B_{\mathrm{cell}}(\delta V/V_0)$ seen in Section 2.2 (eqn (1.33)). Finally, assuming the conservation of solute during the dynamics implies: $\delta c/c_0 = -\delta V/V_0$. At the lowest order, eqn (1.37) then reduces to a linear relaxation equation for the turgor pressure:

$$\frac{d\delta P}{\mathrm{d}t} = -\frac{1}{\tau_{\mathrm{cell}}}\delta P \quad \text{with} \quad \tau_{\mathrm{cell}} = \frac{V_0}{L_{\mathrm{p}}S_0(B_{\mathrm{cell}} + \Pi)}, \tag{1.38}$$

where τ_{cell} is the cell relaxation time and $\Pi = c_0\mathcal{R}T$ is the osmotic pressure. Generally, $B_{\mathrm{cell}} \gg \Pi$; writing $V/S \sim R$ where R is the typical size of the cell thus gives:

$$\tau_{\mathrm{cell}} \sim \frac{R}{L_{\mathrm{p}}B_{\mathrm{cell}}}. \tag{1.39}$$

The cell relaxation time thus increases when the cell size increases or when the membrane permeability or cell elasticity decreases. Typical measurements in giant algal cells of $R = 200$ μm, $B_{\mathrm{cell}} = 30$ MPa, and $L_{\mathrm{p}} = 2\times10^{-12}$ ms^{-1} Pa give $\tau_{\mathrm{cell}} = 3.5$ s.

The cell relaxation time sets the shortest response time of a plant cell to a sudden (small) change of water potential: whatever the timescale of other biochemical or genetic processes, *e.g.* ion flux, aquaporin opening, change of wall mechanical properties, a cell cannot change its volume in a time $\tau < \tau_{\mathrm{cell}}$. The cell relaxation time thus defines the fastest possible swelling/shrinking movement at the cellular level.[39,50] This hydraulic limit is probably reached in some carnivorous fungi (*e.g. Arthrobotrys brochopaga*), which have evolved special mycelial structures to capture nematodes[115] (Figure 1.8b). In these organisms, the trap consists of a constrictive ring composed of three connected cells of size $R \sim 5$ μm. When a nematode enters the ring, the cells rapidly inflate inward, wrapping the nematode in less than 1/10 of second. The current hypothesis for this rapid movement is that the cell wall of the ring is bi-composite, with a very rigid outer wall surrounding a more flexible inner wall. Upon stimulation, the outer wall suddenly breaks, which is equivalent to a sudden drop in the Young's modulus of the wall. To satisfy mechanical equilibrium and Hooke's law, [$\sigma = (R/2h)P = E_{\mathrm{w}}\varepsilon_{\mathrm{el}}$, see Section 2.2], the turgor must therefore drop abruptly, resulting in water being sucked in and the cell swelling very rapidly due to its small size ($\tau_{cell} \propto R$).

1.2.3.2 Water Diffusion in Plant Tissues

We can extend the previous approach to the tissue level. The simplest model of a plant tissue is a one-dimensional chain of identical rectangular cells in which water is transported from cell to cell through the cell plasma membranes, without considering the cell wall between the cells (Philip's model,[122] see Figure 1.8c). Using the water transport law (eqn (1.13)), the mass conservation for the volume V_i of the cell i is given by (assuming $\Psi_{bath} = 0$):

$$\frac{dV_i}{dt} = AL'_p(P_{i+1} - \Pi_{i+1} - P_i + \Pi_i) - AL'_p(P_i - \Pi_i - P_{i-1} + \Pi_{i-1}), \qquad (1.40)$$

where A is the cell area in the direction of the flow, $L'_p = L_p/2$ is the effective membrane hydraulic conductivity (the factor 1/2 comes from the fact that two cell membranes has to be crossed per cell), P_i is the turgor pressure of cell i and $\Pi_i = c_i\mathcal{R}T$ is the osmotic pressure of the cell i. Using again the small perturbation approximation and the initial water potential equilibrium condition, eqn (1.40) reduces to:

$$\frac{d\delta P_i}{dt} = \frac{1}{\tau_{cell}}(\delta P_{i+1} + \delta P_{i-1} - 2\delta P_i). \qquad (1.41)$$

where $\tau_{cell} = V_0/[AL'_p(B_{cell} + \Pi)]$ is the single cell relaxation time. Using the continuum approximation : $\delta P_i(t) \rightarrow \delta P(x,t)$, the set of discrete differential eqn (1.41) becomes a single partial differential equation:

$$\frac{\partial \delta P}{\partial t} = \left(\frac{R^2}{\tau_{cell}}\right)\frac{\partial^2 \delta P}{\partial x^2} = \mathcal{D}_{Philip}\frac{\partial^2 \delta P}{\partial x^2}, \qquad (1.42)$$

with $\mathcal{D}_{Philip} = R^2/\tau_{cell} \sim RL_pB_{cell}$.

Eqn (1.42) corresponds to a diffusion equation for the relaxation of the turgor pressure, with a diffusion coefficient $\mathcal{D}_{Philip}$. The characteristic time for water transport through a tissue of length L, called the *poroelastic time* τ_p, is given by:

$$\tau_p \sim \frac{L^2}{\mathcal{D}_{Philip}} \sim \left(\frac{L}{R}\right)^2 \tau_{cell} \sim \frac{L^2}{RL_pB_{cell}}. \qquad (1.43)$$

The poroelastic time τ_p depends on the cellular hydraulic and mechanical properties and scales with the tissue size as $\propto L^2$.

The Philip's model has been extended to take into account both the cell-to-cell pathway and the apoplast pathway, *i.e.* through the cell wall (see Box 1.1) by Molz and Ikenberry[100] (Figure 1.8d). This model assumes that water can be transported from cell to cell by a gradient of cell water potential Ψ_{cell}, and into the wall by a gradient of hydrostatic pressure P_{wall}. Water can also be exchanged perpendicularly between the wall and the cells. In this framework, the water transport is described by two diffusion equations with a coupling term:

$$\begin{aligned}\frac{\partial \Psi_{\text{cell}}}{\partial t} &= \mathcal{D}_1 \frac{\partial^2 \Psi_{\text{cell}}}{\partial x^2} - C_1(\Psi_{\text{cell}} - P_{\text{wall}}) \\ \frac{\partial P_{\text{wall}}}{\partial t} &= \mathcal{D}_2 \frac{\partial^2 P_{\text{wall}}}{\partial x^2} + C_2(\Psi_{\text{cell}} - P_{\text{wall}})\end{aligned} \tag{1.44}$$

where $\mathcal{D}_1 = \mathcal{D}_{\text{Philip}}$ and $\mathcal{D}_2 = 2B_{\text{wall}}k_{\text{wall}}/2\eta$ is a water diffusion coefficient in the cell wall, with B_{wall} the wall bulk modulus, k_{w} the cell wall (Darcy) permeability and η the water viscosity. The coupling terms involve the properties of the cell wall and cell membrane: $C_1 = 6AL_{\text{p}}(B_{\text{cell}} + \Pi)/V_0$ and $C_2 = 6L_{\text{p}}B_{\text{wall}}/\phi R$, where $\phi \ll 1$ is the wall volume fraction.

In practice, water diffuses much faster through the wall than from cell to cell, because the wall is much more rigid than the cell ($\mathcal{D}_2 \approx 100\ \mathcal{D}_1$). However, the small solid fraction of the wall and the large exchange surface between wall and cells make the coupling between the two paths very strong. In this situation, it can be shown that there is a local quasi-equilibrium between the water potential of the wall and the water potential of the adjacent cell, such that $\Psi_{\text{cell}} \approx P_{\text{wall}} = \Psi_{\text{tissue}}$, where Ψ_{tissue} is the water potential of the tissue. The two coupled equations can then be reduced into a single effective diffusion equation for water transport:

$$\frac{\partial \Psi_{\text{tissue}}}{\partial t} = \mathcal{D}_{\text{tissue}} \frac{\partial^2 \Psi_{\text{tissue}}}{\partial x^2}, \tag{1.45}$$

where $\mathcal{D}_{\text{tissue}}$ is an effective water diffusion coefficient given by:

$$\mathcal{D}_{\text{tissue}} = \frac{\mathcal{D}_1 C_2 + \mathcal{D}_2 C_1}{C_1 + C_2}. \tag{1.46}$$

Typical values for plant tissues (taking $R \sim 10/100\ \mu\text{m}$, $L_{\text{p}} \sim 10^{-12}\ \text{ms}^{-1}\,\text{Pa}^{-1}$, $B_{\text{cell}} \sim 10$ MPa, $\phi \sim 0.1$, $B_{\text{wall}} \sim 100$ MPa, $k_{\text{wall}} \sim 10^{-18}\ \text{m}^2$) gives $\mathcal{D}_{\text{tissue}} \sim 10^{-9}$–$10^{-10}\ \text{m}^2\,\text{s}^{-1}$, in agreement with measurements.[100,148] A three-dimensional extension of this model, taking into account the transport in vascularized tissues, can be found in ref. 130 and 131.

1.2.3.3 *Poroelasticity in Continuum Media*

When the effective medium approximation is valid, one can approach water diffusion in plant tissues using continuum equations and the theory of poroelasticy (see for *e.g* ref. 34). We derive here a simple one-dimensional poroelastic model for a porous material made of an incompressible solid matrix of volume fraction $\phi = \delta V_{\text{S}}/\delta V$, where δV_{S} is the local volume of solid and δV the local total volume, infiltrated by an incompressible liquid of volume fraction $\phi_f = \delta V_f/\delta V = 1 - \phi$, where δV_f is the local volume of fluid (Figure 1.8e). We note $U_s = (1/\delta V_s)\int_{\delta V_s} d\tau U_s^{\text{local}}$ (resp. $U_f = (1/\delta V_f)\int_{\delta V_f} d\tau U_f^{\text{local}}$) the volume averaged velocity of the solid matrix (resp. fluid). The conservation of mass written for the total medium (solid and fluid) and for the solid phase gives two equations:

$$\begin{aligned} &\frac{\partial}{\partial x}\left(\phi U_s + (1-\phi)U_f\right) = 0, \\ &\frac{\partial \phi}{\partial t} + \frac{\partial \phi U_s}{\partial x} = 0. \end{aligned} \tag{1.47}$$

Neglecting fluid and solid's inertia, the conservation of momentum for the total medium and for the fluid phase gives:

$$\begin{aligned} &\frac{\partial}{\partial x}(\sigma_s - P) = 0, \\ &J \equiv (1-\phi)(U_f - U_s) = -\frac{k(\phi)}{\eta}\frac{\partial P}{\partial x}. \end{aligned} \tag{1.48}$$

Here, σ_S is the volume-averaged stress in the solid matrix corrected by a fluid pressure term: $\sigma_s = (1/\delta V)\int_{\delta V_S} d\tau \sigma_s^{\text{local}} + \phi P$, where P is the pressure in the fluid phase (minus the atmospheric pressure by convention). This definition of the solid stress, also called Terzaghi effective stress, ensures that the solid stress is zero when the matrix is immersed in a fluid of uniform pressure without deformation. The second equation for the fluid's momentum balance is Darcy's law, where J is the volume flux of fluid, k is the Darcy permeability of the porous medium and η is the fluid's viscosity.

The system is closed by adding a constitutive law for the mechanical behavior of the solid phase. Assuming an elastic linear relationship between stress and deformation (Hooke's law), we can write:

$$\sigma_s = E\left(\frac{\phi_0 - \phi}{\phi_0}\right), \tag{1.49}$$

where ϕ_0 is the solid fraction of the undeformed matrix, $(\phi_0 - \phi)/\phi_0$ is the one-dimensional elastic strain and $E = -\phi(d\sigma_S/d\phi)$ is a one-dimensional Young's modulus of the solid matrix.

Linearizing eqn (1.47)–(1.49) around the rest state $P = 0$, $U_S = U_f = 0$ and $\phi = \phi_0$ gives:

$$\frac{\partial \phi}{\partial t} = \mathcal{D}\frac{\partial^2 \phi}{\partial x^2} \quad \text{with} \quad \mathcal{D} = \frac{kE}{\eta}. \tag{1.50}$$

One recovers a diffusion equation for the transport of water in the poroelastic medium, as obtained previously using a discrete description of the plant tissue. The effective diffusion coefficient $\mathcal{D}$ is a function of the fluid's viscosity η, the Darcy permeability k and the elastic Young's modulus E of the porous matrix. The poroelastic time of water diffusion across a tissue of size L is given by:

$$\tau_{\mathrm{p}} = \frac{L^2}{\mathcal{D}} = \frac{\eta L^2}{kE}. \tag{1.51}$$

As for the cell relaxation time at the cell scale, the poroelastic time τ_{p} provides a bound for the fastest possible water-driven movements in plants at the tissue and organ level.[146] A plant motion occurring on a timescale $\tau > \tau_{\mathrm{p}}$ can rely on water transport, whereas systems with $\tau < \tau_{\mathrm{p}}$ must use other mechanisms (Figure 1.8f). The strong size dependence of the poroelastic time ($\tau_{\mathrm{p}} \propto L^2$) shows that hydraulic movements are increasingly less efficient in terms of speed as the system size increases. We shall see in Section 2.5 how the use of mechanical instabilities enables plants to overcome this hydraulic limit.

1.2.4 Growth

So far, we have dealt with the small deformation regime of plant tissues, where strains are mainly reversible. We now address the opposite extreme deformation regime corresponding to plant growth. Growth is the irreversible and sustained expansion of cells under physiological conditions.[155] It is obviously a very complex phenomenon that requires a continuous interplay between biochemistry, mechanics and hydraulics to maintain homeostasis. During growth, the cell wall must be continuously synthesized while it expands in order to maintain its mechanical and structural integrity. At the same time, the absorption of water from the surrounding environment must be precisely balanced, in order to allow the increase in cell volume and thus the expansion.

The concept of growth usually encompasses cell division, cell differentiation and the actual growth of cells, *i.e.* their increase in size. Cell division occurs in localized regions called meristems. In plants, a distinction is made between primary growth, responsible for the elongation of organs, and secondary growth, responsible for the radial thickening of organs. Primary growth is driven by meristems located at the tip (apex) of stems and roots, and therefore called apical meristems. The meristem of secondary growth is the cambium, a ring of cells around the axis of stems and roots. We focus here mainly on primary growth and on the mechanical aspects of cell expansion, leaving aside the question of cell division and differentiation. A discussion of growth from a modeling perspective is found in Chapter 4.

1.2.4.1 *Cell Wall Rheology: The Lockhart–Ortega Model*

The basic mechanics of plant growth have been revealed by seminal experiments conducted on giant algal cells.[60,125–127] First, growth requires the existence of a minimum turgor pressure to occur, *i.e.* a minimum stretch or

Box 1.4 Extension of poroelasticity to hygroscopic media.

We have seen how the coupling between flow (Darcy's law) and elastic deformation (Hooke's law) yields a diffusion equation for the transport of water in a deformable porous medium. Similar fluid–solid couplings occur in hygroscopic media like hydrogels, except that in the latter case the strong chemical affinity between the water and the polymer chains generates an additional osmotic driving force for the transport of water. Hydrogels provide a simple model for the hygroscopic behavior of the plant cell wall. We give below a one-dimensional version of the theory developed by Bertrand *et al.* to describe the swelling dynamics of ideal elastomeric gels.[12]

As in a classical poroelastic medium, the total stress in the gel can be written as the sum of a solid stress σ_S and a fluid (pore) pressure P:

$$\sigma = \sigma_S - P. \tag{1.52}$$

For a gel the solid stress is of entropic origin and represents the elastic stretching stress of the polymer chains network (as before, the isotropic atmospheric pressure has been removed from the definition of P and σ, such that for a gel at atmospheric pressure: $P = 0$ and $\sigma = 0$). This elastic contribution for uniaxial deformation can be written as:[12]

$$\sigma_s^{1D} = \frac{\mathcal{R}T}{\nu_p}\left(\frac{1-\phi^2}{\phi}\right), \tag{1.53}$$

where ϕ is the local volume fraction of solid and ν_p is the molar volume of polymers in the dry state. Note that with this convention, the elastic stress is null when the gel is dry ($\phi = 1$, no water inside).

For an ideal elastomeric hydrogels, it is possible to decompose the matrix potential (the water potential inside the gel) between a pressure term and an osmotic term, by analogy with dilute solutions (see ref. 12):

$$\Psi_{gel} = P + \Psi_{os}. \tag{1.54}$$

The osmotic term is the derivative of the mixing Helmholtz free energy of the gel per unit volume, with respect to change of volume. It can be obtained from the Flory–Huggins theory of polymer solutions as:

$$\Psi_{os} = \frac{\mathcal{R}T}{\nu_w}\left(\phi + \ln(1-\phi) - \frac{\phi}{\alpha} + \chi\phi^2\right), \tag{1.55}$$

where ν_w is the molar volume of pure water in the reference state. The first three terms in brackets reflect the entropy of mixing, where α is a measure of the volume per polymer molecule relative to the volume per fluid molecule in the mixture. The fourth term reflects the enthalpy of mixing, where χ is a dimensionless interaction parameter.[12] Once the water potential is known, the generalized Darcy's law is given by:

$$J \equiv (1-\phi)(U_f - U_s) = -\frac{k(\phi)}{\eta}\frac{\partial \Psi_{\text{gel}}}{\partial z}. \tag{1.56}$$

By combining eqn (1.53)–(1.56) with the mass conservation (1.47), and linearizing around a rest state, we recover a similar diffusion equation for the evolution of ϕ as derived previously (eqn (1.50)), except that the diffusion coefficient now incorporates the osmotic contribution:

$$\mathcal{D}_{\text{gel}} = \frac{kE_{\text{gel}}}{\eta} \quad \text{with} \quad E_{\text{gel}} = -\phi\frac{d(\sigma_s^{1D} + \Psi_{\text{os}})}{d\phi}. \tag{1.57}$$

We can apply this theory to estimate the swelling of a "water bead" initially in the dry state with a radius R_{dry}, and immersed in a bath of pure water at atmospheric pressure where it reaches an equilibrium radius R_{eq}. In this situation, swelling occurs in three dimensions, but the above one-dimensional model can be applied to each direction independently, by symmetry of the problem. At thermodynamic equilibrium, the water potential of the gel is equal to that of the bath, such that: $\Psi_{\text{gel}}(\phi_{\text{eq}}) = P + \Psi_{\text{os}}(\phi_{\text{eq}}) = 0$. To be consistent with the one-dimensional theory, ϕ_{eq} must be understood as the *linear* solid fraction, not the volume solid fraction, *i.e.* $\phi_{\text{eq}} = \phi_{\text{dry}} \times (R_{\text{eq}}/R_{\text{dry}})$ with $\phi_{\text{dry}} = 1$. On the other hand, mechanical equilibrium implies that the total stress is null (the bath is at atmospheric pressure), such that: $\sigma^{1D}(\phi_{\text{eq}}) - P = 0$. The linear solid fraction of the water bead at equilibrium thus satisfies:

$$\frac{\mathcal{R}T}{v_{\text{p}}}\left(\frac{1-\phi_{\text{eq}}^2}{\phi_{\text{eq}}}\right) + \frac{\mathcal{R}T}{v_{\text{w}}}\left(\phi_{\text{eq}} + \ln(1-\phi_{\text{eq}}) - \frac{\phi_{\text{eq}}}{\alpha} + \chi\phi_{\text{eq}}^2\right) = 0. \tag{1.58}$$

In practice, ϕ_{eq} is small and $v_p/(\alpha v_w) \gg 1$. At the lowest order, the swelling ratio is thus given by:

$$\frac{R_{\text{final}}}{R_{\text{dry}}} = \frac{\phi_{\text{dry}} = 1}{\phi_{\text{eq}}} \approx \sqrt{\frac{v_p}{\alpha v_{\text{w}}}} \approx 6, \tag{1.59}$$

with typical values for hydrogels: $\alpha \approx 250$ and $v_{\text{w}}/(\alpha v_{\text{p}}) \approx 10^{-4}$ (see ref. 12).

stress for the cell wall to yield. Second, above this yield pressure, the rate of cell expansion, or growth rate, depends on the pressure difference between the turgor pressure and the yield pressure. Third, when applying increments of turgor pressure, the growth rate stabilizes at its steady value only after a given relaxation time.

These observations are reminiscent of the behavior of yield-stress fluids in soft matter, such as emulsions, polymeric gels or clays, which exhibit elastic behavior below a yield-stress and viscous behavior above.[6] From the mechanical standpoint, the simplest constitutive law of the plant cell wall is therefore that of a Bingham fluid or an elasto-viscoplastic fluid. Such Bingham model was first proposed by Lockhart in 1965[92] and then extended by Ortega in 1985[119] to take into account elasticity. For uniaxial deformation, the Lockhart–Ortega rheology of the cell wall can be written as:

$$\sigma_{\text{wall}} = E_{\text{wall}}\varepsilon_{\text{el}} \quad \text{and} \quad \varepsilon = \varepsilon_{\text{el}} + \varepsilon_{\text{ir}},$$
$$\text{with} \begin{cases} \dot{\varepsilon}_{\text{ir}} = 0 \quad \text{if} \quad \sigma_{\text{wall}} < \sigma_{\text{Y}}, \\ \dot{\varepsilon}_{\text{ir}} = \Phi_{\text{wall}}(\sigma_{\text{wall}} - \sigma_{\text{Y}}) \quad \text{if} \quad \sigma_{\text{wall}} > \sigma_{\text{Y}}. \end{cases} \tag{1.60}$$

In this model, the total strain in the cell wall ε is decomposed into an elastic part, ε_{el}, and an irreversible or plastic part, ε_{ir}, which represents growth (Figure 1.9a). Below a critical stress σ_{Y}, the cell wall behaves as an elastic solid ($\varepsilon = \varepsilon_{\text{el}}$) and follows Hooke's law with a Young's modulus E_{wall}. Above σ_{Y}, the cell wall deforms irreversibly and flows with a strain rate $\dot{\varepsilon}_{\text{ir}}$ proportional to the excess stress ($\sigma_{\text{wall}} - \sigma_{\text{Y}}$). The proportionality coefficient Φ_{wall} between the strain rate and the excess stress is called the extensibility of the cell wall and is associated with the irreversible creep of the wall under stress, also called wall-loosening.[166] Dimensionally, the cell wall extensibility is an inverse viscosity [unit (Pa s)$^{-1}$]. However, it should be kept in mind that the extensibility reflects a complex process at the microscopic level, including the addition of mass and a chemically mediated remodeling of wall material. The extensibility strongly depends on the temperature – a sign that it is under the control of cell metabolism[127] – and can be modified under the action of pH-dependent wall-loosening enzymes, such as expansins.[25,32] Several molecular-scale[2,42,132,167] or thermodynamic[7] models have been proposed to describe this active process, some of which recovering a Bingham-like rheology at the macroscopic level (see Chapter 4 for a detailed presentation of one of these microscopic models).

The simple uniaxial Lockhart–Ortega model (1.60) has been extended to anisotropic cell wall under multiaxial stress, either by assuming an anisotropic extensibility[40] or an anisotropic elasticity by analogy with fiber-reinforced composite materials.[15,18] A more complex constitutive law explicitly accounting for the reorientation of microfibrils under flow has been also developed, by analogy with liquid crystals.[43] For such tensorial laws, the question of whether the cell wall yields above a critical strain ε_{Y} or a critical stress σ_{Y}, which is equivalent in the one-dimensional Lockhart model, is of importance, as the principal direction of strain may not coincide with the principal direction of stress.[117] For elongated cells and organs for which $\sigma_{\perp} > \sigma_{\parallel}$ (see Section 1.2.1), the direction of growth is mainly longitudinal and thus perpendicular to the maximal stress, suggesting that the yield condition in terms of strain is more relevant.

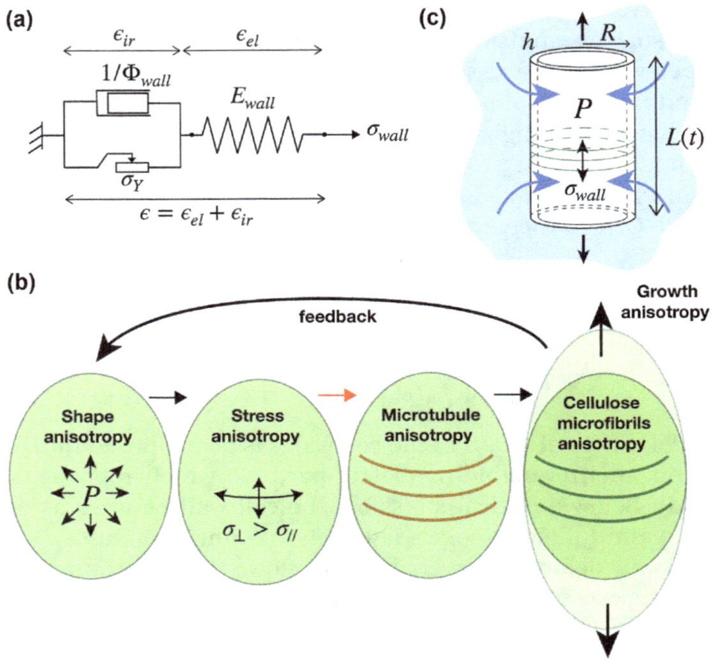

Figure 1.9 Growth and cell wall rheology. (a) One-dimensional mechanical representation of the elasto-viscoplastic rheology of the cell wall. (b) Mechanosensing (red arrow) of the wall stress by the cortical microtubules (brown) fixing the orientation of cellulose microfibrils (green) generates a mechanical feedback loop between shape and growth. (c) Unidirectional growth of a single cell immersed in a water bath (Lockhart's model).

Finally, a growing body of evidence suggests that the cell wall rheology is regulated not only by molecular signals, but also by the mechanical state of the wall itself.[159] A key molecular actor of this feedback is the cortical microtubules, whose orientation determines the orientation of the cellulose microfibrils in the wall, and thus wall anisotropy (see Box 1.1. and Figure 1.1b). Experiments show that the orientation of the microtubules aligns with the direction of maximal stress in the wall, generating a feedback loop between the organ's shape (determining stress direction) and material anisotropy (determining growth direction and thus shape)[63] (Figure 1.9b). This mechanical feedback is taken into account in the most recent rheological models[118] but the precise transduction mechanism is still debated.[64,117] More generally, how mechanics modified growth, either through direct effect on the wall stress or through the perception and transduction of internal and external mechanical signals – a process known as thigmomorphogenesis – is a very active area of research in plant biophysics. A discussion of these aspects in the context of root growth in given in Chapter 5.

1.2.4.2 Growth of a Single Plant Cell

The Lockhart–Ortega rheology can be used to describe the unidirectional and uniform steady growth of an isolated cell, modeled as a cylinder of fixed radius R and length $L(t)$, immersed in an external bath of water potential Ψ_{bath} (Figure 1.9c). This situation typically corresponds to the growth of the giant internode cells of the green algae *Chara* or *Nitella*. For these cells, the wall expansion is evenly distributed over the wall surface and growth is uniform along the cell, a mode of growth called diffuse growth§§. Moreover, the cellulose microfibrils in these cells run mainly circumferentially, restricting growth in the radial direction and promoting growth in length.

In this cylindrical geometry, the balance between the turgor pressure and the wall stress implies that the longitudinal wall stress is given by $\sigma_{wall} = PR/(2h)$, where h is the wall thickness and P the turgor pressure. For a steady state of growth the elastic strain ε_{el} is constant. The growth strain rate, also termed relative elongation rate (RER) or relative elementary growth rate (REGR) in the plant science community, is then given by $\dot{\varepsilon} = \dot{\varepsilon}_{ir} = (1/L)(\mathrm{d}L/\mathrm{d}t)$. Combining the force balance and the Lockhart–Ortega wall rheology (1.60) thus implies:

$$\frac{1}{L}\frac{\mathrm{d}L}{\mathrm{d}t} = \Phi(P - P_Y), \tag{1.61}$$

where $\Phi = (\Phi_{wall}R/2h)$ is an effective extensibility defined at the cellular level and $P_Y = 2h\sigma_Y/R$ is the yield turgor pressure for growth. This first Lockhart's equation must be complemented by a second equation for water transport. Indeed, for cell expansion to be possible, an inward flow of water must compensate for the increase in volume, which requires a difference in water potential between the inside and outside of the cell. Using the water transport eqn (1.13), the second Lockhart equation is:

$$\frac{\mathrm{d}L}{\mathrm{d}t} = -\frac{2L}{R}L_p(P - c\mathcal{R}T - \Psi_{bath}), \tag{1.62}$$

where L_p is the hydraulic conductivity of the membrane and c the cell solute concentration. To close the system, a third Lockhart equation should be added to describe the temporal evolution of the solute concentration.[105] We assume here that growth is slow enough that osmotic balance is maintained, so that solute concentration remains constant. The solution of (1.61–1.62) is therefore:

$$\frac{1}{L}\frac{\mathrm{d}L}{\mathrm{d}t} = \frac{2L_p}{R}\frac{c\mathcal{R}T - P_Y}{1 + \dfrac{2L_p}{R\Phi}} \quad \text{and} \quad P = \frac{P_Y + \dfrac{2L_p}{R\Phi}c\mathcal{R}T}{1 + \dfrac{2L_p}{R\Phi}}. \tag{1.63}$$

§§Diffuse growth is opposed to tip growth, characterized by a highly localized wall expansion at the tip of the cells. Tip growth is the mode of growth of invading cells like root hairs or pollen tubes and is discussed in Chapter 7.

Two limiting regimes can be considered depending on the value of the Lockhart's parameter: $\mathcal{L}=2L_p/R\Phi$. For $\mathcal{L}\gg1$ (small extensibility, large membrane permeability), eqn (1.63) becomes:

$$\frac{1}{L}\frac{\mathrm{d}L}{\mathrm{d}t}\approx\Phi(c\mathcal{R}T-P_Y)\quad\text{and}\quad P\approx c\mathcal{R}T. \tag{1.64}$$

This regime is called the extensibility-limited regime. In this case the growth rate of the cell is set by how fast the cell can extend its cell wall (embedded in the parameter Φ), not by its ability to take up water across the plasma membrane. The water potential balance is then almost satisfied and the turgor pressure is simply given by the osmotic pressure $c\mathcal{R}T$. Growth of most plant organs is assumed to operate in this extensibility-limited regime. From typical growth rate in young shoots $\dot{\varepsilon}\sim0.1-1\ \mathrm{h}^{-1}$ (see ref. 144) and using standard values $c\mathcal{R}T\sim0.5$ MPa, $P_Y\sim0.3$ MPa[31] and $R/2h\sim10$, we find that the cell wall "viscosity" is $1/\Phi_{\mathrm{wall}}\sim10$–$100$ GPas – a value close to the viscosity of ice![54]

For the opposite regime $\mathcal{L}\ll1$ (large extensibility, small membrane permeability), the Lockhart eqn (1.63) becomes:

$$\frac{1}{L}\frac{\mathrm{d}L}{\mathrm{d}t}\approx\frac{2L_p}{R}(c\mathcal{R}T-P_Y)\quad\text{and}\quad P\approx P_Y. \tag{1.65}$$

In this conductivity-limited regime, the extensibility of the cell is so large that the growth rate is limited, not by the cell-wall extensibility, but by the water uptake resistance set by the plasma membrane. The turgor pressure is then poised at the yield pressure P_Y, not at the osmotic pressure $c\mathcal{R}T$, meaning that a significant water potential gradient exists between the inside and outside of the cell.

1.2.4.3 *Extension of Lockhart's Model to Tissues*

Since the pioneering work of Molz and Boyer,[99] the Lockhart equations have been extended to multicellular tissues using cell-based models to couple water transport and growth.[26] A scaling analysis of the problem can also be made using the same kind of continuum approach as we used for poroelasticty (Section 2.3). Consider a tissue of size L growing at a constant rate $\dot{\varepsilon}$. Volume conservation requires a water flux to sustain growth, which is given by: $J=(1/S)\mathrm{d}V/\mathrm{d}t=\dot{\varepsilon}L$, where V is the volume of the growing tissue and S the cross-section perpendicular to the flow. From Darcy's law, this water flux must be induced by a turgor pressure gradient across the tissue (assuming uniform solute concentration) given by: $(\Delta P/L)=(\eta/k)J\sim(\eta/k)\dot{\varepsilon}L$, such that:

$$\dot{\varepsilon}\sim\frac{k}{\eta}\frac{\Delta P}{L^2}. \tag{1.66}$$

This relationship is the tissue version of the second Lockhart equation (1.62) on water transport. On the other hand, the force balance between the internal turgor pressure and the cell wall still holds at the tissue level, such that $\sigma \sim (R/h)P$. Using the Lockhart–Ortega rheology for the wall in steady state, we recover the same relationship as the first Lockhart's equation:

$$\dot{\varepsilon} \sim \Phi(P - P_Y). \tag{1.67}$$

This analysis shows that the gradient of turgor pressure induced by tissue growth is negligible if $\Delta P \ll P - P_Y$, which using eqn (1.66) gives:

$$\mathcal{L}_{\text{tissue}} = \frac{k}{\Phi \eta L^2} \gg 1. \tag{1.68}$$

The dimensionless number $\mathcal{L}_{\text{tissue}}$ is the tissue equivalent of the Lockhart number $\mathcal{L}$ defined at the cellular level. For $\mathcal{L}_{\text{tissue}} \gg 1$, the tissue growth rate is limited by cell-wall extensibility while for $\mathcal{L}_{\text{tissue}} \ll 1$, it is limited by water conductivity. Interestingly, this "tissue Lockhart number" depends not only on the mechanics and hydraulic properties of the tissue but also on the system size, with a strong size dependence $\mathcal{L}_{\text{tissue}} \propto L^{-2}$. It has long been assumed that growth in plants operates in the extensibility-limited regime and that growth-associated water potential gradients were small, with the exception of fast-growing tissues such as coleoptiles or roots.[16,17,30,31,99,148] However, recent measurements of turgor pressure at cellular resolution in the shoot apical meristem of *Arabidopsis* revealed high cell-to-cell pressure heterogeneity, which can only be predicted by taken into account water transport in addition to wall mechanics.[93] Water conductivity was also shown to strongly influence the emergence of lateral roots in *Arabidopsis*.[121] Therefore, the role of water conductivity in the control of growth and development might be more important than previously anticipated, even at small scales.

1.2.4.4 Differential Growth in Rod-like Organs: Plant Tropisms

In the previous section, we considered uniform and unidirectional growth. When growth varies spatially, kinematic constraints may lead to a change in growth direction and thus in organ shape. We consider here the important case of differential growth in a slender, rod-like body, such as a plant shoot or stem, which leads to the bending of the organ.[9,105,144]

To address this question, we model the shoot by a thin rod of length $L(t)$ and constant radius $R \ll L$, restricting its shape to a curve in two dimensions for simplicity. The spatio-temporal shape of the rod is then fully described by the angle $\theta(s,t)$ made by the rod with the vertical axis, where s is the arc length from the fixed base and t is time (Figure 1.10a). By definition, the local curvature of the rod is: $C(s,t) = -\partial\theta/\partial s$. Here the convention is such that curvature is positive when the shoot angle decreases away from the base, as in Figure 1.10a. The growth kinematics are characterized by the velocity field $v(s, t)$ of the different material points along the rod. A portion $\mathrm{d}s$ of the rod at position s and time t increases to a length $(1+\dot{\varepsilon}\,\mathrm{d}t)\mathrm{d}s$ at time $t+\mathrm{d}t$, where $\dot{\varepsilon}=\partial v/\partial s$ is the local growth rate. Similarly, the Lagrangian change of curvature during $\mathrm{d}t$, *i.e.* following the material displacement, is:

$$\frac{\mathrm{d}C}{\mathrm{d}t}=\frac{\partial C}{\partial t}+v(s,t)\frac{\partial C}{\partial s}. \tag{1.69}$$

We now assume that the growth rate not only varies along the rod but also across the thickness of the rod, from a value $\dot{\varepsilon}_2$ at the bottom side of the rod to a value $\dot{\varepsilon}_1<\dot{\varepsilon}_2$ at the upper side (Figure 1.10a). During $\mathrm{d}t$, the lower side length thus expands to $(1+\dot{\varepsilon}_2\mathrm{d}t)\mathrm{d}s$ while the upper side expands to $(1+\dot{\varepsilon}_1\mathrm{d}t)\mathrm{d}s$. As shown in Figure 1.10a, this differential expansion must induce a small change of rod curvature $\mathrm{d}C$. Geometry imposes that: $[(1+\dot{\varepsilon}_2\mathrm{d}t)\mathrm{d}s]/(r+R)=[(1+\dot{\varepsilon}_1\mathrm{d}t)\mathrm{d}s]/(r-R)$, where $r=(1/\mathrm{d}C)\gg R$. At first order, the curvature change induced by differential growth is then given by:

$$R\frac{\mathrm{d}C}{\mathrm{d}t}\approx\frac{\dot{\varepsilon}_2-\dot{\varepsilon}_1}{2}. \tag{1.70}$$

Combining this kinematic growth relation with the expression (1.69) of the Lagrangian derivative gives:

$$R\frac{\mathrm{d}C}{\mathrm{d}t}=R\left(\frac{\partial C}{\partial t}+v(s,t)\frac{\partial C}{\partial s}\right)=\dot{\varepsilon}\Delta, \quad \text{with } \Delta=\frac{\dot{\varepsilon}_2-\dot{\varepsilon}_1}{\dot{\varepsilon}_1+\dot{\varepsilon}_2} \text{ and } \dot{\varepsilon}=\frac{\dot{\varepsilon}_1+\dot{\varepsilon}_2}{2}. \tag{1.71}$$

This equation shows that change of curvature in a thin elongated organ is driven by differential growth, here expressed in terms of the relative growth asymmetry $\Delta=(\dot{\varepsilon}_2-\dot{\varepsilon}_1)/(\dot{\varepsilon}_2+\dot{\varepsilon}_1)$. For a given Δ, the rate of change of curvature is set by the mean growth rate $\dot{\varepsilon}=(\dot{\varepsilon}_2+\dot{\varepsilon}_1)/2=\partial v(s,t)\partial s$. Note that local change in curvature impacts the entire organ orientation because the relationship between angle and curvature is non-local:

$$\theta(s,t)=-\int_0^s C(s',t)\mathrm{d}s'+\theta_0. \tag{1.72}$$

where θ_0 is the angle of the shoot at the base.

Eqn (1.71) shows that maintaining a steady curved shape in a growing organ ($\partial C/\partial t=0$) requires a subtle spatio-temporal regulation of differential growth to satisfy: $v(s,t)R(\partial C/\partial s)=\dot{\varepsilon}(s,t)\Delta(s,t)$ – an idea first put forward by Wendy Silk (see ref. 145). Such steady growing shapes are found in many seedlings[145] and also in some compound leaves.[129] In these examples, the

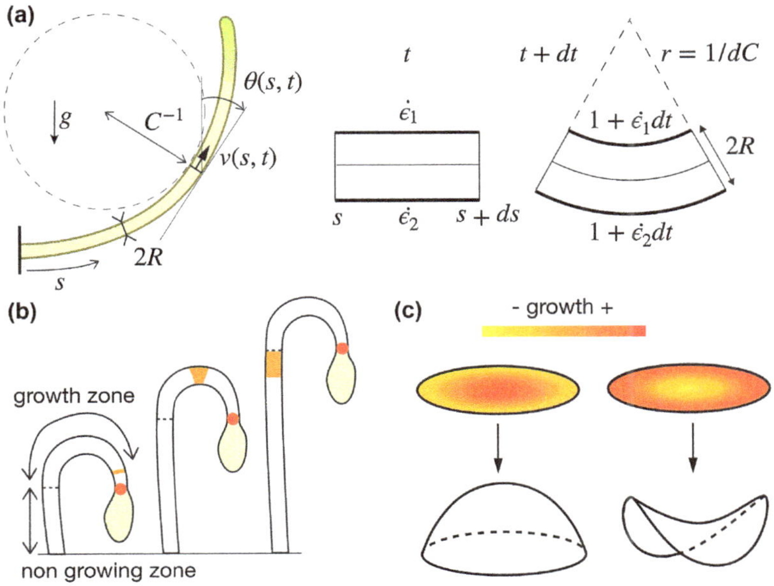

Figure 1.10 Differential growth in slender organs. (a) Kinematic description of a growing rod-like organ and bending induced by differential growth across the thickness. (b) Steady hook shape in a growing shoot, showing the growth and spatio-temporal path of a group of cells (orange region) produced at the apical meristem (red spot) and advected outside the growth zone. (c) Differential growth on a two-dimensional, leaf-like organ induces the generation of a positive (left) or a negative (right) Gauss curvature depending on the growth gradient.

end of the shoot or stem exhibits a hook shape that remains steady and at a fixed distance from the plant apex, all along the growth. Although the hook shape appears steady, each tissue element must follow a complex bending and unbending growth pattern as it moves from the tip, in order to comply with kinematics (see Figure 1.10b).

Bending induced by differential growth is also the basis of most tropic movements in plants,[58] *i.e.* the directed motion in response to external stimuli such as light (phototropism), touch (thigmotropism) or gravity (gravitropism), which we discuss below. Plants detect gravity using specialized cells (called statocytes), in which starch-rich particles (called amyloplasts or statoliths) sediment under gravity to form miniature "granular" piles at the bottom of the cell[101,103,109,158] (see Figure 1.11 and Box 1.5). When the cell is inclined, the statoliths move toward the side of the cell even at a small inclination, unlike sand grains, because statoliths are fluidized by the cytoskeleton's activity (see ref. 13 and Box 1.5). The resulting asymmetric position of statoliths within the cell triggers a complex molecular pathway, leading to the generation of a growth-hormone gradient across the thickness (the growth hormone in plants is called auxin) and thus to differential growth and organ curvature.

The gravitropic response of plant shoots is inclination-dependent and follows a sine-like shape, such that:

$$\Delta_{\text{gravity}} = \left(\frac{\dot{\varepsilon}_2 - \dot{\varepsilon}_1}{\dot{\varepsilon}_1 - \dot{\varepsilon}_2}\right)_{\text{gravity}} = \beta \sin\theta, \tag{1.73}$$

where β is a dimensionless parameter characterizing the gravitropic sensitivity of the plant. This sine-law has classically been interpreted as the consequence of a force-sensing mechanism at the cellular level, the projection of the weight of the statoliths along the side of the cell being proportional to $\sin\theta$. However, recent experiments have shown that the response is independent of the gravity intensity¶¶, suggesting that the gravitropic stimulus is the position of the statoliths within the cell, not their weight.[24,124]

Surprisingly, when the expression (1.74) of the gravitropic response is introduced in eqn (1.71)–(1.72) (without the advective term) to predict the gravitropic bending of an inclined stem, an unrealistic spatio-temporal dynamics is found. Instead of converging toward the vertical, the tip of the stem keeps oscillating back and forth, with regions of increasing curvature accumulating close to the base of the stem.[8] This paradox was solved by adding to the gravitropic law (1.74) an additional term proportional to the local curvature, which counteracts the generation of curvature induced by gravity sensing:[8]

$$\Delta_{\text{prop}} = \left(\frac{\dot{\varepsilon}_2 - \dot{\varepsilon}_1}{\dot{\varepsilon}_1 + \dot{\varepsilon}_2}\right)_{\text{prop}} = -\gamma RC, \tag{1.74}$$

where γ is a dimensionless parameter. Such tendency of a stem or a shoot to perceive its own curvature and straighten was coined "autotropism" in early studies (see ref. 21 and 48 and older references therein) and latterly "proprioception", by analogy with the sense of body perception in vertebrates.[8] In plants, the mechanism of proprioception is still debated[104,116] but its reality is well supported, in particular by experiments in microgravity environments which show that an initially curved stem spontaneously straightens during growth. Writing for simplicity $\sin\theta \approx \theta$ and neglecting the advection term in eqn (1.71) gives for the evolution of the curvature:

$$R\frac{\partial C}{\partial t} = \dot{\varepsilon}(\beta\theta - \gamma RC). \tag{1.75}$$

¶¶This observation is valid as long as gravity intensity is high enough. In very low gravity environments (typically $g < 10^{-2}$ m s^{-2}), the avalanche time of the statoliths may become longer than the growth timescale $\dot{\varepsilon}^{-1}$ so that statoliths have not the time to reach their equilibrium position before the plant bends. The gravitropic response then decreases and becomes null without gravity, as expected.

This mathematical model of gravitropism, which includes both graviperception and proprioception, was first introduced by Bastien *et al.*[8] and coined the "AC model" (for Angle–Curvature). Unlike the purely gravitropic model, the AC model exhibits a steady solution ($\partial_t C = 0$) given by $\theta(s) = \theta_0 \exp(-s/L_c)$, where θ_0 is the inclination of the stem at the base and $L_c = \gamma R/\beta$ is a bending length scale. This steady shape is controlled by the balance number $\mathcal{B}$:

$$\mathcal{B} = \frac{L}{L_c} = \frac{L\beta}{\gamma R} \approx \mathcal{O}\left(\frac{\Delta_{\text{gravity}}}{\Delta_{\text{prop}}}\right), \tag{1.76}$$

which quantifies the ratio of the gravitropic term to the proprioceptive term. For $\mathcal{B} \gg 1$ ($L_c \ll L$), graviperception is large compare to proprioception and the plant rapidly recovers the vertical after a bending length L_c. In the opposite situation $\mathcal{B} \ll 1$ ($L_c \gg L$), proprioception overcomes graviperception and the plant does not recover the vertical over its length L. Note that the balance number $\mathcal{B}$ not only controls the final shape of the plant but also its dynamics toward vertical: for large $\mathcal{B}$, the tip of the stem oscillates several times around the vertical before converging, which is not the case for small $\mathcal{B}$.[8]

The previous discussion shows that the perception of gravity alone through the sine law (1.74) does not enable a proper posture control in plants, and that an additional straightening or proprioceptive mechanism is needed. The AC model constitutes the building block from which additional effects may be considered. For instance, the gravitropic response exhibits several timescales not included in the simple sine law, such as a delay time before the beginning of bending and a memory time that filters rapid changes in inclination.[23] The purely kinematic model presented here can also be extended to add mechanical effects, such as elasticity and stem sagging under its own weight.[27] In this case, the actual (or observed) curvature C contains an elastic contribution due to the self-weight. It must thus be distinguished from the natural[‖‖] curvature C^0 induced by differential growth and described by eqn (1.71). Finally, other kinds of tropisms and plant movements may be included in this framework, such as phototropism, thigmotropism or circumnutation – the spontaneous oscillation and circular movements of the tip of stems.[1,10,106] Extension to three-dimensional growing rods, including torsion and helicity, are also available (see ref. 106 and references therein).

1.2.4.5 *Growth in Thin Sheets and Morphogenesis*

We have seen how differential growth across the thickness of a rod-shaped organ like a stem induces a change in natural curvature. When the organ

‖‖The natural curvature is also called the intrinsic, or the spontaneous, curvature.

Box 1.5 Gravity sensing in plants: an active granular flow problem?

From tiny shoots to large trees, all plants are able to sense gravity and reorient their growth toward the vertical direction set by the gravitational field. This ability is important at early stages of development for roots to anchor in the soil and shoots to find light. It is also key throughout the plant's life, for the plant to maintain its upright position and not fall against its own weight. The detection of gravity in plants is known to originate in specialized cells (statocytes) containing dense starch grains (amyloplasts or statoliths); but how cells detect the statoliths and how this sensing is converted into a bending growth response at the organ level are still largely open questions (see, for example, the reviews[101,103,105,109,158]).

Recent experiments studying the gravitropic response of shoots to hyper- and hypo-gravity conditions have shown that plants are not sensitive to the intensity of the gravity field, but only to the inclination against the direction of the gravity vector.[24,124] The gravisensor in plants is thus a position sensor, not a force sensor. This finding is surprising because it implies that the pile of statoliths at the bottom of the cell move and respond to even the tiniest tilt. At first sight, such a behavior contradicts our knowledge of the physics of granular media, which stipulates that an assembly of grains cannot move below a critical avalanche angle set by friction and steric constrains between particles.[3]

The solution to this conundrum comes from the *in-situ* visualization of statolith movement in response to a large cell tilt[13] (Figure 1.11b). Initially, statoliths flow in bulk like a granular "avalanche", with a pile angle that rapidly relaxes toward a critical angle θ_c, as expected for a classical granular medium. However, over a long time the behavior strongly contrasts with that of a classical granular medium. Instead of being stuck at θ_c, the statolith pile keeps evolving and slowly creeps until its free surface recovers the horizontal, as a liquid would do. Investigation of statolith motion at the particle level reveals that this liquid-like behavior comes from the agitation of the statoliths, which helps the grains to unjam and flow even for very small inclinations. This agitation comes from cellular activity and not thermal agitation, as confirmed by comparison with the behavior of passive Brownian particles in the same geometry. It is likely to involve the dynamics of the actin–myosin network in the cytoskeleton, and its interaction with statoliths.

The remarkable sensitivity of plants to gravity therefore relies on an active granular material at the cellular level.[13] This strategy, which combines local active noise and signal integration of statolith position, shares similarities with other biological sensors, such as hair cells[108] or tactile

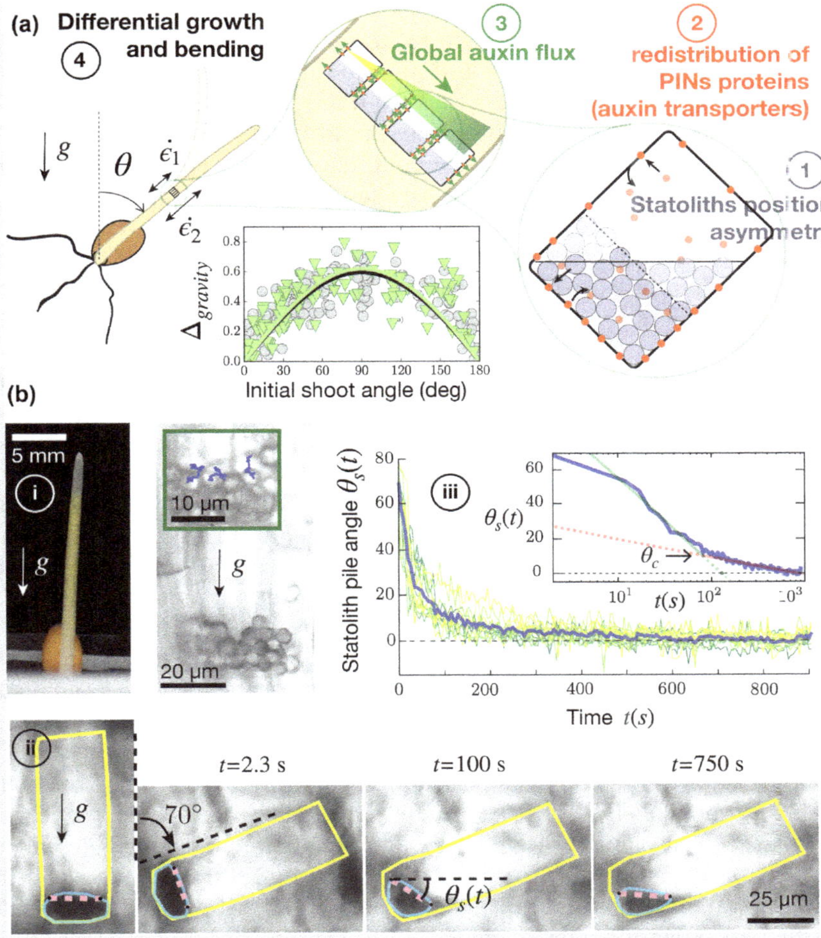

Figure 1.11 (a) Signaling pathway of plant gravitropism (adapted from ref. 90). The plot shows the gravitropic response of wheat coleoptiles to steady inclination[24] (sine law). (b) Flowing behavior of the plant gravisensors (statoliths). (i) Gravisensing cells of wheat coleoptile. (ii) In-situ visualization of statolith motion in response to a large tilt. The angle of the free-surface of the statolith pile (θ_s) first decreases rapidly toward a critical angle θ_c, then slowly creeps in order to recover the horizontal, as a liquid would do (iii, plot in lin-lin and log-lin scale). This pseudoliquid behavior comes from the agitation of statoliths by cell activity (blue trajectories in top-left image). Adapted from ref. 13, https://doi.org/10.1073/pnas.1801895115, under the terms of the CC BY 4.0 licence, https://creativecommons.org/licenses/by/4.0/.

whiskers.[28] Understanding the physics of such active particulate media is an exciting soft matter topic in itself, in addition to the biological motivations.

remains thin but extends in two directions, such as in a flat leaf, two different types of differential growth-induced shapes can occur***.

The first mode of deformation is similar to the one we just saw in the case of a one-dimensional stem and comes from a differential growth across the *thickness* of the sheet. In this situation, a natural curvature††† C^0 in the direction of differential growth is generated, causing the sheet to curl into a cylindrical shape. Many plant leaves, for example grass blades or corn leaves, exhibit such rolled shapes induced by a differential expansion across the leaf thickness. However, when the leaf is curved in two perpendicular directions like a shell, bending in one direction is coupled to the other direction through the generation of in-plane tissue stresses, causing subtle rolling/unrolling mechanisms.[102]

The second type of deformation is associated with differential *surface* growth: different regions of the sheet grow at different rates (Figure 1.10c). In general, these spatial variations in growth cannot happen without generating in-plane compressive stresses. For a thin sheet of lateral size L and thickness $h \ll L$, the energy cost of such compressive deformation is very high and the system prefers to bulge out of the plane. A non-zero natural Gauss curvature κ_G^0 is then created‡‡‡. When the growth is larger in the center of the sheet than at the edges, the generated Gauss curvature is positive $\kappa_G^0 > 0$, *i.e.* the sheet deforms into a dome shape. In contrast, when the edges grow faster than the center, negative Gauss curvature $\kappa_G^0 < 0$ is created, resulting in rippling edges and saddle-like shapes. Many flat organs in plants and algae, such as leaves, blades or flowers, show this type of rippling at their edges, induced by faster growth at their periphery.[29,83,142,143]

Overall, geometry and mechanics, in addition to genetics, play a primary role in morphogenesis and development – an idea promoted by pioneering scientists such as d'Arcy Thompson[156] and Paul Green.[61] This role can be purely "passive", as when differential growth generates out-of-plane bending to satisfy geometric compatibility. It can also be "active" and directly influence the molecular and genetic machinery of the plant, as we have seen in the case of the cell wall, where the stress in the wall controls the orientation of the cortical microtubules, which in turn affects wall anisotropy[63] (Figure 1.9b). Understanding such feedback between mechanical signals and biology is one of the main current challenges in plant morphogenesis.[139]

***For general textbooks on the elasticity of plates and shells, see for example[5,95,157] A mechanical description of thin growing sheets in the context of differential geometry can be found in ref. 44.

†††The natural (or intrinsic, or spontaneous) curvature is the curvature that remains when all external forces, geometrical frustrations or in-plane stresses are released. Experimentally, the natural curvature along a given direction in a plate or shell is found by cutting a thin strip in this direction and measuring the resulting curvature.

‡‡‡The Gauss curvature of a surface is defined by $\kappa_G = C_1 \times C_2$, where C_1 and C_2 are the two principal curvatures of the surface (*i.e.* the two extrema curvatures at the surface point considered). In practice, the natural Gauss curvature is found by cutting small discs onto the surface and measuring the two principal curvatures.

1.2.5 Mechanical Instabilities and Fast Movements

Throughout the chapter we have seen how plants, although lacking muscle, use gradients of water potential to transport water and perform various swelling or growth movements. The timescale of these water-driven movements is, however, constrained by a physical limit: for a cell or a tissue to swell, water must be transported from one place to another within the soft, porous plant material. We have seen that the shortest response time of this process is set by the cell relaxation time at the cellular level, and by the poroelastic time at the tissue level (Section 1.2.3). These timescales depend on the hydraulic and elastic properties of the medium and, more importantly, on the size of the system: the larger the organ, the greater the time required to produce purely hydraulic motion.

Plants have developed a simple and elegant strategy to overcome this hydraulic limit: the use of mechanical instabilities.[50,52] The general principle is as follows: during a first "slow" phase, elastic deformation is stored in the cell wall due to a slow modification of the hydraulic or elastic properties (*e.g.* a water movement due to evaporation or active solute transport, a change of the mechanical properties of the wall, *etc*). However, this elastic energy is not released because there exists an "energy barrier" in the system. When the stored elastic energy reaches a critical point, the energy barrier is crossed and a second "fast" phase occurs where the deformation in the wall is rapidly released and converted into a fast movement.

The existence of an energy barrier is the key ingredient for developing a mechanical instability. This barrier can be of molecular origin, as for bubble nucleation in a liquid under negative pressure or crack nucleation in a solid under tension. In this case, the cohesion between molecules is associated with an energy cost to create a new surface – the bubble or crack – in the medium. When the size of the bubble (or length of the crack) is large enough, the release of bulk energy is larger than the cost of surface energy; the bubble (or crack) then expands suddenly. Fracture propagation and its associated explosive elastic energy release is used by many plants to disperse their seeds or spores.[37,46,74,76,133] The cavitation of bubbles in a liquid under tension, so harmful for trees, has been harnessed by some species of ferns to catapult their spores and thereby generate the fastest movement in the plant kingdom.[81,113] The rich physics of cavitation in the context of plants is discussed in Chapter 4.

Mechanisms based on fracture propagation or bubble cavitation are one-shot, *irreversible* fast movements: once the stretched tissue is torn or the metastable liquid transformed into vapor, no resetting of motion is possible. A more flexible strategy to actuate *reversible* rapid movements consists of using an elastic or *snap-buckling instability*. A snap-buckling instability is the discontinuous transition between two states of minimal elastic energy in a slender body. In this case, the energy barrier comes from the existence of a geometrical constraint: the system cannot transit from one state to the other without passing through an intermediate shape that generates in-plane stresses. We have seen that for thin bodies these in-plane stresses are associated with an important elastic energy cost. Elastic energy is then accumulated until it is so large that the barrier is crossed and the shape suddenly snaps through.

This principle is used by the carnivorous plant *Dionaea muscipula* (best known as the Venus flytrap) to trap insects[53,123,136] (Figure 1.12a). The trap of the Venus flytrap consists of two doubly curved, shell-like lobes, which are convex (curved outward) in the initial state. Because bending such a shell requires crossing a stretching energy barrier (see below), the trap can store elastic energy without closing. However, when the inner hairs are triggered, the plant actively changes the natural curvature of its lobes in one direction. This is enough to cross the elastic energy barrier, leading to a sudden change of curvature and closure.

It is possible to quantify this mechanism using a minimal one-dimensional elastic energy model and scaling arguments.[53] Let consider a shell of typical size L, thickness h and Young's modulus E. From Hooke's law, the total elastic energy of the shell scales as: $\mathcal{E}_{\rm el} \sim \frac{1}{2} E V \varepsilon^2$, where $V \sim hL^2$ is the volume of the shell and ε is the characteristic elastic strain associated to the deformation of the shell. When the shell is bent, two different modes of elastic deformation occur.[87] The first one is associated to pure bending, *i.e.* the fact that changing the curvature of a plate with respect to its natural curvature induces a gradient of elastic deformation within the *thickness* of the shell. The elastic strain associated to this bending mode is: $\varepsilon_{\rm bend} \sim h(C - C^0)$, where C is the actual curvature and C^0 is the natural curvature. The corresponding bending elastic energy is thus:

$$\mathcal{E}_{\rm bend} \sim E h^3 L^2 (C - C^0)^2. \tag{1.77}$$

The second mode of deformation is related to geometrical incompatibility: it is not possible to bend and reverse the curvature of a shell without stretching or compressing its surface. This is a direct consequence of the Gauss' Theorema egregium,[5] which implies that a surface cannot change its Gauss curvature without changing the natural length, or metric, of its surface. To estimate the stretching deformation $\varepsilon_{\rm stretch}$ induced by such a change of Gauss curvature, let consider a shallow shell of radius of curvature $r \gg L$, which is deformed into a flat disc (Fig. 1.12b). Assuming that the length L along the diameter of the shell is conserved, the shell's perimeter must increase from πL_0, where L_0 is the initial projected diameter, to πL, so that $\varepsilon_{\rm stretch} \sim (L - L_0)/L_0$. Pythagoras's theorem implies that $\Delta^2 + (L_0^2/4) = L^2/4$ and $(r - \Delta)^2 + (L_0^2/4) = r^2$. Since $\Delta \ll r$ and $\Delta \ll L$, we have at first order: $\varepsilon_{\rm stretch} \sim 2\Delta^2/L^2 \sim L^2/(32r^2) \propto L^2 \kappa_G^0$, where $\kappa_G^0 = 1/r^2$ corresponds to the natural Gauss curvature of the shell. Here the Gauss curvature of the deformed state is zero since the disc is flat. This relationship can be generalized to a deformed state of arbitrary non-zero Gauss curvature as: $\varepsilon_{\rm stretch} \sim L^2(\kappa - \kappa_G^0)$, where $\kappa_G = C^2$ is the Gauss curvature of the deformed state. Therefore, the elastic energy associated to the stretching mode is given by:

$$\mathcal{E}_{\rm stretch} \sim E h L^6 (\kappa_G - \kappa_G^0)^2 \sim E h L^6 (C^2 - \kappa_G^0)^2. \tag{1.78}$$

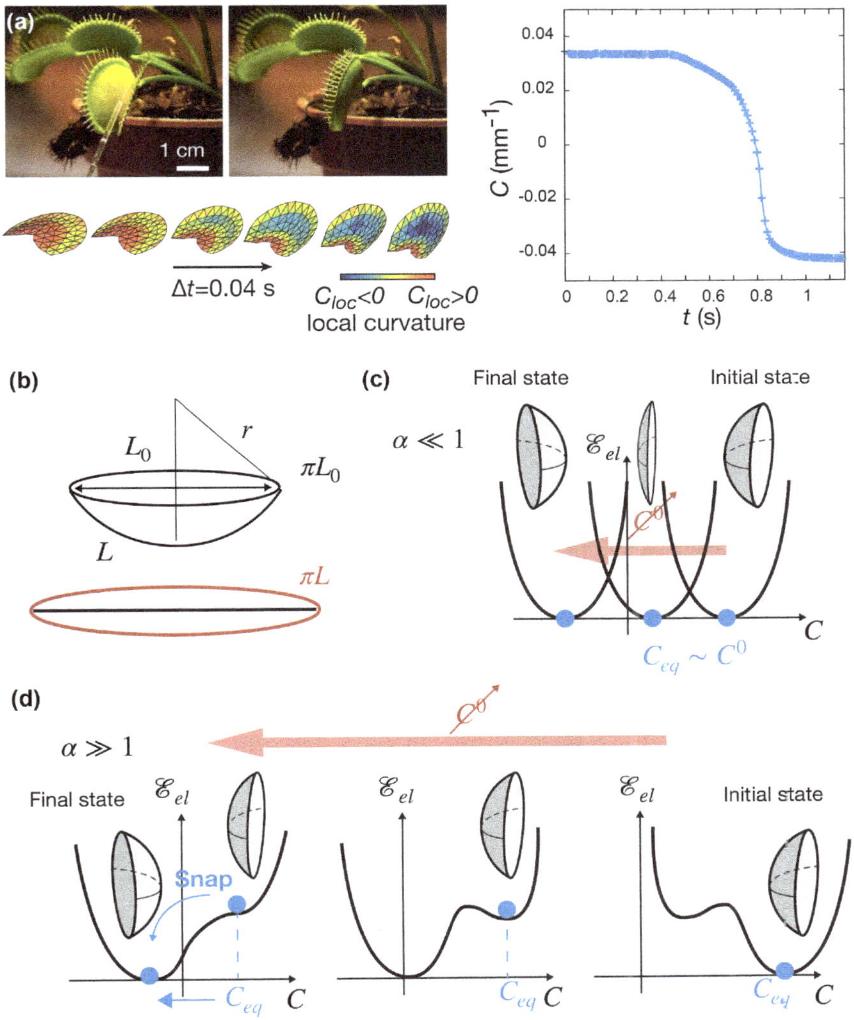

Figure 1.12 Rapid plant movements induced by a snap-buckling instability. (a) The carnivorous plant Venus flytrap (*Dionaea muscipula*) in the open (left image) and closed (right image) states. Bottom panel: three-dimensional shape reconstruction of the lobe during closure. Left panel: spatially averaged mean curvature of one lobe as function of time, showing a sudden transition from convex ($C>0$) to concave ($C<0$) (adapted from ref. 53). (b) Sketch of a shallow shell stretched into a flat disc, in which the shell's diameter L is conserved but the perimeter increases. (c) Elastic energy landscape for a thin shell when the bending energy dominates ($\alpha \ll 1$), showing a smooth transition from convex to concave as the natural curvature C^0 changes. C_{eq} is the equilibrium curvature that minimizes the energy. (d) When the stretching energy dominates ($\alpha > 1$), the energy landscape exhibits two local minima and the equilibrium shape suddenly snaps from convex to concave.

The total elastic energy of the shell is the sum of the bending energy (1.77) and stretching energy (1.78): $\mathcal{E}_{\mathrm{el}} = \mathcal{E}_{\mathrm{bend}} + \mathcal{E}_{\mathrm{stretch}}$. It is convenient to make the curvatures dimensionless using the characteristic curvature $\sqrt{\kappa_G^0}$, such that: $\tilde{C} = C/\sqrt{\kappa_G^0}$, $\tilde{C}^0 = C^0/\sqrt{\kappa_G^0}$. Using eqn (1.77)–(1.78), the dimensionless total elastic energy of the shell $\tilde{\mathcal{E}}_{\mathrm{el}} = \mathcal{E}_{\mathrm{el}}/(EV(\kappa_G^0 h)^2)$ is then:

$$\tilde{\mathcal{E}}_{\mathrm{el}} \sim \left(\tilde{C} - \tilde{C}^0\right)^2 + \alpha\left(\tilde{C}^2 - 1\right)^2,$$

with

$$\alpha \sim \frac{L^4 \kappa_G^0}{h^2} = \mathcal{O}\left(\frac{\mathcal{E}_{\mathrm{stretching}}}{\mathcal{E}_{\mathrm{bending}}}\right). \tag{1.80}$$

The parameter α quantifies the ratio of the stretching elastic energy to the bending elastic energy and depends only on the geometry of the shell: the thinner, wider or curved the shell, the greater the stretching energy compared to the bending energy. We see below that α determines how the shell flips from convex to concave when the natural curvature C^0 is gradually changed. To this end, we note that, for a given value of $\tilde{C}^0$ and α, the equilibrium shape of the shell $\tilde{C}_{\mathrm{eq}}$ is obtained by minimizing the total elastic energy with respect to $\tilde{C}$:

$$\left(\frac{\partial \tilde{\mathcal{E}}_{\mathrm{el}}}{\partial \tilde{C}}\right)_{\tilde{C} = \tilde{C}_{\mathrm{eq}}} = 0. \tag{1.81}$$

Figure 1.12c,d show qualitatively the shape of the total elastic energy as function of $\tilde{C}$ when the natural curvature of the shell $\tilde{C}^0$ goes continuously from convex ($\tilde{C}^0 > 0$) to concave ($\tilde{C}^0 < 0$). For $\alpha \ll 1$, the bending energy dominates: the elastic energy exhibits only one minimum and the curvature of the shell at equilibrium is always very close to the natural curvature $C_{\mathrm{eq}} \approx C^0$. Therefore, when the natural curvature changes from positive to negative, the shell shape follows the natural curvature and smoothly changes from convex to concave (Figure 1.12c). The situation is markedly different for $\alpha \gtrsim 1$ (Figure 1.12d). In this case, the stretching energy is large and the total elastic energy (1.79) may exhibit two local minima separated by an energy barrier, depending on the value of the natural curvature. As shown in Fig. 1.12d, when the sign of the natural curvature changes, the system initially remains stuck in the first local minimum of convex shape, until the energy barrier disappears and the shell suddenly flips to the second minimum of energy, corresponding to the concave shape. Therefore, a small change of natural curvature (the internal motor of the motion) may lead to a very large change of actual curvature from convex to concave. This is the amplification mechanism used by the Venus flytrap and other carnivorous plants to speed up their motion and engulf their preys.[53,162]

The active process by which plants like the Venus flytrap trigger the instability and overcome the energy barrier is still not fully elucidated, raising interesting questions about signaling and fast mechanical actuation at the molecular level.[50,66,154] However, the general principle of using an elastic instability to overcome the hydraulic limit and amplify the speed of motion is robust and weakly depends on these microscopic details.[146] This strategy has already inspired several artificial devices, such as fast soft actuators and jumping robots.[73,79,94,168] More generally, the non-muscular movements of plants offer a wealth of mechanisms worth studying from a physics and engineering perspective.[19,22,140] Biomimetic applications of plant movements are further discussed in Chapter 8.

Acknowledgements

I would like to thank the whole scientific community of Plant Biomech' and GDR PhyP ("Plant Biophysics and biomechanics", CNRS no 2007), from whom I learned most of the ideas and concepts discussed in this chapter. I particularly thank Jacques Dumais and Bruno Moulia who introduced me to the fascinating field of plant biomechanics almost 20 years ago. This work was supported by the European Research Council (ERC) under the European Union's Horizon 2020 research and innovation program (Grant 647384).

References

1. D. Agostinelli, A. Lucantonio, G. Noselli and A. DeSimone, Nutations in growing plant shoots: The role of elastic deformations due to gravity loading, *J. Mech. Phys. Solids*, 2020, **136**, 103702.
2. O. Ali and J. Traas, Force-driven polymerization and turgor-induced wall expansion, *Trends Plant Sci.*, 2016, **21**(5), 398–409.
3. B. Andreotti, Y. Forterre and O. Pouliquen, *Granular Media: Between Fluid and Solid*, Cambridge University Press, 2013.
4. S. Armon, E. Efrati, R. Kupferman and E. Sharon, Geometry and mechanics in the opening of chiral seed pods, *Science*, 2011, **333**(6050), 1726–1730.
5. B. Audoly and Y. Pomeau. *Elasticity and Geometry: From Hair Curls to the Non-linear Response of Shells*, Oxford university press, 2010.
6. N. J. Balmforth, I. A. Frigaard and G. Ovarlez, Yielding to stress: recent developments in viscoplastic fluid mechanics, *Annu. Rev. Fluid Mech.*, 2014, **46**, 121–146.
7. A. Barbacci, M. Lahaye and V. Magnenet, Another brick in the cell wall: biosynthesis dependent growth model, *PLoS One*, 2013, **8**(9), e74400.
8. R. Bastien, T. Bohr, B. Moulia and S. Douady, Unifying model of shoot gravitropism reveals proprioception as a central feature of posture control in plants, *Proc. Natl. Acad. Sci.*, 2013, **110**(2), 755–760.
9. R. Bastien, S. Douady and B. Moulia, A unifying modeling of plant shoot gravitropism with an explicit account of the effects of growth, *Front. Plant Sci.*, 2014, **5**, 136.

10. R. Bastien, S. Douady and B. Moulia, A unified model of shoot tropism in plants: photo-, gravi-and propio-ception, *PLoS Comput. Biol.*, 2015, **11**(2), e1004037.
11. L. Beauzamy, J. Derr and A. Boudaoud, Quantifying hydrostatic pressure in plant cells by using indentation with an atomic force microscope, *Biophys. J.*, 2015, **108**(10), 2448–2456.
12. T. Bertrand, J. Peixinho, S. Mukhopadhyay and C. W. MacMinn, Dynamics of swelling and drying in a spherical gel, *Phys. Rev. Appl.*, 2016, **6**(6), 064010.
13. A. Bérut, H. Chauvet, V. Legué, B. Moulia, O. Pouliquen and Y. Forterre, Gravisensors in plant cells behave like an active granular liquid, *Proc. Natl. Acad. Sci.*, 2018, **115**(20), 5123–5128.
14. A. Boudaoud, An introduction to the mechanics of morphogenesis for plant biologists, *Trends Plant Sci.*, 2010, **15**(6), 353–360.
15. F. Boudon, J. Chopard, O. Ali, B. Gilles, O. Hamant, A. Boudaoud, J. Traas and C. Godin, A computational framework for 3d mechanical modeling of plant morphogenesis with cellular resolution, *PLoS Comput. Biol.*, 2015, **11**(1), e1003950.
16. J. S. Boyer, A. Cavalieri and E.-D. Schulze, Control of the rate of cell enlargement: excision, wall relaxation, and growth-induced water potentials, *Planta*, 1985, **163**(4), 527–543.
17. J. S. Boyer and W. K. Silk, Hydraulics of plant growth, *Funct. Plant Biol.*, 2004, **31**(8), 761–773.
18. B. Bozorg, P. Krupinski and H. Jönsson, A continuous growth model for plant tissue, *Phys. Biol.*, 2016, **13**(6), 065002.
19. I. Burgert and P. Fratzl, Actuation systems in plants as prototypes for bioinspired devices, *Philos. Trans. R. Soc., A*, 2009, **367**(1893), 1541–1557.
20. N. Carpita, D. Sabularse, D. Montezinos and D. P. Delmer, Determination of the pore size of cell walls of living plant cells, *Science*, 1979, **205**(4411), 1144–1147.
21. D. K. Chapman, A. Johnsson, C. Karlsson, A. Brown and D. Heathcote, Gravitropically-stimulated seedlings show autotropism in weightlessness, *Physiol. Plant.*, 1994, **90**(1), 157–162.
22. V. Charpentier, P. Hannequart, S. Adriaenssens, O. Baverel, E. Viglino and S. Eisenman, Kinematic amplification strategies in plants and engineering, *Smart Mater. Struct.*, 2017, **26**(6), 063002.
23. H. Chauvet, B. Moulia, V. Legué, Y. Forterre and O. Pouliquen, Revealing the hierarchy of processes and time-scales that control the tropic response of shoots to gravi-stimulations, *J. Exp. Bot.*, 2019, **70**(6), 1955–1967.
24. H. Chauvet, O. Pouliquen, Y. Forterre, V. Legué and B. Moulia, Inclination not force is sensed by plants during shoot gravitropism, *Sci. Rep.*, 2016, **6**(1), 1–8.
25. Y. Chebli and A. Geitmann, Cellular growth in plants requires regulation of cell wall biochemistry, *Curr. Opin. Cell Biol.*, 2017, **44**, 28–35.

26. I. Cheddadi, M. Génard, N. Bertin and C. Godin, Coupling water fluxes with cell wall mechanics in a multicellular model of plant development, *PLoS Comput. Biol.*, 2019, **15**(6), e1007121.
27. R. Chelakkot and L. Mahadevan, On the growth and form of shoots, *J. R. Soc., Interface*, 2017, **14**(128), 20170001.
28. L. N. Claverie, Y. Boubenec, G. Debrégeas, A. M. Prevost and E. Wandersman, Whisker contact detection of rodents based on slow and fast mechanical inputs, *Front. Behav. Neurosci.*, 2017, **10**, 251.
29. E. Coen, A.-G. Rolland-Lagan, M. Matthews, J. A. Bangham and P. Prusinkiewicz, The genetics of geometry, *Proc. Natl. Acad. Sci.*, 2004, **101**(14), 4728–4735.
30. D. Cosgrove and E. Steudle, Water relations of growing pea epicotyl segments, *Planta*, 1981, **153**(4), 343–350.
31. D. J. Cosgrove, Cell wall yield properties of growing tissue: evaluation by in vivo stress relaxation, *Plant Physiol.*, 1985, **78**(2), 347–356.
32. D. J. Cosgrove, Growth of the plant cell wall, *Nat. Rev. Mol. Cell Biol.*, 2005, **6**(11), 850–861.
33. D. J. Cosgrove, Nanoscale structure, mechanics and growth of epidermal cell walls, *Curr. Opin. Plant Biol.*, 2018, **46**, 77–86.
34. O. Coussy, *Poromechanics*, John Wiley & Sons, 2004.
35. J. Dainty, Water relations of plant cells, *Advances in Botanical Research*, Elsevier, 1963, vol. 1, pp. 279–326.
36. C. Dawson, J. F. Vincent and A.-M. Rocca, How pine cones open, *Nature*, 1997, **390**(6661), 668.
37. R. D. Deegan, Finessing the fracture energy barrier in ballistic seed dispersal, *Proc. Natl. Acad. Sci.*, 2012, **109**(14), 5166–5169.
38. H. H. Dixon and J. Joly, *Xii. on the ascent of sap. Philos. Trans. R. Soc., B*, 1895, **186**, 563–576.
39. J. Dumais and Y. Forterre, "vegetable dynamicks": the role of water in plant movements, *Annu. Rev. Fluid Mech.*, 2012, **44**, 453–478.
40. J. Dumais, S. L. Shaw, C. R. Steele, S. R. Long and P. M. Ray, An anisotropic-viscoplastic model of plant cell morphogenesis by tip growth, *Int. J. Dev. Biol.*, 2006, **50**, 209–222.
41. P. Durand-Smet, E. Gauquelin, N. Chastrette, A. Boudaoud and A. Asnacios, Estimation of turgor pressure through comparison between single plant cell and pressurized shell mechanics, *Phys. Biol.*, 2017, **14**(5), 055002.
42. R. Dyson, L. Band and O. Jensen, A model of crosslink kinetics in the expanding plant cell wall: yield stress and enzyme action, *J. Theor. Biol.*, 2012, **307**, 125–136.
43. R. Dyson and O. Jensen, A fibre-reinforced fluid model of anisotropic plant cell growth, *J. Fluid Mech.*, 2010, **655**, 472.
44. E. Efrati, E. Sharon and R. Kupferman, The metric description of elasticity in residually stressed soft materials, *Soft Matter*, 2013, **9**(34), 8187–8197.

45. R. Elbaum, L. Zaltzman, I. Burgert and P. Fratzl, The role of wheat awns in the seed dispersal unit, *Science*, 2007, **316**(5826), 884–886.
46. D. Evangelista, S. Hotton and J. Dumais, The mechanics of explosive dispersal and self-burial in the seeds of the filaree, erodium cicutarium (geraniaceae), *J. Exp. Biol.*, 2011, **214**(4), 521–529.
47. A. Finkelstein, *Water Movement Through Lipid Bilayers, Pores, and Plasma Membranes: Theory and Reality*, John Wiley and Sons, 1987.
48. R. Firn and J. Digby, A study of the autotropic straightening reaction of a shoot previously curved during geotropism, *Plant, Cell Environ.*, 1979, **2**(2), 149–154.
49. A. Fleischer, M. A. O'Neill and R. Ehwald, The pore size of non-graminaceous plant cell walls is rapidly decreased by borate ester cross-linking of the pectic polysaccharide rhamnogalacturonan ii, *Plant Physiol.*, 1999, **121**(3), 829–838.
50. Y. Forterre, Slow, fast and furious: understanding the physics of plant movements, *J. Exp. Bot.*, 2013, **64**(15), 4745–4760.
51. Y. Forterre and J. Dumais, Generating helices in nature, *Science*, 2011, **333**(6050), 1715–1716.
52. Y. Forterre, P. Marmottant, C. Quilliet and X. Noblin, Physics of rapid movements in plants, *Europhys. News*, 2016, **47**(1), 27–30.
53. Y. Forterre, J. M. Skotheim, J. Dumais and L. Mahadevan, How the venus flytrap snaps, *Nature*, 2005, **433**(7024), 421–425.
54. A. Fowler, Glaciers and ice sheets, *The Mathematics of Models for Climatology and Environment*, Springer, 1997, pp. 301–336.
55. P. J. Franks, T. N. Buckley, J. C. Shope and K. A. Mott, Guard cell volume and pressure measured concurrently by confocal microscopy and the cell pressure probe, *Plant Physiol.*, 2001, **125**(4), 1577–1584.
56. A. Geitmann, Experimental approaches used to quantify physical parameters at cellular and subcellular levels, *Am. J. Bot.*, 2006, **93**(10), 1380–1390.
57. L. Gibson and M. Ashby, *Cellular Solids: Structure and Properties*, cambridge univ press, *Cambridge, UK*, 1999.
58. S. Gilroy and P. Masson. *Plant Tropisms*, Blackwell Publishing, Oxford, UK, 2008.
59. S. Gravelle, L. Joly, F. Detcheverry, C. Ybert, C. Cottin-Bizonne and L. Bocquet. Optimizing water permeability through the hourglass shape of aquaporins. *Proc. Natl. Acad. Sci.*, 2013, 110(41):16367–16372.
60. P. Green, R. Erickson and J. Buggy, Metabolic and physical control of cell elongation rate: in vivo studies in nitella, *Plant Physiol.*, 1971, **47**(3), 423–430.
61. P. B. Green, Organogenesis-a biophysical view, *Annu. Rev. Plant Physiol.*, 1980, **31**(1), 51–82.
62. P. B. Green and F. W. Stanton, Turgor pressure: direct manometric measurement in single cells of nitella, *Science*, 1967, **155**(3770), 1675–1676.

63. O. Hamant, M. G. Heisler, H. Jönsson, P. Krupinski, M. Uyttewaal, P. Bokov, F. Corson, P. Sahlin, A. Boudaoud and E. M. Meyerowitz, *et al.*, Developmental patterning by mechanical signals in arabidopsis, *Science*, 2008, **322**(5908), 1650–1655.
64. O. Hamant, D. Inoue, D. Bouchez, J. Dumais and E. Mjolsness, Are microtubules tension sensors?, *Nat. Commun.*, 2019, **10**(1), 1–12.
65. E. S. Haswell, R. Peyronnet, H. Barbier-Brygoo, E. M. Meyerowitz and J.-M. Frachisse, Two mscs homologs provide mechanosensitive channel activities in the arabidopsis root, *Curr. Biol.*, 2008, **18**(10), 730–734.
66. R. Hedrich and E. Neher, Venus flytrap: how an excitable, carnivorous plant works, *Trends Plant Sci.*, 2018, **23**(3), 220–234.
67. Z. Hejnowicz and A. Sievers, Tissue stresses in organs of herbaceous plants: I. poisson ratios of tissues and their role in determination of stresses, *J. Exp. Bot.*, 1995, **46**, 1035–1043.
68. V. Hernández-Hernández, M. Bentez and A. Boudaoud, Interplay between turgor pressure and plasmodesmata during plant development, *J. Exp. Bot.*, 2020, **71**(3), 768–777.
69. N. Hervieux, M. Dumond, A. Sapala, A.-L. Routier-Kierzkowska, D. Kierzkowski, A. H. Roeder, R. S. Smith, A. Boudaoud and O. Hamant, A mechanical feedback restricts sepal growth and shape in arabidopsis, *Curr. Biol.*, 2016, **26**(8), 1019–1028.
70. A. N. Heyn, Further investigations on the mechanism of cell elongation and the properties of the cell wall in connection with elongation, *Protoplasma*, 1933, **19**(1), 78–96.
71. B. S. Hill and G. P. Findlay, The power of movement in plants: the role of osmotic machines, *Q. Rev. Biophys.*, 1981, **14**(2), 173–222.
72. H. Höfte, A. Peaucelle and S. Braybrook, Cell wall mechanics and growth control in plants: the role of pectins revisited, *Front. Plant Sci.*, 2012, **3**, 121.
73. D. P. Holmes and A. J. Crosby, Snapping surfaces, *Adv. Mater.*, 2007, **19**(21), 3589–3593.
74. C. T. Ingold, *et al.*, Spore liberation, *Spore liberation*, 1965.
75. K. H. Jensen, K. Berg-Sørensen, H. Bruus, N. M. Holbrook, J. Liesche, A. Schulz, M. A. Zwieniecki and T. Bohr, Sap flow and sugar transport in plants, *Rev. Mod. Phys.*, 2016, **88**(3), 035007.
76. L. Jost and R. J. Harvey-Gibson, *Lectures on Plant Physiology*, Clarendon Press, 1907.
77. O. Kedem and A. Katchalsky, Thermodynamic analysis of the permeability of biological membranes to non-electrolytes, *Biochem. Biophys. Acta*, 1958, **27**, 229–246.
78. D. Kierzkowski, N. Nakayama, A.-L. Routier-Kierzkowska, A. Weber, E. Bayer, M. Schorderet, D. Reinhardt, C. Kuhlemeier and R. S. Smith, Elastic domains regulate growth and organogenesis in the plant shoot apical meristem, *Science*, 2012, **335**(6072), 1096–1099.
79. S.-W. Kim, J.-S. Koh, J.-G. Lee, J. Ryu, M. Cho and K.-J. Cho, Flytrap-inspired robot using structurally integrated actuation based on

bistability and a developable surface, *Bioinspiration Biomimetics*, 2014, **9**(3), 036004.
80. Y. X. Kim and E. Steudle, Light and turgor affect the water permeability (aquaporins) of parenchyma cells in the midrib of leaves of zea mays, *J. Exp. Bot.*, 2007, **58**(15–16), 4119–4129.
81. A. L. King, The spore discharge mechanism of common ferns, *Proc. Natl. Acad. Sci. U. S. A.*, 1944, **30**(7), 155.
82. J. Knoblauch, D. L. Mullendore, K. H. Jensen and M. Knoblauch, Pico gauges for minimally invasive intracellular hydrostatic pressure measurements, *Plant Physiol.*, 2014, **166**(3), 1271–1279.
83. M. A. Koehl and W. K. Silk, How kelp in drag lose their ruffles: environmental cues, growth kinematics, and mechanical constraints govern curvature, *J. Exp. Bot.*, 2021, **72**(10), 3677–3687.
84. E. M. Kramer and D. R. Myers, Osmosis is not driven by water dilution, *Trends Plant Sci.*, 2013, **18**(4), 195–197.
85. P. J. Kramer and J. S. Boyer, *Water Relations of Plants and Soils*, Academic press, 1995.
86. U. Kutschera, Regulation of cell expansion, *The Cytoskeletal Basis of Plant Growth and Form*, 1991, pp. 85–99.
87. L. D. Landau and E. M. Lifshitz, *et al.*, *Theory of Elasticity*, Pergamon Press, Oxford New York, 1986, vol. 7.
88. T. Lawson and J. Matthews, Guard cell metabolism and stomatal function, *Annu. Rev. Plant Biol.*, 2020, **71**, 273–302.
89. P. Lehmann and D. Or, Effects of stomata clustering on leaf gas exchange, *New Phytol.*, 2015, **207**(4), 1015–1025.
90. N. Levernier, O. Pouliquen and Y. Forterre, An integrative model of plant gravitropism linking statoliths position and auxin transport, *Front. Plant Sci.*, 2021, **12**, 474.
91. P. M. Lintilhac, C. Wei, J. J. Tanguay and J. O. Outwater, Ball tonometry: a rapid, nondestructive method for measuring cell turgor pressure in thin-walled plant cells, *J. Plant Growth Regul.*, 2000, **19**(1), 90–97.
92. J. A. Lockhart, An analysis of irreversible plant cell elongation, *J. Theor. Biol.*, 1965, **8**(2), 264–275.
93. Y. Long, I. Cheddadi, G. Mosca, V. Mirabet, M. Dumond, A. Kiss, J. Traas, C. Godin and A. Boudaoud, Cellular heterogeneity in pressure and growth emerges from tissue topology and geometry, *Curr. Biol.*, 2020, **30**(8), 1504–1516.
94. D. Lunni, M. Cianchetti, C. Filippeschi, E. Sinibaldi and B. Mazzolai, Plant-inspired soft bistable structures based on hygroscopic electrospun nanofibers, *Adv. Mater. Interfaces*, 2020, **7**(4), 1901310.
95. E. H. Mansfield, *The Bending and Stretching of Plates*, Cambridge university press, 1989.
96. C. Maurel, L. Verdoucq, D.-T. Luu and V. Santoni, Plant aquaporins: membrane channels with multiple integrated functions, *Annu. Rev. Plant Biol.*, 2008, **59**, 595–624.

97. P. Milani, S. A. Braybrook and A. Boudaoud, Shrinking the hammer: micromechanical approaches to morphogenesis, *J. Exp. Bot.*, 2013, **64**(15), 4651–4662.
98. P. Milani, M. Gholamirad, J. Traas, A. Arnéodo, A. Boudaoud, F. Argoul and O. Hamant, In vivo analysis of local wall stiffness at the shoot apical meristem in arabidopsis using atomic force microscopy, *Plant J.*, 2011, **67**(6), 1116–1123.
99. F. J. Molz and J. S. Boyer, Growth-induced water potentials in plant cells and tissues, *Plant Physiol.*, 1978, **62**(3), 423–429.
100. F. J. Molz and E. Ikenberry, Water transport through plant cells and cell walls: theoretical development, *Soil Sci. Soc. Am. J.*, 1974, **38**(5), 699–704.
101. M. T. Morita, Directional gravity sensing in gravitropism, *Annu. Rev. Plant Biol.*, 2010, **61**, 705–720.
102. B. Moulia, Leaves as shell structures: double curvature, auto-stresses, and minimal mechanical energy constraints on leaf rolling in grasses, *J. Plant Growth Regul.*, 2000, **19**(1), 19–30.
103. B. Moulia, R. Bastien, H. Chauvet-Thiry and N. Leblanc-Fournier, Posture control in land plants: growth, position sensing, proprioception, balance, and elasticity, *J. Exp. Bot.*, 2019, **70**(14), 3467–3494.
104. B. Moulia, S. Douady and O. Hamant, Fluctuations shape plants through proprioception, *Science*, 2021, **372**(6540), eabc6868.
105. B. Moulia and M. Fournier, The power and control of gravitropic movements in plants: a biomechanical and systems biology view, *J. Exp. Bot.*, 2009, **60**(2), 461–486.
106. D. E. Moulton, H. Oliveri and A. Goriely, Multiscale integration of environmental stimuli in plant tropism produces complex behaviors, *Proc. Natl. Acad. Sci.*, 2020, **117**(51), 32226–32237.
107. E. Münch, Die stoffbewegung in der pflanze, fischer, jena, 1930.
108. B. Nadrowski, P. Martin and F. Jülicher, Active hair-bundle motility harnesses noise to operate near an optimum of mechanosensitivity, *Proc. Natl. Acad. Sci.*, 2004, **101**(33), 12195–12200.
109. M. Nakamura, T. Nishimura and M. T. Morita, Gravity sensing and signal conversion in plant gravitropism, *J. Exp. Bot.*, 2019, **70**(14), 3495–3506.
110. K. J. Niklas and H.-C. Spatz, *Plant Physics*, University of Chicago Press, 2012.
111. S. B. Nilsson, C. H. Hertz and S. Falk, On the relation between turgor pressure and tissue rigidity. ii: Theoretical calculations on model systems, *Physiol. Plant.*, 1958, **11**(4), 818–837.
112. P. S. Nobel *et al.*, *Physicochemical & Environmental Plant Physiology*, Academic press, 1999.
113. X. Noblin, N. Rojas, J. Westbrook, C. Llorens, M. Argentina and J. Dumais, The fern sporangium: a unique catapult, *Science*, 2012, **335**(6074), 1322.
114. B. Nordbring-Hertz, H.-B. Jansson, Y. Persson, E. Friman and C. Dackman, Nematophagous fungi, In TIB AV-PORTAL, https://av.tib.eu/media/14259.

115. B. Nordbring-Hertz, H.-B. Jansson and A. Tunlid, Nematophagous fungi, *Encyclopedia of Life Sciences*, Wiley Online Library, 2006, pp. 173–204.
116. K. Okamoto, H. Ueda, T. Shimada, K. Tamura, T. Kato, M. Tasaka, M. T. Morita and I. Hara-Nishimura, Regulation of organ straightening and plant posture by an actin -myosin xi cytoskeleton, *Nat. Plants*, 2015, **1**(4), 1–7.
117. H. Oliveri, On the role of mechanical feedback in plant morphogenesis, PhD thesis, Université Montpellier, 2019.
118. H. Oliveri, J. Traas, C. Godin and O. Ali, Regulation of plant cell wall stiffness by mechanical stress: a mesoscale physical model, *J. Math. Biol.*, 2019, **78**(3), 625–653.
119. J. K. Ortega, Augmented growth equation for cell wall expansion, *Plant Physiol.*, 1985, **79**(1), 318–320.
120. T. Pedley, Calculation of unstirred layer thickness in membrane transport experiments: a survey, *Q. Rev. Biophys.*, 1983, **16**(2), 115–150.
121. B. Péret, G. Li, J. Zhao, L. R. Band, U. Voß, O. Postaire, D.-T. Luu, O. Da Ines, I. Casimiro, M. Lucas, D. M. Wells, L. Lazzerini, P. Nacry, J. R. King, O. E. Jensen, A. R. SchÃòner, C. Maurel and M. J. Bennett, Auxin regulates aquaporin function to facilitate lateral root emergence, *Nat. Cell Biol.*, 2012, **14**(10), 991–998.
122. J. Philip, Propagation of turgor and other properties through cell aggregations, *Plant Physiol.*, 1958, **33**(4), 271.
123. S. Poppinga and M. Joyeux, Different mechanics of snap-trapping in the two closely related carnivorous plants dionaea muscipula and aldrovanda vesiculosa, *Phys. Rev. E*, 2011, **84**(4), 041928.
124. O. Pouliquen, Y. Forterre, A. Bérut, H. Chauvet, F. Bizet, V. Legue and B. Moulia, A new scenario for gravity detection in plants: the position sensor hypothesis, *Phys. Biol.*, 2017, **14**(3), 035005.
125. M. Probine and R. Preston, Cell growth and the structure and mechanical properties of the wall in internodal cells of nitella opaca: Ii. mechanical properties of the walls, *J. Exp. Bot.*, 1962, **13**(1), 111–127.
126. T. E. Proseus, J. K. Ortega and J. S. Boyer, Separating growth from elastic deformation during cell enlargement, *Plant Physiol.*, 1999, **119**(2), 775–784.
127. T. E. Proseus, G.-L. Zhu and J. S. Boyer, Turgor, temperature and the growth of plant cells: using chara corallina as a model system, *J. Exp. Bot.*, 2000, **51**(350), 1481–1494.
128. P. S. Raux, S. Gravelle and J. Dumais, Design of a unidirectional water valve in tillandsia, *Nat. Commun.*, 2020, **11**(1), 1–7.
129. M. Rivière, Y. Corre, A. Peaucelle, J. Derr and S. Douady, The hook shape of growing leaves results from an active regulatory process, *J. Exp. Bot.*, 2020, **71**(20), 6408–6417.
130. F. E. Rockwell, N. M. Holbrook and A. D. Stroock, Leaf hydraulics i: Scaling transport properties from single cells to tissues, *J. Theor. Biol.*, 2014, **340**, 251–266.

131. F. E. Rockwell, N. M. Holbrook and A. D. Stroock, Leaf hydraulics ii: vascularized tissues, *J. Theor. Biol.*, 2014, **340**, 267–284.
132. E. R. Rojas, S. Hotton and J. Dumais, Chemically mediated mechanical expansion of the pollen tube cell wall, *Biophys. J.*, 2011, **101**(8), 1844–1853.
133. A. Rolena, M. Paetkau, K. A. Ross, D. V. Godfrey, J. S. Church and C. R. Friedman, Thermogenesis-triggered seed dispersal in dwarf mistletoe, *Nat. Commun.*, 2015, **6**(1), 1–5.
134. J.-P. Rospars and N. Meyer-Vernet, Force per cross-sectional area from molecules to muscles: a general property of biological motors, *R. Soc., Open Sci.*, 2016, **3**(7), 160313.
135. A.-L. Routier-Kierzkowska, A. Weber, P. Kochova, D. Felekis, B. J. Nelson, C. Kuhlemeier and R. S. Smith, Cellular force microscopy for in vivo measurements of plant tissue mechanics, *Plant Physiol.*, 2012, **158**(4), 1514–1522.
136. R. Sachse, A. Westermeier, M. Mylo, J. Nadasdi, M. Bischoff, T. Speck and S. Poppinga, Snapping mechanics of the venus flytrap (dionaea muscipula), *Proc. Natl. Acad. Sci.*, 2020, **117**(27), 16035–16042.
137. L. Sack and N. M. Holbrook, Leaf hydraulics, *Annu. Rev. Plant Biol.*, 2006, **57**, 361–381.
138. A. Sampathkumar, P. Krupinski, R. Wightman, P. Milani, A. Berquand, A. Boudaoud, O. Hamant, H. Jönsson and E. M. Meyerowitz, Subcellular and supracellular mechanical stress prescribes cytoskeleton behavior in arabidopsis cotyledon pavement cells, *eLife*, 2014, **3**, e01967.
139. A. Sampathkumar, A. Yan, P. Krupinski and E. M. Meyerowitz, Physical forces regulate plant development and morphogenesis, *Curr. Biol.*, 2014, **24**(10), R475–R483.
140. S. Schleicher, J. Lienhard, S. Poppinga, T. Speck and J. Knippers, A methodology for transferring principles of plant movements to elastic systems in architecture, *Comput. -Aided Des.*, 2015, **60**, 105–117.
141. J. Schroeder, R. Hedrich and J. Fernandez, Potassium-selective single channels in guard cell protoplasts of vicia faba, *Nature*, 1984, **312**(5992), 361–362.
142. J. M. Selker, G. L. Steucek and P. B. Green, Biophysical mechanisms for morphogenetic progressions at the shoot apex, *Dev. Biol.*, 1992, **153**(1), 29–43.
143. E. Sharon, B. Roman, M. Marder, G.-S. Shin and H. L. Swinney, Buckling cascades in free sheets, *Nature*, 2002, **419**(6907), 579.
144. W. K. Silk, Quantitative descriptions of development, *Annu. Rev. Plant Physiol.*, 1984, **35**(1), 479–518.
145. W. K. Silk and R. O. Erickson, Kinematics of hypocotyl curvature, *Am. J. Bot.*, 1978, **65**(3), 310–319.
146. J. M. Skotheim and L. Mahadevan, Physical limits and design principles for plant and fungal movements, *Science*, 2005, **308**(5726), 1308–1310.

147. E. Steudle, Water flow in plants and its coupling to other processes: An overview, *Methods Enzymol.*, 1989, **174**, 183–225.
148. E. Steudle, The biophysics of plant water: compartmentation, coupling with metabolic processes, and flow of water in plant roots, *Water and life*, Springer, 1992, pp. 173–204.
149. E. Steudle. Pressure probe technique: basic prinples and application to studies of water and solute relations at the cell, tissue and organ level, *Water deficits. Plant responses from cell to community*, 1993, pp. 5–36.
150. E. Steudle and U. Zimmermann, Determination of the hydraulic conductivity and of reflection coefficients in nitella flexilis by means of direct cell-turgor pressure measurements, *Biochim. Biophys. Acta, Biomembr.*, 1974, **332**(3), 399–412.
151. E. Steudle, U. Zimmermann and U. Lüttge, Effect of turgor pressure and cell size on the wall elasticity of plant cells, *Plant Physiol.*, 1977, **59**(2), 285–289.
152. H. Stoeckel and K. Takeda, Calcium-sensitivity of the plasmalemmal delayed rectifier potassium current suggests that calcium influx in pulvinar protoplasts from mimosa pudica l. can be revealed by hyperpolarization, *J. Membr. Biol.*, 1995, **146**(2), 201–209.
153. A. D. Stroock, V. V. Pagay, M. A. Zwieniecki and N. Michele Holbrook, The physicochemical hydrodynamics of vascular plants, *Annu. Rev. Fluid Mech.*, 2014, **46**, 615–642.
154. H. Suda, H. Mano, M. Toyota, K. Fukushima, T. Mimura, I. Tsutsui, R. Hedrich, Y. Tamada and M. Hasebe, Calcium dynamics during trap closure visualized in transgenic venus flytrap, *Nat. Plants*, 2020, **6**(10), 1219–1224.
155. L. Taiz and E. Zeiger, *Plant Physiology*, *Sinauer Associates, Inc.* Publishers, sunderland, ma, 2002, 3:484.
156. D. W. Thompson *et al.*, On growth and form, *On Growth and Form*, 1942.
157. S. P. Timoshenko and S. Woinowsky-Krieger, *Theory of Plates and Shells*, McGraw-hill, 1959.
158. M. Toyota and S. Gilroy, Gravitropism and mechanical signaling in plants, *Am. J. Bot.*, 2013, **100**(1), 111–125.
159. D.-C. Trinh, J. Alonso-Serra, M. Asaoka, L. Colin, M. Cortes, A. Malivert, S. Takatani, F. Zhao, J. Traas, C. Trehin and O. Hamant, How mechanical forces shape plant organs, *Curr. Biol.*, 2021, **31**(3), R143–R159.
160. M. T. Tyree and M. H. Zimmermann, *Xylem Structure and the Ascent of Sap*, Springer, 2002.
161. D. Vella, A. Ajdari, A. Vaziri and A. Boudaoud, Indentation of ellipsoidal and cylindrical elastic shells, *Phys. Rev. Lett.*, 2012, **109**(14), 144302.
162. O. Vincent, C. Weißkopf, S. Poppinga, T. Masselter, T. Speck, M. Joyeux, C. Quilliet and P. Marmottant, Ultra-fast underwater suction traps, *Proc. R. Soc. B*, 2011, **278**(1720), 2909–2914.

163. M. Warner, B. Thiel and A. Donald, The elasticity and failure of fluid-filled cellular solids: theory and experiment, *Proc. Natl. Acad. Sci.*, 2000, **97**(4), 1370–1375.
164. A. Weber, S. Braybrook, M. Huflejt, G. Mosca, A.-L. Routier-Kierzkowska and R. S. Smith, Measuring the mechanical properties of plant cells by combining micro-indentation with osmotic treatments, *J. Exp. Bot.*, 2015, **66**(11), 3229–3241.
165. C. Wei, E. Steudle and M. T. Tyree, Water ascent in plants: do ongoing controversies have a sound basis?, *Trends Plant Sci.*, 1999, **4**(9), 372–375.
166. T. Zhang, H. Tang, D. Vavylonis and D. J. Cosgrove, Disentangling loosening from softening: insights into primary cell wall structure, *Plant J.*, 2019, **100**(6), 1101–1117.
167. Y. Zhang, J. Yu, X. Wang, D. M. Durachko, S. Zhang and D. J. Cosgrove, Molecular insights into the complex mechanics of plant epidermal cell walls, *Science*, 2021, **372**(6543), 706–711.
168. Q. Zhao, X. Yang, C. Ma, D. Chen, H. Bai, T. Li, W. Yang and T. Xie, A bioinspired reversible snapping hydrogel assembly, *Mater. Horiz.*, 2016, **3**(5), 422–428.

CHAPTER 2

Fluid–Structure Interactions in Plant Vascular Flows

KAARE H. JENSEN

Department of Physics, Technical University of Denmark, DK-2800 Kgs. Lyngby, Denmark
Email: khjensen@fysik.dtu.dk

2.1 Introduction

Transport of matter from one part of the body to another in vascular plants generally involves fluid flow along or across the walls of conduits formed by cells. The most commonly studied tube systems are the xylem (water), phloem (products of photosynthesis), and aerenchyma (gases). There are many similarities between the fluid physics of plants and animals, the latter of which has received a great deal of attention in the context of, for instance, cardiovascular transport,[2,15] cerebrospinal flows,[44] and drinking.[21] There are, however, characteristics which appear unique to plants. First, vegetal vascular flows occur predominantly inside individual either living (phloem) or dead (xylem) cells. (An exception is arenchyma, where gases flow in the intercellular space.) Another prominent feature of plant vascular tissues is the presence of a rigid walls, which allow cells to withstand substantial pressures, often in the ~1 MPa-range. This exceeds the typical pressure in animal tissues (10 kPa[21,24]) by one to two orders of magnitude. Remarkably, the hydrostatic pressure in plant cells can take positive as well as negative values. Positive pressures are generated by osmosis, while evaporation can create tension. (The physical origin of both effects is discussed in greater detail in Chapters 1 and 4.)

Soft Matter Series No. 15
Soft Matter in Plants: From Biophysics to Biomimetics
Edited by Kaare H. Jensen and Yoël Forterre

Published by the Royal Society of Chemistry, www.rsc.org

Tolerating extreme mechanical stress is associated with substantial expense related to tissue construction and maintenance. If the channels are too soft, they may burst or collapse, and will thus no longer be able to carry flow. In contrast, building rigid pipes represents an unnecessary investment with little or no functional benefit. Consequently, the channel's mechanical properties are constrained by trade-offs between functional advantage and costs. The central physical question is therefore whether the conduits' stress tolerance has been subject to natural selection and if it represents an optimal solution to the coupled flow-mechanics problem.

Motivated by these questions, this chapter explores the interplay between fluid flow and elasticity in plants subjected to relatively strong environmental perturbations. We begin by considering a simple example: the bending of a branch where the macroscopic strain induces an excess pressure in the fluid phase. This is followed by an analysis of the link between conduit deformation and flow in the xylem when under tension. We then progress to a discussion of cell-to-cell flows *via* intercellular pores that contain soft elements. Finally, we briefly discuss osmosis and cytoplasmic streaming. Although the reciprocal fluid–structure coupling is less clear in the last two cases, their inclusion is merited by modeling approaches which treat them as moving boundaries. In each case, perspectives and open questions are discussed.

2.2 From Bending a Branch to Increasing Cell Pressure

Our journey starts at the tissue scale and we begin by considering a process familiar to all: the bending of a branch. The terminal branches of most plants can be bent with little effort, and offshoots often do not break even when subjected to relatively large loads due to, *e.g.*, a climbing animal, wind, rain, or snow.[27,36] There are some exceptions to this, *e.g.*, in the context of nests.[47] However, it is well established that the remarkable elasticity of plant tissue follows from the material properties of cellulose, hemicellulose, lignin, and pectin, and from the ordered cellular structure.[10,50] It is therefore interesting to consider how externally imposed macroscopic deformations feed back on physiological conditions inside cells. This question is foundational to the studies of plant biomechanics and slow growth (see, *e.g.*, ref. 9 for a recent review). However, relatively few studies have addressed how external perturbations impact cellular conditions on short timescales.

In what is arguably the simplest experiment, Louf *et al.*[23] measured the hydrostatic cellular pressure in tree and shrub branches subjected to cantilever loads (Figure 2.1A). The pressure Δp was measured in the xylem, the main vasculature responsible for water transport. (Here, Δp represents the excess pressure measured relative to the relaxed state.) The deflection of the branch itself is an approximately linear function of the load (see, *e.g.*, ref. 36). Classical linear beam theory predicts that the cell volume V_c in a

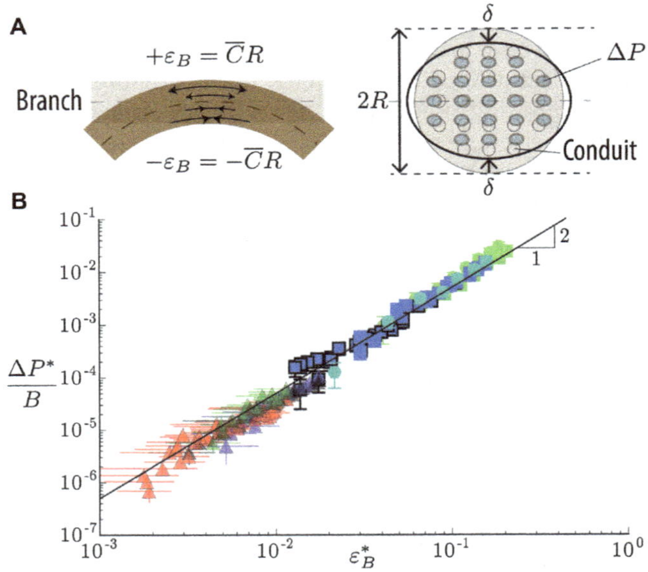

Figure 2.1 From bending a branch to increasing cell pressure. (A) When a branch of radius R is bent it generates a maximal longitudinal strain $\varepsilon_B = \bar{C}R$ in the beam, where $\bar{C}$ is the mean curvature. The cross-section of the beam is squeezed by an amount δ to reduce the elastic bending energy. (B) This leads to an increase in the xylem cell pressure $\Delta p \sim B\varepsilon^2$, where B is the bulk modulus. Note that panel (B) includes data from both real and synthetic branches. Adapted from ref. 23, https://doi.org/10.1073/pnas.1707675114, under the terms of the CC BY 4.0 license https://creativecommons.org/licenses/by/4.0/.

symmetrical bent beam should remain constant, regardless of its elastic properties. This would predict a null pressure variation $\Delta p \sim -BV_c = 0$ according to Hooke's law in a closed beam with interconnected channels. (Here, B is the bulk modulus.) Surprisingly, Louf *et al.* found a non-linear relationship (Figure 2.1B) between the between cell pressure $\Delta p \sim B\varepsilon_B^2$ and bending strain $\varepsilon_B = \bar{C}R$, where $\bar{C}$ is the mean curvature, R the radius of the branch, and $B \approx 0.1$ GPa is the bulk modulus. This contradicts the simplest theory cited above ($\Delta p = 0$). When an elastic beam is bent, however, the longitudinal elastic strain (extension or compression) increases linearly with the distance to the neutral surface (Figure 2.1A). This induces a bending elastic energy per unit volume $U_b \sim E\varepsilon_B^2 \sim E\bar{C}^2R^2$ that varies quadratically with the radius of the beam. (E is Young's modulus of the beam.) Therefore, a way for the system to lower the bending elastic energy is to squeeze its cross-section, hence decreasing the beam radius by a quantity δ in the transverse direction. This lowers the bending energy to $U_b \sim E\bar{C}^2(R-\delta)^2$ and is associated with a reduction in the channel volume $\Delta V_c/V_c \sim -\delta/R$. However, the transverse compression δ/R comes at the cost of a positive compressive energy in the transverse direction $U_c \sim E(\delta/R)$. Minimizing the total energy

$U_b + U_c$ with respect to the deformation δ leads to a specific relation between the transverse and longitudinal strain: $\delta/R \sim \varepsilon_B^2$. This implies that the pressure in the cellular conduits scale as

$$\Delta p = -B\frac{\Delta V_c}{V_c} \sim B\varepsilon_B^2, \tag{2.1}$$

in accord with experimental data (Figure 2.1B). For moderately large deformations ($\varepsilon_B = 0.01$) bending thus induces an excess pressure of $\Delta p \approx 10^4$ Pa. Precisely how this overpressure may be involved in signalling, or indeed sensed, is still under debate.[8,13,23]

2.3 Xylem Flow Under Tension and the Effects of Conduit Collapse

Having considered the effects of macroscopic organ-scale deformations on flow, we continue our discussion of elastic phenomena at tissue scales. Xylem transport is driven by evaporation from the leaf surface. (This is discussed in Chapter 4, and we shall not provide further details of the mechanism here.) The suction process reduces the pressure in the vascular conduits to values lower than measured in the surrounding tissue, and in the atmosphere. A pressure drop across the xylem tube boundaries therefore develops that tends to pull the cell wall inwards. The Hagen–Poiseuille law (eqn (1.9)) tells us that a reduction in the conduit radius leads to a substantial reduction in flow capacity because the flow rate Q scales as the radius a raised to the fourth power. Deformations of the xylem cell wall could therefore be of physiological significance.

Complete conduit collapse is believed to be a rare event, because most xylem conduits are appropriately reinforced and able to handle all but the most extreme levels of water stress. We will return to this bracing problem in the next section. Nevertheless, a few instances have been reported in which vascular conduits have caved in.[3,6,40,53] For instance, Bouche *et al.*[3] reported conduit collapse during dehydration in pine needles, mostly in the soft tracheids surrounding the xylem (Figure 2.2). The tissue-scale deformation in a seedling is illustrated in Figure 2.2A: the needle retains its shape at low-to-moderate water stress but begins to collapse when the pressure in the conduits is sufficiently negative. The physiological impact of dehydration on transport is clear: the hydraulic conductance decreases rapidly as the tension increases from the relaxed state ($\Delta p = 0.1$ MPa) to the closed state [$\Delta p \approx 4$ MPa (Figure 2.2C)]. The origin of this loss of conductivity is still debated and could originate from biological responses within the living tissue surrounding the vessels.[40] However, it is instructive to examine the simplest hypothesis of a purely mechanical effect due to the collapse of individual vascular conduits upon tissue contraction (Figure 2.2B).

Describing in detail the transition from an open to a closed conduit is a complex mechanical problem, possibly involving buckling and contact

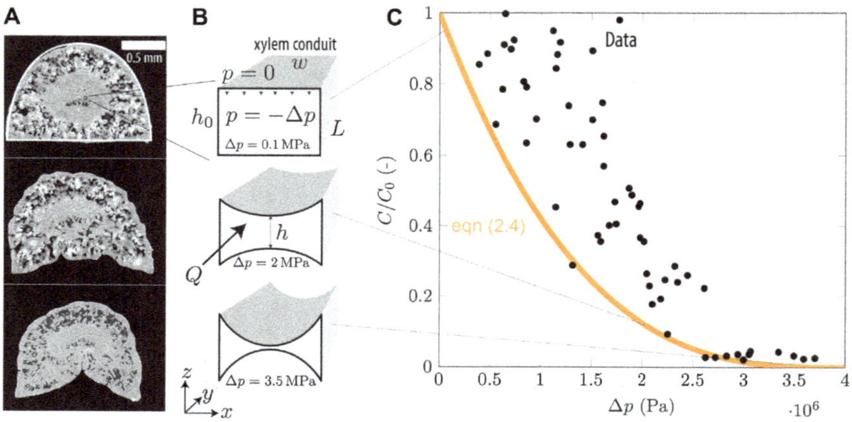

Figure 2.2 Xylem flow under tension and the effects of conduit collapse. (A-B) When subjected to dry conditions, plant tissue deforms. The images show transverse slices of pine needles based on X-ray computed tomography. The tissue is compressed as the difference Δp between the internal and ambient pressure increases from $\Delta p = 0.1$ MPa to 3.5 MPa. This translates to the cellular scale (see also ref. 53). Therefore, the xylem conduits, which carry a water current Q driven by the same evaporative pressure drop Δp, loose the capacity to carry flow. Here, we approximate the xylem tubes by rectangular channels of with w, height h and length L. (C) Experimental conductivity C plotted as function of the pressure drop Δp (data taken from ref. 3). The data are normalized by the stress-free conductivity, C_0. The solid line represents the theoretical prediction in eqn (2.4) with $p_c = 4$ MPa. Panel A and data in C adapted from ref. 3 with permission from John Wiley & Sons, Copyright © 2015 John Wiley & Sons Ltd.

mechanics, as well as interactions with the surrounding living tissue. The process, however, does appear to be reversible.[53] It is therefore conceivable that the full coupled fluid-elasticity problem could be analyzed using, for instance, a numerical finite-element scheme combined with a detailed tissue model. To our knowledge this has not yet been attempted, but Cochard *et al.*[6] pointed out that channel occlusion has a strong geometric impact on flow.

To get a first insight into how the closing process could impact the hydraulic resistance, and how it depends on the geometric and mechanical parameters, we will briefly discuss a simple model of the process. It is not meant to be a fully comprehensive description, but we hope it might spur the interest of researchers in the field. When setting up the problem, we will consider the flow rate Q through a tube of length L driven by the pressure drop Δp (Figure 2.2B). The channels in question are reasonably small, typically measuring approximately $a \approx 10$ μm in radius, and the fluid motion is dominated by viscous forces, *i.e.*, the Reynolds number $\mathrm{Re} \ll 1$. The cross-sectional shape of the conduits is complex and difficult to describe in a precise mathematical form. To simplify matters here we assume that the channels are approximately rectangular of width w and height h, chosen such that $h<w$ (Figure 2.2). (Both h and w, however, are of the same order of magnitude as the aforementioned conduits, *i.e.*, $h \sim w \sim 10$ μm.)

Because the flow is dominated by viscous forces and the geometry is shallow, we can use Stokes' equation, and the lubrication approximation thereof to examine the link between flow and deformation. The flow rate Q can therefore be expressed as

$$Q = \frac{h^3 w}{12\eta} \frac{\Delta p}{L}, \tag{2.2}$$

where η is the viscosity and Δp is the pressure drop along the length L of the channel. An important feature of eqn (2.2) is the strong geometric dependence in the h^3w-factor that links the pressure gradient $\Delta p/L$ and the flow rate Q. Fluid–solid interactions that modify the channel height h *via* the pressure p thus have the potential to strongly impact the flow characteristics.

To model the channel deformation we consider the transverse pressure drop $\delta p = p - p_e = p$ across the cell wall. Here, $p_e = 0$ is the external tissue pressure and $p<0$ is the pressure in the conduit itself, which tends to pull the walls inwards. When δp is (numerically) relatively small the channel is unaffected and the height $h \approx h_0$ is close to the relaxed state (Figure 2.2B). However, as δp increases the channel closes and $h \rightarrow 0$. In in general, the pressure $p(x)$, and hence deformation $h(x)$, will vary along the length of the channel according to the lubrication equation, where $\partial_x p \sim h^{-3} w^{-1} \eta Q$. In the limit of a completely blocked channel $(h \rightarrow 0)$ the pressure gradient $\partial_x p$ grows and the majority of the viscous dissipation will therefore be confined to a relatively narrow region. To avoid dealing with the mathematical complexities associated with this self-consistent problem, we will assume that the typical transverse deformation is controlled by the pressure drop Δp, *i.e.*, the maximum available pressure difference across the cell wall. The primary deformation induced by this pressure imbalance across the cell wall will occur along the z direction, *i.e.*, perpendicular to the widest axis of the tube (Figure 2.2B). The precise magnitude of the change in the channel height h depends on the detailed boundary conditions, and on the neighbouring tissue. Putting these details aside, and focusing only on the deflection of the cell wall, we can estimate the local channel height from linear plate theory $h \approx h_0 - (w^4/t^3)\Delta p/E = h_0(1 - \Delta p/p_c)$, where t is the thickness and E is the Young's modulus of the cell wall. The parameter $p_c = Et^3 h_0/w^4$ is the closing pressure, *i.e.*, the approximate pressure at which the two widest opposing cell wall segments will meet near the centre of the channel, thus blocking, or at least severely restricting, flow. Finally, using this height profile in eqn (2.2) yields the flow rate relation

$$Q = \frac{h_0^3 w}{12\eta L} \Delta p \left(1 - \frac{\Delta p}{p_c}\right)^3. \tag{2.3}$$

To assess the ability of the conduit to carry flow in a deformed channel, we consider the hydraulic conductance $C = Q/\Delta p$ defined as the ratio of the flow rate Q to the applied pressure Δp. When the applied pressure drop $\Delta p \ll p_c$ is substantially smaller than the closing pressure p_c the conductance $C = C_0$ is equal to that of an undeformed rectangular channel $C_0 = h_0^3 w/(12\eta L)$. The pressure-dependent conductance can be expressed as

$$\frac{C}{C_0} = \left(1 - \frac{\Delta p}{p_c}\right)^3 \tag{2.4}$$

which is gradually reduced as the applied pressure Δp approaches the closing pressure p_c, underscoring conduit collapse as a suitable mechanism for controlling xylem flow. The simple model eqn (2.4) is in rough accord with data (Figure 2.2C). Moreover, as already pointed out by Cochard *et al.*,[6] it highlights the clear impetus for preventing deformation to maintain unimpeded flow. This process is discussed further in the next section. For comprehensive reviews of related flow control and fluid–structure interaction problems in animal cardiovascular flows, see, *e.g.*, ref. 2 and 15.

2.4 Resistance to Collapse

Having outlined the potentially detrimental impact of channel collapse on flow, we proceed with a brief analysis of the structural features employed by plants to prevent, or at least control, cell wall deformations. The focus of this analysis is again xylem cells under tension, and to set up this problem we closely follow the anaysis by G. N. Karam.[20] Tyree and Zimmermann[46] also provides an excellent and accessible introduction to the remarkable complexity of the xylem.

For simplicity, we return to the picture of a xylem conduit as a slender cylindrical tube of radius a and length $L \gg a$ (Figure 2.3A). The trans-wall pressure difference Δp causes hoop and longitudinal stresses, with the hoop stresses being twice the longitudinal stresses. The forces per unit length in the hoop and longitudinal directions, F_h and F_l, respectively, that the tube has to resist can be written as:

$$F_h = \Delta p a, \tag{2.5}$$

$$F_l = \frac{\Delta p a}{2}, \tag{2.6}$$

where we note that a positive pressure Δp leads to tensile stresses in the cell wall while a negative pressure leads to compressive stresses. (The latter case is relevant to xylem flows.)

Plant cell walls are made from cellulose, hemicellulose, lignin, and pectin.[10,50] If the maximum allowable stress in the wall is σ^*, then the available resisting hoop force per unit length will be $F_{max} = \sigma^* t$, where t is the cell wall

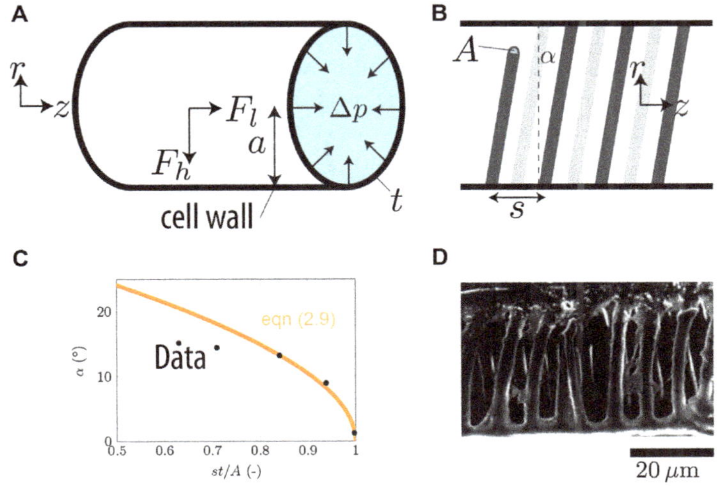

Figure 2.3 Resistance to collapse. (A) Xylem conduits are approximately cylindrical and often function under tension (magnitude Δp). This result in hoop, F_{h}, and longitudinal, F_{l}, forces as the cell wall is pulled inwards. (B) Helical thickenings of cross-section area A, spacing s and pitch angle α are frequently observed. They act to strengthen the pressure vessel. (C) Pitch angle α plotted as function of non-dimensional spacing st/A. The solid line represents the optimal configuration in eqn (2.9). (D) Micrograph of thickening patterns in xylem of the grass species *Elytrigia repens*. Panel (C) and (D) adapted from ref. 20.

thickness. If a numerically small (but negative) internal pressure p is applied and $\Delta pa < \sigma^{\star}t$ the tube can readily resist the hoop and longitudinal stresses caused by the internal pressure. However, if the internal pressure is sufficiently great, specifically when $\Delta pa < \sigma^{\star}t$, the tube wall needs reinforcing in the longitudinal as well as the hoop direction to resist internal pressures.

In the xylem, the reinforcement is often in the form of a helical-coil thickening of secondary wall material with cross-section area A and spacing s at an angle α (Figure 2.3B). Equating the resisting forces to the applied ones in the hoop and longitudinal directions, respectively, leads to

$$\sigma^{\star}(t + A/s\cos^2\alpha) = \Delta pa, \tag{2.7}$$

$$\sigma^{\star}(t + A/s\sin^2\alpha) = \frac{\Delta pa}{2}. \tag{2.8}$$

For a given amount of cell wall and reinforcement material, the angle must therefore fulfil

$$\cos^2\alpha - 2\sin^2\alpha = \frac{st}{A}, \tag{2.9}$$

in reasonable agreement with data where angles in the range $\alpha \approx 10$–20° are most frequently observed (Figure 2.3C). It has been speculated that soft xylem-inspired conduits based on this principle could provide a robust alternative to rigid pipes.[1] Notably, the designs used by plants are reasonably far from the situation in which the primary wall is neglected, *i.e.*, when $t \ll A/s$. In this case eqn (2.7) and eqn (2.8) yield $\alpha \approx 35°$, which is the optimal winding angle often used in fibre-reinforced pressure vessels.

2.5 Fluids and Elasticity in Intercellular Flows

Having explored the physics of flow at organ and tissue scales, we now focus our attention on intercelluar transport processes. A unique feature of plants is their relatively rigid compartmentalization in which cells remain fixed in space throughout their existence. The organism as a whole can, of course, morph, grow, move, and so on, but the relative position of a mature cell and its immediate surroundings remains approximately constant throughout its life.[43]

A consequence of the inherent architectural rigidity is that many plants cells establish permanent channels that link them to their neighbours (Figure 2.4A). There are many different types of cell–cell pores, *e.g.*, plasmodesmata (PD),[37] sieve pores,[25] and pit pores[4] (some of which contain PD sub-channels[17]), and we will not discuss all of them in detail here. The strong integration, however, implies that the cellular cytoplasmic liquid is more or less continuous across all cells.[22] This stands in stark contrast to animal cells, where individual cytoplasmic connections mediated by, for instance, gap junctions, most often are limited in time and space.[41] It is, however, important to stress that the cytoplasmic continuity does not imply that transport is unregulated.[22] Finally, we note that transport across the cell membrane plays a major role in cell–cell communication,[43] but we will not discuss that in detail here.

Direct cell–cell trafficking of small molecules in plants is mediated by PD nanopores.[37] These cylindrical nanopores (typically several hundred nanometres long and tens of nanometres wide) cross the wall between plant cells. The pores are open, *i.e.*, the plasma membranes of adjacent cells meet inside the pore. The endoplasmic reticulum (ER) permeates each pore, and the gap between the cylindrical desmotubule and the membrane forms an annular cytoplasmic sleeve through which water and solutes move. The ER–desmotubule complex is anchored by filamentous protein tethers.[7] Interestingly, transport across PD is sensitive to the difference in pressure between two neighbouring cells. For instance, a difference in cell turgor of 200 kPa instantly reduces PD transport by 50% between adjoining trichome cells of *Nicotiana clevelandii*,[28] and turgor differences that arise during cell growth have also been associated with reduced permeability.[38,49,51,52] The rapid reduction in transport cannot be explained by standard models of PD transport, which assume that cell–cell movement occurs by a combination of molecular diffusion and bulk flow in static PD geometries. The physical mechanism of pressure regulation of the permeability is a topic of active research.[16,30]

The precise size of the PD aperture is still under debate, but it is believed to be of order O(1 nm).[26,32] This is sufficiently large to select and

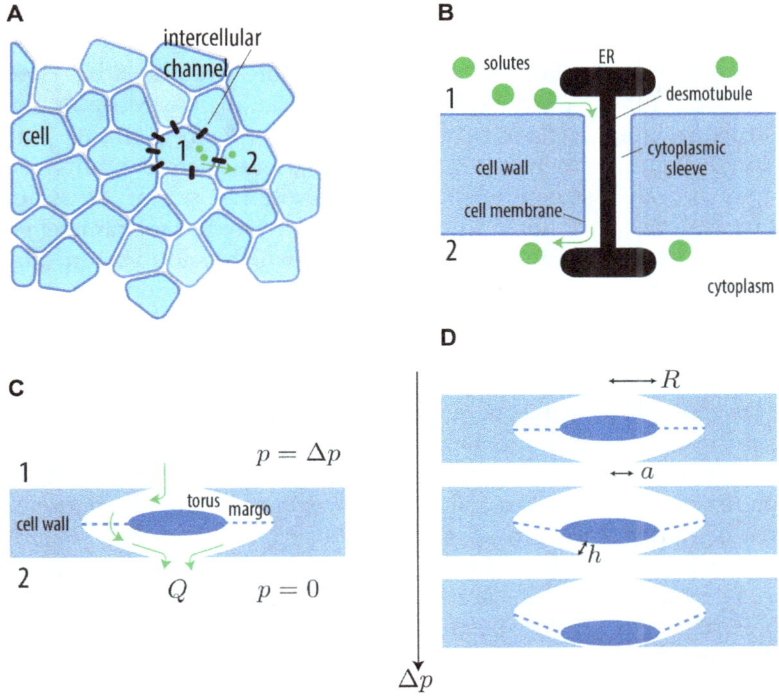

Figure 2.4 (A) Most plant cells are connected to their neighbours by small pores or channels which facilitate intercellular transport. (B) One example is plasmodesmata, nanopores that directly link the cytoplasm of adjacent cells (1 and 2). Transport across the cell wall occurs in a concentric cytoplasmic sleeve formed in the narrow gap between the cell membrane and the desmotubule (an organelle which links the ER of adjacent cells). (C) Another case is pit pores, which connect neighbouring xylem vascular conduits and limit the ability of air pockets to spread. The cell–cell flow rate Q of water is driven by the pressure difference Δp across the pore. A unique feature of pit pores in conifers (and related plant families) is the presence of soft cellulose filaments (the margo), which deform when subjected to mechanical stress. This enables the central solid region (the torus) to gradually block flow across the pore as the pressure drop increases. (D) The actuation is small when a negligible pressure drop Δp is applied. However, as Δp increases, the gap height h diminishes, thus increasing the resistance to flow. See additional details in the text.

mediate diffusive transport of individual molecules (*e.g.*, ions, sugars, hormones), but not effective at facilitating bulk fluid flow. Indeed, it would appear that bigger holes are needed. This is the case, for instance, in the vasculatures formed by phloem and xylem cells. (Here, we remind ourselves that water flows from roots to leaves in the xylem, while the products of photosynthesis move in the phloem). In these tissues, the constituent cells are linked by large open pores which permit liquids to move more or less freely. Another notable feature of plant vascular cells is that they are slender. The diameter-to-length aspect ratio is often of order O(0.1) or less.

We will not discuss all the different vascular pore types here, but choose to focus on torus margo pit pores, as these provide an illustrative example of fluid–structure interactions in plant vascular flows (Figure 2.4B). Pit pores are small intercellular microchannels that link adjacent xylem conduits. (For detailed reviews of their structure, function, and evolution, see ref. 14, 39 and 46.) A unique feature of pit pores in conifers (and related plant families) is the presence of soft cellulose filaments (the margo) which deform when subjected to mechanical stress. This enables the central solid region (the torus) to gradually block flow across the pore as the pressure drop increases. The key function of pit pores is to prevent gas pockets from migrating between cells.[5,29] At the interface between two liquid-filled neighbours, however, the pressure difference could, in principle, be sufficiently great for the soft valve to actuate in repose to strong localized transpiration stream fluctuations or tissue damage.

To understand how fluid–solid interactions impact the intercellular flow characteristics, we proceed to consider the system sketched in Figure 2.4C: a flat disk of radius $R \approx 5$ μm suspended above a small circular opening of radius $a \approx 1$ μm. Liquid flows in the shallow gap of height $h < 1$ μm between the torus and the cell wall. A pressure drop Δp drives the flow rate Q across the pores. The typical flow speed is moderate ($Q/R^2 \approx 1$ mm s^{-1}), hence the Reynolds number $\mathrm{Re} = \rho v h \eta \approx 10^{-1}$ is reasonably small. It is therefore reasonable to assume that the system is not too far removed from ideal Stokes' flow conditions. Moreover, because the geometry is reasonably shallow we can deploy the lubrication approximation thereof to obtain a first approximate to the link between pressure drop and flow rate. With these assumptions, the pressure drop Δp across the pore can be expressed as

$$\Delta p = -\int_a^R \partial_p \mathrm{d}r = \frac{6\eta Q}{\pi}\int_a^R \frac{1}{rh^3}\mathrm{d}r. \tag{2.10}$$

The flow rate Q is thus a linear function of the pressure drop

$$Q = \frac{\pi}{6\eta}\frac{h^3}{\ln(R/a)}\Delta p, \tag{2.11}$$

although the margo deformation will reduce the channel height h as pressure increases. To estimate this effect, we think of the torus–margo complex as a thin membrane of thickness t, radius R, and Young's modulus E. Using linear thin-plate theory (see *e.g.*, ref. 45), we can to a first approximation express the gap height as $h(\Delta p) = h_0\ (1 - \Delta p/p_c)$, where $p_c = Eh_0t^3/R^4$ is the pressure required to close the valve. Using this expression in eqn (2.11) leads to a non-linear pressure-flow relationship:

$$Q = \frac{\pi}{6\eta}\frac{h_0^3}{\ln(R/a)}\Delta p\left(1 - \frac{\Delta p}{p_c}\right)^3. \tag{2.12}$$

The pressure-dependent conductance can thus be expressed as

$$\frac{C}{C_0} = \left(1 - \frac{\Delta p}{p_c}\right)^3, \tag{2.13}$$

which, notably, is identical to eqn (2.4), which describes the hydraulic conductance of a collapsing channel (plotted in Figure 2.2C). Again, we note the strong geometric impact on on flow. The conductance is reduced by almost 50% when $\Delta p = 0.2p_c$ and by nearly 90% when the pressure drop $\Delta p = 0.5p_c$ reaches half the closing pressure p_c. The predicted flow control profile in eqn (2.13) was first obtained by Park *et al.*[31] and is still awaiting experimental validation.

2.6 Intracellular Flows and Cytoplasmic Streaming

We end this chapter by discussing fluid–solid interactions at sub-cellular scales. The interior of a plant cell is a complex physical environment dominated by the presence of numerous organelles suspended in the membrane-enclosed cytoplasmic fluid. This cytosol also contains a myriad of smaller particles, from large proteins to biomolecules and ions.

Intracellular flows are important for many phenomena in plant cells. Indeed, when seen under a microscope, bulk fluid motion is commonly observed in many cell types. These intracellular flows are (primarily) driven by two effects: osmosis and cytoplasmic streaming. Osmosis is the transport of water across the semipermeable cell membrane in response to solute concentration differences between the cell and its surroundings. (For a detailed discussion, see Chapter 1 and ref. 18 and 42.) One interesting question is how the cell radius a and typical flow speed V are related. In an osmotic process, the transmembrane flux j of the fluid entering (or exiting) the cell is proportional to potential difference across the membrane

$$j = L_p(RT\Delta c - \Delta p) \tag{2.14}$$

where L_p is the permeability, R is the gas constant, T is temperature, and c is the concentration. The first term in the bracket corresponds to the van't Hoff approximation of the osmotic pressure, while the second term is the hydrostatic pressure difference across the membrane. Volume accumulation will continue until a balance between pressure buildup and concentration difference has been reached and $j = 0$. In phloem cells, which form a long file of axially connected tubes, of, say, length L, the typical flow speed V measured along the cell axis is

$$V = \frac{2\pi a L j}{\pi a^2} = \frac{2Lj}{a}, \tag{2.15}$$

which, interestingly scales inversely with cell size. This means that the flow is comparatively faster in a smaller cell. If speed is important, as is the case, for instance, in signal transduction, one would expect that the tubes would be relatively small. Consistent with this concept, phloem cells, which carry both photoassimilates and signals, are among the smallest plant vascular conduits with radius $a \approx 10$ μm.[19]

While osmotic flows are suitable for driving transport in relatively small cells, it is unlikely to be a particularly useful strategy for mixing or transport in larger cells. Instead, many cells use cytoplasmic streaming to facilitate intracellular circulation (Figure 2.5A). This process drives fluid flow using molecular motors called myosin.[12,33] These proteins burn energy (ATP) while sliding along actin filaments (Figure 2.5B). The myosin is in turn attached to organelles (*e.g.*, vesicles), which are dragged along the actin cables, thus entraining the cellular fluids which move in the same direction. One can think of cytoplasmic streaming as a special case of the no-slip boundary

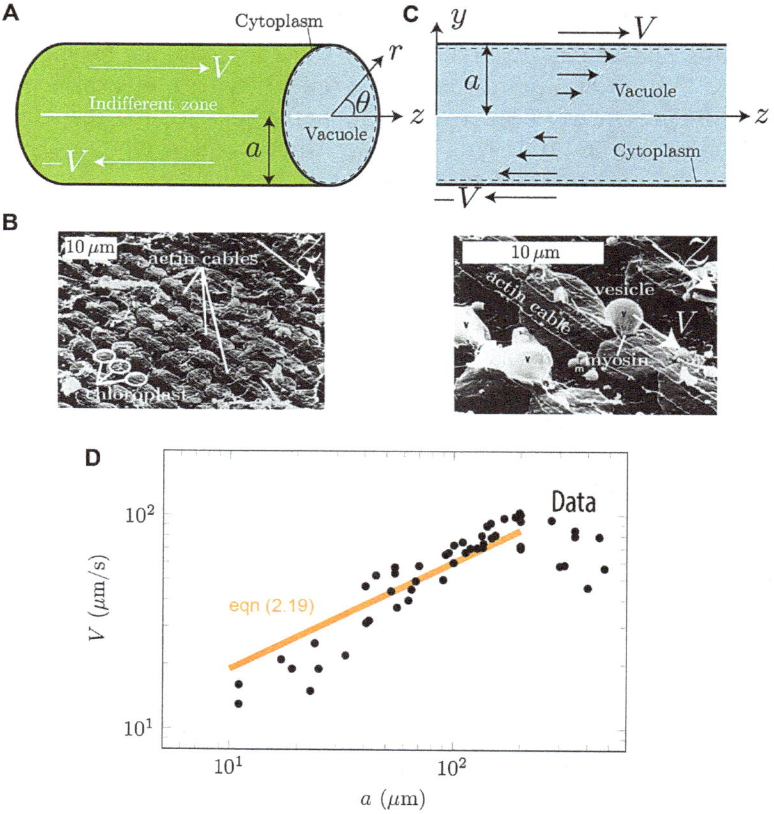

Figure 2.5 Cytoplasmic flows. (A) Cylindrical geometry analogue to *Chara* cells. (B) Microscopic arrangement of actin cables, myosin, and vesicles involved in dragging the cytoplasm. Adapted from ref. 54 with permission from Rockefeller University Press, Copyright 1976. (C) Simplified counter-moving parallel plate geometry to estimate the energetic cost of driving. (D) Typical cytoplasmic speed as function of the cell radius. Data taken from ref. 35. See also ref. 55–69.

condition, in which the channel wall, and hence fluid, is moving in the laboratory frame.

Cytoplasmic streaming is readily observable in many plant cells (you may have seen it in onion epidermis cells), and is especially striking in the large cells of the *Chara* algae. Algae and land plants are both eukaryotes, perform photosynthesis, and share many other features. Note, however, that algae lack true roots, stems, and leaves.

Streaming flow in *Chara* can reach speeds of up to $V = 100\ \mu\text{m s}^{-1}$ in the largest cells, which are up to $2a = 1$ mm in diameter. Interestingly, the typical flow speed V first increases with cell radius a before stabilizing at the plateau value $V \approx 100\ \mu\text{m s}^{-1}$ (Figure 2.5D). In contrast, osmotic flows were found to decrease with increasing cell size (eqn (2.15)). The key question is, What physical effects drive this difference?

To our knowledge, W. F. Pickard was the first to notice the accelerating flow in relatively small cells, and we follow his analysis here.[35] To get a better understanding of the problem, we begin by providing a more detailed account of the microscopic conditions. The actin filaments in *Chara* are located at the periphery ($r = a$) of the cylindrical cells, which can reach lengths of $L = 10$ cm and radii of $a = 0.5$ mm (Figure 2.5A). One half of the cell periphery (angles $0 < \theta < \pi$) is covered by actin cables whose polarity determines that myosin proteins move at the speed $u(a, \theta) = V$ along the $+z$ direction. The other half (angles $\pi < \theta < 2\pi$) moves at the same speed but in the opposite direction. The regions where the actin polarity changes sign (at $\theta = 0$ and $\theta = \pi$) are known as indifferent zones. This arrangement of actin filaments induces a bi-directional pipe flow with a zero net current.[48] To characterize the flow conditions in this relatively complex geometry, we notice that the Reynolds number $\text{Re} = \rho Va/\eta \approx 0.1$ is reasonably small, which allows us to ignore inertial effects and use Stokes' equation in the analysis. This leads to the velocity profile

$$u(r, \theta) = \frac{2}{\pi} V \arctan\left(\frac{2(r/a)\sin\theta}{1 - (r/a)^2}\right), \tag{2.16}$$

which, to our knowledge, was first discussed by Pickard,[34] and later refined by R. E. Goldstein and co-workers to include the effect of non-zero indifferent zone angles.[11] When averaged over the top half of the cell, the mean flow speed is $\langle u \rangle = 8V/\pi^2$. The Pickard flow profile in eqn (2.16) has stood up to detailed experimental scrutiny.[11,34]

Returning to the link between cell size and flow speed, we note that because the observed speed V enters into the flow problem as a boundary condition, eqn (2.16) does not permit us to justify the observed relation between cell radius a and V itself. In a subsequent paper, however, Pickard proposed a solution to this problem:[35] the relation is determined by a balance between viscous energy dissipation and metabolic cost. To avoid unnecessary computational complexities, we divert from exact geometry and consider for a moment the flow between two counter-moving parallel plates (Figure 2.5C). This case is known as Couette flow. In this configuration, the velocity varies linearly between the two edges

$$u = V\frac{z}{a} \tag{2.17}$$

where V is again the speed of the moving boundary and the z direction is perpendicular to the axis of motion ($-a<z<a$). This flow profile is reasonably consistent with eqn (2.16) near the cell mid-plane $\theta = \pi/2$, but deviates near the indifferent zones. The shear force which must be applied to, say, the top surface at $z = a$, is

$$F = \frac{\eta V A}{a}, \tag{2.18}$$

where A is half the cell surface area. The energetic cost of driving the flow can be computed by considering the work performed when displacing a section of the wall. The viscous power dissipation is the product of the shear force and the boundary speed, *i.e.*, $P_{\text{visc}} = FV$. The energy input required to drive the flow is delivered by the the actin cables, which consume ATP. Assuming that the cable density and the power output per cable is independent of cell size, the power delivered P_{ATP} is simply proportional to the cell surface area A, *i.e.*, $P_{\text{ATP}} = kA$. Equating the viscous power dissipation P_{visc} and the input power P_{ATP} leads to an equation that relates the cell radius a and the streaming velocity V:

$$V \sim \left(\frac{ka}{\eta}\right)^{1/2}. \tag{2.19}$$

Based on this energy model, eqn (2.19) predicts that the speed V grows with the square root of the cell radius a. The predictions of eqn (2.19) are in rough accord with data (Figure 2.5D). Notably, however, the flow speed peaks at or near the cell radius of $a \approx 200$ μm. To our knowledge, the reason for this optimum is unknown. Note that the exact calculation in cylindrical coordinates yields the same scaling, so it is unlikely to be a purely geometric effect.[35]

References

1. A. P. Almeida, J. P. Canejo, S. N. Fernandes, C. Echeverria, P. L. Almeida and M. H. Godinho, Cellulose-based biomimetics and their applications, *Adv. Mater.*, 2018, **30**(19), 1703655.
2. G. Batchelor, H. Moffatt, M. Worster, and T. Osborn, *Perspectives in Fluid Dynamics*, 2000.
3. P. S. Bouche, S. Delzon, B. Choat, E. Badel, T. J. Brodribb, R. Burlett, H. Cochard, K. Charra-Vaskou, B. Lavigne and S. Li, *et al.*, Are needles of

pinus pinaster more vulnerable to xylem embolism than branches? new insights from x-ray computed tomography, *Plant Cell Environ.*, 2016, **39**(4), 860–870.
4. B. Choat, A. R. Cobb and S. Jansen, Structure and function of bordered pits: new discoveries and impacts on whole-plant hydraulic function, *New Phytol.*, 2008, **177**(3), 608–626.
5. B. Choat, S. Jansen, T. J. Brodribb, H. Cochard, S. Delzon, R. Bhaskar, S. J. Bucci, T. S. Feild, S. M. Gleason and U. G. Hacke, *et al.*, Global convergence in the vulnerability of forests to drought, *Nature*, 2012, **491**(7426), 752–755.
6. H. Cochard, F. Froux, S. Mayr and C. Coutand, Xylem wall collapse in water-stressed pine needles, *Plant Physiol.*, 2004, **134**(1), 401–408.
7. B. Ding, R. Turgeon and M. V. Parthasarathy, Substructure of freeze-substituted plasmodesmata, *Protoplasma*, 1992, **169**(1–2), 28–41.
8. H. Fromm and Y. Fichman, Water sensing in plants, in *Sensory Biology of Plants*, Springer, 2019, pp. 79–94.
9. A. Geitmann and J. Gril, *Plant Biomechanics*, Springer International Publishing, Cham, Switzerland, 2018.
10. L. J. Gibson, The hierarchical structure and mechanics of plant materials, *J. R. Soc., Interface*, 2012, **9**(76), 2749–2766.
11. R. E. Goldstein, I. Tuval and J.-W. van de Meent, Microfluidics of cytoplasmic streaming and its implications for intracellular transport, *Proc. Natl. Acad. Sci.*, 2008, **105**(10), 3663–3667.
12. R. E. Goldstein and J.-W. van de Meent, A physical perspective on cytoplasmic streaming, *Interface Focus*, 2015, **5**(4), 20150030.
13. Y. Guerringue, S. Thomine and J.-M. Frachisse, Sensing and transducing forces in plants with msl10 and dek1 mechanosensors, *FEBS Lett.*, 2018, **592**(12), 1968–1979.
14. U. G. Hacke, B. Lachenbruch, J. Pittermann, S. Mayr, J.-C. Domec and P. J. Schulte The hydraulic architecture of conifers, in *Functional and Ecological Xylem Anatomy*, Springer, 2015, pp. 39–75.
15. M. Heil and A. L. Hazel, Fluid-structure interaction in internal physiological flows, *Annu. Rev. Fluid Mech.*, 2011, **43**, 141–162.
16. V. Hernández-Hernández, M. Bentez and A. Boudaoud, Interplay between turgor pressure and plasmodesmata during plant development, *J. Exp. Bot.*, 2020, **71**(3), 768–777.
17. S. Jansen, J.-B. Lamy, R. Burlett, H. Cochard, P. Gasson and S. Delzon, Plasmodesmatal pores in the torus of bordered pit membranes affect cavitation resistance of conifer xylem, *Plant Cell Environ.*, 2012, **35**(6), 1109–1120.
18. K. H. Jensen, K. Berg-Sørensen, H. Bruus, N. M. Holbrook, J. Liesche, A. Schulz, M. A. Zwieniecki and T. Bohr, Sap flow and sugar transport in plants, *Rev. Mod. Phys.*, 2016, **88**(3), 035007.
19. K. H. Jensen, J. Liesche, T. Bohr and A. Schulz, Universality of phloem transport in seed plants, *Plant Cell Environ.*, 2012, **35**(6), 1065–1076.

20. G. N. Karam, Biomechanical model of the xylem vessels in vascular plants, *Ann. Bot.*, 2005, **95**(7), 1179–1186.
21. W. Kim and J. W. Bush, Natural drinking strategies, *J. Fluid Mech.*, 2012, **705**, 7–25.
22. Z. P. Li, A. Paterlini, M. Glavier and E. M. Bayer, Intercellular trafficking via plasmodesmata: molecular layers of complexity, *Cell. Mol. Life Sci.*, 2020, 1–18.
23. J.-F. Louf, G. Guéna, E. Badel and Y. Forterre, Universal poroelastic mechanism for hydraulic signals in biomimetic and natural branches, *Proc. Natl. Acad. Sci.*, 2017, **114**(42), 11034–11039.
24. T. A. McMahon and J. T. Bonner, *On Size and Life*, 1983. No. Sirsi i9780716750000.
25. D. L. Mullendore, C. W. Windt, H. Van As and M. Knoblauch, Sieve tube geometry in relation to phloem flow, *Plant Cell*, 2010, **22**(3), 579–593.
26. W. J. Nicolas, M. S. Grison, S. Trépout, A. Gaston, M. Fouché, F. P. Cordelières, K. Oparka, J. Tilsner, L. Brocard and E. M. Bayer, Architecture and permeability of post-cytokinesis plasmodesmata lacking cytoplasmic sleeves, *Nat. Plants*, 2017, **3**(7), 1–11.
27. K. J. Niklas and H.-C. Spatz, *Plant Physics*. University of Chicago Press, 2012.
28. K. Oparka and D. Prior, Direct evidence for pressure-generated closure of plasmodesmata, *Plant J.*, 1992, **2**(5), 741–750.
29. J. Park, J. Ryu, S. H. Park and S. J. Lee, Air spread through a wetted deformable membrane: Implications for the mechanism of soft valves in plants, *Phys. Rev. E*, 2021, **103**(6), 062407.
30. K. Park, J. Knoblauch, K. Oparka and K. H. Jensen, Controlling intercellular flow through mechanosensitive plasmodesmata nanopores, *Nat. Commun.*, 2019, **10**(1), 1–7.
31. K. Park, A. Tixier, A. Christensen, S. Arnbjerg-Nielsen, M. Zwieniecki and K. Jensen, Viscous flow in a soft valve, *J. Fluid Mech.*, 2018, **836**, R3.
32. W. S. Peters, K. H. Jensen, H. A. Stone and M. Knoblauch, Plasmodesmata and the problems with size: interpreting the confusion, *J. Plant Physiol.*, 2020, 153341.
33. W. Pickard, The role of cytoplasmic streaming in symplastic transport, *Plant Cell Environ.*, 2003, **26**(1), 1–15.
34. W. F. Pickard, Further observations on cytoplasmic streaming in chara braunii, *Can. J. Bot.*, 1972, **50**(4), 703–711.
35. W. F. Pickard, Hydrodynamic aspects of protoplasmic streaming inchara braunii, *Protoplasma*, 1974, **82**(4), 321–339.
36. P. M. Ray and M. S. Bret-Harte, Elastic and irreversible bending of tree and shrub branches under cantilever loads, *Front. Plant Sci.*, 2019, **10**, 59.
37. A. Robards and W. Lucas, Plasmodesmata, *Annu. Rev. Plant Biol.*, 1990, **41**(1), 369–419.
38. Y.-L. Ruan, D. J. Llewellyn and R. T. Furbank, The control of single-celled cotton fiber elongation by developmentally reversible gating of

plasmodesmata and coordinated expression of sucrose and k+ transporters and expansin, *Plant Cell*, 2001, **13**(1), 47–60.
39. P. J. Schulte and U. G. Hacke, Solid mechanics of the torus–margo in conifer intertracheid bordered pits, *New Phytol.*, 2021, **229**(3), 1431–1439.
40. C. Scoffoni, L. Sack and D. Ort, The causes and consequences of leaf hydraulic decline with dehydration, *J. Exp. Bot.*, 2017, **68**(16), 4479–4496.
41. J. W. Smyth and R. M. Shaw, The gap junction life cycle, *Heart Rhythm*, 2012, **9**(1), 151–153.
42. A. D. Stroock, V. V. Pagay, M. A. Zwieniecki and N. Michele Holbrook, The physicochemical hydrodynamics of vascular plants, *Annu. Rev. Fluid Mech.*, 2014, **46**, 615–642.
43. L. Taiz, E. Zeiger, I. M. Møller and A. Murphy, *et al.*, *Plant Physiology and Development*, Sinauer Associates Incorporated, 6th edn, 2015.
44. J. H Thomas, Fluid dynamics of cerebrospinal fluid flow in perivascular spaces, *J. R. Soc., Interface*, 2019, **16**(159), 20190572.
45. S. Timoshenko and J. Goodier, *Theory of Elasticity*, Mc Graw-Hill, New York, 1970.
46. M. T. Tyree and M. H. Zimmermann, *Xylem Structure and the Ascent of Sap*, Springer Science & Business Media, 2013.
47. A. van Casteren, W. I. Sellers, S. K. Thorpe, S. Coward, R. H. Crompton, J. P. Myatt and A. R. Ennos, Nest-building orangutans demonstrate engineering know-how to produce safe, comfortable beds, *Proc. Natl. Acad. Sci.*, 2012, **109**(18), 6873–6877.
48. J.-W. van de Meent, A. J. Sederman, L. F. Gladden and R. E. Goldstein, Measurement of cytoplasmic streaming in single plant cells by magnetic resonance velocimetry, *J. Fluid Mech.*, 2010, **642**, 5–14.
49. A. Vatén, J. Dettmer, S. Wu, Y.-D. Stierhof, S. Miyashima, S. R. Yadav, C. J. Roberts, A. Campilho, V. Bulone and R. Lichtenberger, *et al.*, Callose biosynthesis regulates symplastic trafficking during root development, *Dev. Cell*, 2011, **21**(6), 1144–1155.
50. U. G. Wegst, H. Bai, E. Saiz, A. P. Tomsia and R. O. Ritchie, Bioinspired structural materials, *Nat. Mater.*, 2015, **14**(1), 23–36.
51. M. Xu, E. Cho, T. M. Burch-Smith and P. C. Zambryski, Plasmodesmata formation and cell-to-cell transport are reduced in decreased size exclusion limit 1 during embryogenesis in arabidopsis, *Proc. Natl. Acad. Sci. U. S. A.*, 2012, 201202919.
52. S. R. Yadav, D. Yan, I. Sevilem and Y. Helariutta, Plasmodesmata-mediated intercellular signaling during plant growth and development, *Front. Plant Sci.*, 2014, **5**, 44.
53. Y.-J. Zhang, F. E. Rockwell, A. C. Graham, T. Alexander and N. M. Holbrook, Reversible leaf xylem collapse: a potential "circuit breaker" against cavitation, *Plant Physiol.*, 2016, **172**(4), 2261–2274.
54. Y. M. Kersey and N. K. Wessells, Localization of actin filaments in internodal cells of characean algae. A scanning and transmission electron microscope study, *J. Cell Biol.*, 1976, **68**(2), 264–275.

55. J.-W. van de Meent, A. J. Sederman, L. F. Gladden and R. E. Goldstein, Measurement of cytoplasmic streaming in single plant cells by magnetic resonance velocimetry, *J. Fluid Mech.*, 2010, **642**, 5–14.
56. W. F. Pickard, Further observations on cytoplasmic streaming in Chara braunii, *Can. J. Bot.*, 1972, **50**(4), 703–711.
57. M. Yoneda and R. Nagai, Structural basis of cytoplasmic streaming in Characean internodal cells. A hydrodynamic analysis, *Protoplasma*, 1988, **147**(1), 64–76.
58. R. V. Mustacich and B. R. Ware, Velocity distributions of the streaming protoplasm in Nitella flexilis, *Biophys. J.*, 1977, **17**(3), 229–241.
59. Y. Tsuchiya, H. Yamazaki and T. Aoki, Steady and transient behaviors of protoplasmic streaming in Nitella internodal cell, *Biophys. J.*, 1991, **59**(1), 249–251.
60. W. F. Pickard, Hydrodynamic aspects of protoplasmic streaming in Chara braunii, *Protoplasma*, 1974, **82**(4), 321–339.
61. E. Kamitsubo, Y. Ohashi and M. Kikuyama, Cytoplasmic streaming in internodal cells of Nitella under centrifugal acceleration: a study done with a newly constructed centrifuge microscope, *Protoplasma*, 1989, **152**(2), 148–155.
62. M. Tominaga, S. Sonobe and T. Shimmen, Mechanism of inhibition of cytoplasmic streaming by auxin in root hair cells of Hydrocharis, *Plant Cell Physiol.*, 1998, **39**(12), 1342–1349.
63. R. V. Mustacich and B. R. Ware, A study of protoplasmic streaming in Nitella by laser Doppler spectroscopy, *Biophys. J.*, 1976, **16**(5), 373–388.
64. T. Hayama, T. Shimmen and M. Tazawa, Participation of Ca^{2+} in cessation of cytoplasmic streaming induced by membrane excitation in Characeae internodal cells, *Protoplasma*, 1979, **99**(4), 305–321.
65. P. R. Andjus and D. Vučelić, D2O-induced cell excitation, *J. Membr. Biol.*, 1990, **115**(2), 123–127.
66. T. E. Bostrom and N. A. Walker, Intercellular Transport in Plants: II. Cyclosis and the Rate of Intercellular Transport of Chloride in Chara, *J. Exp. Bot.*, 1976, **27**(2), 347–357.
67. K. Ogata and U. Kishimoto, Rhythmic change of membrane potential and cyclosis of Nitella internode, *Plant Cell Physiol.*, 1976, **17**(2), 201–207.
68. T. Hayashi, Some aspects of behavior of the protoplasmic streaming in plant cells, *Bot. Mag., Tokyo*, 1952, **65**, 51–55.
69. A. A. Bulychev, A. V. Alova and A. B. Rubin, Propagation of photoinduced signals with the cytoplasmic flow along Characean internodes: evidence from changes in chloroplast fluorescence and surface pH, *Eur. Biophys. J.*, 2013, **42**(6), 441–453.

CHAPTER 3

Theoretical Tools and Concepts for Modelling Growing Plant Tissues

OLIVER E. JENSEN

Department of Mathematics, University of Manchester, UK
Email: Oliver.Jensen@manchester.ac.uk

Theory provides an important complement to experiment in understanding the role of mechanical forces in plant growth and development, and the associated concepts underpin our interpretation of observations. In an area where biology, engineering, physics, chemistry and mathematics converge, the individual researcher is likely to be challenged to step outside their area of primary expertise. To help open a few doors between disciplines, this chapter aims to show readers who may have some familiarity with basic mechanics how traditional approaches can be adapted to develop new theoretical models describing growing multicellular plant tissues. More mathematical material appears within a handful of *Technical Comments* that can be ignored by readers uncomfortable with unfamiliar notation. There is not space here to be fully comprehensive, and the topics presented are deliberately selective, influenced by the author's experience working on root growth.[1] A wider view of mechanical modelling of growing tissues is provided by Goriely;[2] a number of plant-specific reviews are also available.[3–6]

Soft Matter Series No. 15
Soft Matter in Plants: From Biophysics to Biomimetics
Edited by Kaare H. Jensen and Yoël Forterre

Published by the Royal Society of Chemistry, www.rsc.org

3.1 An Introduction to the Mechanics of Multicellular Materials

Deformations of plant tissues may be passive (a stem bending in the wind or a leaf hanging under gravity) or active (involving growth, through cell division and cell elongation). Setting aside cell division for the time being, a crucial feature of growing plant tissues is that cells generally remain tightly adhered to their neighbours (in so-called *symplastic growth*). In the absence of dramatic deformations that lead to fracture, strains must then be reasonably smooth functions of position, varying slowly from cell to cell in a tissue. This does not preclude the plant from undergoing large deformations, as small strain gradients can accumulate along slender structures to allow substantial deformations. This is of critical importance in allowing a plant to accommodate and exploit its environment. For example, differential expansion drives the bending of a gravitropic root[7] or the unfurling of an anther.[8] Likewise a long slender leaf can be strongly deformed in a breeze[9] without the strains on its tissues being large enough to cause irreversible deformations.

While deformations may generally be smoothly varying functions of position, the resulting stress fields may not be.[10] This is because plant cells are heterogeneous structures, with soft pressurised interiors and relatively stiff cell walls. In general, stress in a biological material can be very sensitive to the size and location of the domain over which it is measured, as explained in the *Technical Comment.* This raises interesting questions about the heterogeneous distribution of stress within a cell, and how stress (or strain) is sensed within a cell. Cosgrove[11] discusses some of the challenges of measuring material properties of plant tissues at small scales.

Technical Comment. The notion of stress heterogeneity can be expressed more precisely as follows. The (Cauchy) stress $\boldsymbol{\sigma}$ of a material can be interpreted with respect to a sample area A with unit normal $\hat{\mathbf{n}}$, such that the force exerted on A is $\int_A \boldsymbol{\sigma}\cdot\hat{\mathbf{n}}\mathrm{d}A$. That is to say, on each small element $\mathrm{d}A$ of the surface A, the tensor $\boldsymbol{\sigma}$ maps the vector $\hat{\mathbf{n}}$ to a new vector (a stress) $\boldsymbol{\sigma}\cdot\hat{\mathbf{n}}$, leading to a force $\boldsymbol{\sigma}\cdot\hat{\mathbf{n}}\mathrm{d}A$. The total force on A is an integral (a sum) over all such elements. If $\hat{\mathbf{n}}$ is uniform over A, we may therefore define the *average stress* over the sample area as $\langle\boldsymbol{\sigma}\rangle \equiv (1/A)\int \boldsymbol{\sigma}\,\mathrm{d}A$. For a material with microstructure, we expect the average stress to be well defined (*i.e.* to have a robust statistical average) for sufficiently large A. However $\langle\boldsymbol{\sigma}\rangle$ is likely to be highly variable for small A, depending on which component of the microstructure is being sampled.

A cartoon visualising stress across a primary root or stem is shown in Figure 3.1. Each cell contains a vacuole that draws in water from its environment, generating a high osmotic pressure (of approximate magnitude RTC, where $R \approx 8.3\ \mathrm{J\,K^{-1}\,mol^{-1}}$ is the ideal gas constant, T the absolute temperature and C the solute concentration in the vacuole). The tendency of the cell to inflate is restricted by the stiff cell wall, generating a tension in the cell wall. The balance between expansive forces in the cytoplasm, and tensile forces in the cell periphery, give the cell an intrinsic rigidity (like an inflated balloon). Strong adhesion forces between neighbouring cells reinforce this effect, rigidifying the tissue. (Secondary mechanisms of stiffening through deposit of woody material are not considered here.) Across the cross-section of a stem, therefore, we expect to see alternating regions of tensile and compressive tissue (Figure 3.1). When measured with respect to sufficiently small areas A, the stress field fluctuates dramatically across individual cells. However, once integrated over multiple cells, we can consider a field that varies more smoothly. Figure 3.1 illustrates a postulated stress field whereby cells at the periphery of the stem are under net tension, while those nearer the centre are under net compression (mimicking the stress field within individual cells), even though the stress averaged across the whole stem cross-section may vanish. This particular distribution of so-called *residual stress* helps give the stem rigidity.[12] (Similarly, the shoot apical meristem has a stiff epidermal layer, which can be modelled as a pressurised shell.[13,14]) Stress is hard to measure non-invasively but the effects of residual stress are revealed when the stem is cut, allowing central cells to elongate and peripheral cells to shrink, causing the cross-section to bulge.

The stress field in Figure 3.1 illustrates the mathematical idea of *homogenisation*, whereby a fluctuating field is approximated by its spatial

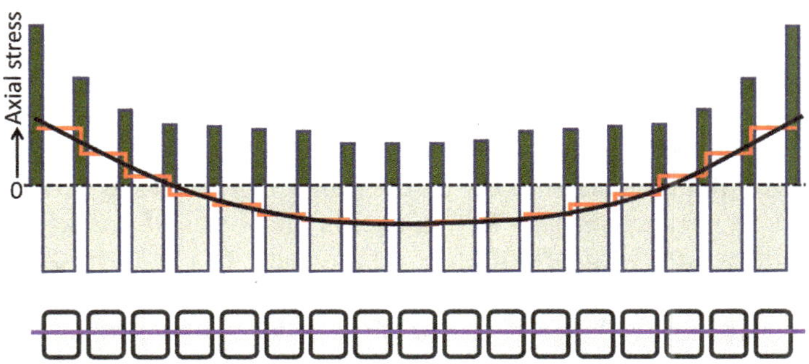

Figure 3.1 A sketch of the axial component of stress (top) in a layer of cells (below) lying across the cross-section of a root or stem, indicating the inhomogeneous nature of residual stress (adapted from[10]). Stress is sampled along the purple line that intersects cell walls (under tension) and cytoplasm (under compression). The stress field therefore fluctuates from positive (dark green) to negative (pale green) values with respect to a reference level. Averaging the stress across each cell (red line) indicates how cells in the centre of the stem may be under net compression, while those near the periphery may be under net tension. The black curve shows the smoothly varying stress field when it is averaged over multiple cells. When averaged over the whole stem, the stress field may vanish (dotted line), even though individual components of the tissue may be subject to large tensile or compressive stresses.

average, which may vary smoothly over long lengthscales. For plant tissues, we might then wish to treat the tissue as a continuum (smoothing out the variations over individual cells) and seek a mechanical model formulated in terms of partial differential equations (PDEs) relating the slowly varying stress field to strain or strain rate, to model deformation under loading or growth. A *constitutive assumption* is required, capturing the mechanical properties of the tissue (as linearly or nonlinearly elastic, viscoelastic, *etc.*) and possibly including a framework for modelling growth (*e.g.* the popular Skalak–Rodriguez formulation[15,16]). We review a set of such models below, starting from simple spatially one-dimensional (1D) descriptions (which are very useful for slender structures) before moving to 3D.

A key question in this approach is to define the appropriate constitutive model. One then asks: how are the macroscopic parameters describing a tissue related to the geometric and physical properties of its microstructure? This question motivates models that resolve individual cells, or components within them. In principle, a cell-scale model can be carefully averaged to derive bulk tissue properties. When cells have a highly organised arrangement (for example a periodic lattice), then the averaging can be performed systematically (*via asymptotic* or *two-scale* homogenisation[17]). More generally, however, there will be a degree of spatial disorder in the tissue, which raises interesting questions about the validity of averaging procedures. The inherently discrete nature of a multicellular tissue may even present mechanical features (such as force chains[18]) that are lost *via* spatial averaging.

Even in upscaling from cells to tissues, a constitutive assumption is needed to describe cell properties, which in turn raises questions about cell microstructure. For plant cells, attention turns immediately to the cell wall, a composite material in which stiff cross-linked cellulose fibres are embedded in a pectin matrix. Plant cells regulate the material properties of the wall, and in particular the orientation of the fibres within it, to influence cell, tissue and organ properties. Key concepts here are *anisotropy* (material properties that vary with orientation with respect to the fibre distribution) and *plasticity* (irreversible deformations that arise for materials that are subject to sufficiently large stress or strain). Plants present a fascinating challenge in that the molecular microstructure (fibre orientation in cell walls, at the sub-micron scale) can influence organ shape (up to metre scales). Alongside the 'bottom-up' question of how microstructure influences macroscopic properties, we must also consider the 'top-down' question of how the plant senses its environment (*via* signals associated with gravity, light, water, nutrients, *etc.*) and responds at the molecular level. Hierarchies of mathematical and computational models are needed that can efficiently communicate information up and down scales, exemplifying the challenge of multiscale modelling.

This endeavour is complemented by the increasing availability of 3D imaging of plant tissues, allowing detailed geometric data to be integrated with models of mechanical and biological function. To some extent this is beguiling: while shape (and the distribution of specific molecules) can be studied in exquisite detail, measurement of mechanical properties and particularly of stress (as mentioned above) remains much more challenging. Theoretical and computational models are essential in inferring mechanical properties from imaging data and stress assays.

3.2 Simple Constitutive Models and Lockhart's Equation

In nonlinear elasticity, different representations of stress can be used, defined with respect to the deformed or undeformed configurations of a material. For small deformations, however, these distinctions can be ignored, and for slender structures it is often sufficient to focus on a single dominant (axial, scalar) stress component σ (taken from a tensor $\boldsymbol{\sigma}$), with dimensions force/area. When integrated over the thickness of structure, the resulting force is sometimes called a *stress resultant* (with dimensions force/length).

Likewise, large deformations of materials in 3D need careful definition of strain. For small (axial) deformations of slender structures, however, it is sufficient to represent the strain as the *relative elongation* $e = \delta l/l$ for an object of length l deforming by extension δl. When length $l(t)$ varies with respect to time t, the *strain rate* $\dot{e} = (1/l)\mathrm{d}l/\mathrm{d}t$, is termed the *relative elongation rate* (RER) or *relative elementary growth rate* (REGR) in the plant science community and is a central quantity in modelling growth by elongation.

We now revisit some standard constitutive descriptions of materials, moving toward the *elasto-viscoplastic* model that is commonly used to model growing plant tissues. Cosgrove[19] makes the important point about accuracy in terminology: wall 'softness', for example, is vague, as it does not distinguish *compliance* (an elastic property) from *extensibility* (a viscous property).

3.2.1 Simple Constitutive Laws in One Dimension

For an *elastic* material, deformation is fully reversible: a load induces a strain and removal of the load causes the material to recover its initial state. At a molecular level, it is assumed there is no breakage of molecular bonds. In 1D, we write $\sigma = Ee$ for some stiffness E (associated with the Young's modulus). This linear relation is a statement of Robert Hooke's famous 'law'. There is an associated mechanical *strain energy* $W = \frac{1}{2}Ee^2$, so that $\sigma = \mathrm{d}W/\mathrm{d}e$ and $E = \mathrm{d}^2W/\mathrm{d}e^2$.

For a *viscous* material, deformation is irreversible (implying breakage of molecular bonds), and the material does not return to its original configuration after a load is removed. In 1D, we write $\sigma = \eta\dot{e}$ for some viscosity η, where a dot denotes a time derivative. This can be inverted to read $\dot{e} = \phi\sigma$ where $\phi = \eta^{-1}$ is called an *extensibility* in the plant science community. The quantity $\sigma\dot{e} = \eta\dot{e}^2$ is the rate at which mechanical energy is dissipated by bond breakage.

For *viscoelastic* materials, two basic formulations are normally adopted, with viscous and elastic elements assumed to act in series (the so-called Maxwell model) or in parallel (Kelvin), representing predominantly fluid-like or solid-like materials, respectively. When in parallel (so that the stresses of each element contribute to the total stress), we write $\sigma = Ee + \eta\dot{e}$. When in series (so that the strains of each element contribute to the total strain), we write

$$\dot{e} = \phi\sigma + E^{-1}\dot{\sigma}. \tag{3.1}$$

These expressions reveal a *relaxation time* $\eta/E \equiv E^{-1}/\phi$ over which the material responds to a change in loading conditions. For example, the Maxwell model (eqn (3.1)) describes how, under a step-change in deformation, there is a rapid elastic response, followed by *relaxation* of the stress to zero. Alternatively, the Kelvin model describes *creep*, the time-dependent deformation arising in response to a step-change in stress. More elaborate combinations of elements can be used to model more complex responses.

To describe a *plastic* material we must introduce the concept of a *yield stress* Y (or possibly a *yield strain*), below which the material remains (largely) undeformed but above which there is irreversible deformation. The viscous model is adapted to read

$$\dot{e} = \phi[\sigma - Y]_{+} \tag{3.2}$$

where the + subscript denotes (for a scalar argument) a *Heaviside function*: $[x]_{+} = x$ if $x > 0$ and $[x]_{+} = 0$ otherwise. This innocuous constitutive law in eqn (3.2) (sometimes called a *Bingham model*) is piecewise linear, but the discontinuity makes eqn (3.2) strongly nonlinear overall (not even everywhere differentiable). Yield is a complex topic, reviewed elsewhere.[20]

We can combine eqn (3.1) and (3.2) to form a (Maxwell) *elasto-viscoplastic* material, satisfying

$$\dot{e} = \phi[\sigma - Y]_{+} + E^{-1}\dot{\sigma}. \tag{3.3}$$

Weak loads, for which σ remains below Y, lead to reversible elastic deformation with $\sigma = Ee$. Stronger loads irreversibly deform the material. This model can be used to explain the way in which a ribbon (or a strip of paper, *i.e.* cellulose) can be curled by running it over a blade, provided it is under sufficient load.[21]

It is important to remember that all of these constitutive models are based on very restrictive assumptions. A real material is likely to have complex behaviour that falls outside these standard descriptions. Obvious missing features are dimensionality (considering 2D or 3D rather than 1D deformations, implicating additional components of stress and strain), nonlinearity, anisotropy, heterogeneity and, for plants, active processes that lead to growth.

3.2.2 Lockhart's Model

Lockhart[22] introduced an influential model for the expansion of a single plant cell. Suppose that an isolated thin-walled cylindrical cell elongates along its axis. Its volume is $V = Al$, where A is its cross-section. As long as A is constant, then the volumetric strain rate is equivalent to the linear strain rate,

$$\frac{1}{V}\frac{\mathrm{d}V}{\mathrm{d}t} = \frac{1}{l}\frac{\mathrm{d}l}{\mathrm{d}t} = \dot{e}. \tag{3.4}$$

If P is the pressure within the cell, then a force balance on the end plate (assuming the cell is under no external load) gives

$$PA = Tp \tag{3.5}$$

where T is the axial stress resultant (axial tension) in the cell wall and p the length of the perimeter. (For a circular cross-section, $A = \pi R^2$ and $p = 2\pi R$ where R is the cell radius, so that $T = PR/2$. Away from the ends of the cell, there is an additional transverse (or hoop) stress resultant in the curved cell wall, given by the *Young–Laplace* condition as PR, exactly double the axial stress resultant.) If we adopt a 1D viscoplastic model for the cell wall (eqn (3.2)), then

$$\frac{1}{l}\frac{\mathrm{d}l}{\mathrm{d}t} = \phi[\sigma - Y]_+ \tag{3.6}$$

where $\sigma = T/h$ and Y is a yield stress. (Here σ represents the axial stress in the wall, averaged over the wall thickness h.) Then eqn (3.4) and (3.5) give

$$\frac{1}{V}\frac{\mathrm{d}V}{\mathrm{d}t} = \theta\phi\left[P - \frac{Y}{\theta}\right]_+ \quad \text{where } \theta \equiv \frac{A}{ph}. \tag{3.7}$$

The geometric structure of the cell, captured through the dimensionless ratio θ, converts the wall extensibility ϕ and wall yield stress Y to an effective cell extensibility $\theta\phi$ and cell yield stress Y/θ. We will shortly see how the viscoplastic structure of the model is preserved as we move up to tissue level.

A plant cell regulates its turgor pressure using solutes that draw in water from its surroundings by osmosis. The flux per unit area across the plasma membrane (adjacent to the cell wall) is given by a Starling relation, of the form $J = k(\Delta P - \Delta\Pi)$, where ΔP is a hydrostatic pressure difference and $\Delta\Pi$ an osmotic pressure difference. k is a *membrane permeability*. For simplicity, let us assume that the hydrostatic and osmotic pressures are zero outside the cell, and that the flux takes place across the curved wall of the cell, of area pl. Then mass conservations demands that

$$\frac{\mathrm{d}V}{\mathrm{d}t} = plk(\Pi - P). \tag{3.8}$$

If we assume that P is large enough to allow cell expansion, we can eliminate P between (3.7) and (3.8) to obtain

$$\frac{1}{V}\frac{\mathrm{d}V}{\mathrm{d}t}=\frac{\phi\theta[\Pi-Y/\theta]}{1+(\phi\theta^2 h/k)} \tag{3.9}$$

if $\Pi - Y/\theta$ is sufficiently large to induce yielding. The parameter $\phi\theta^2 h/k$ represents a ratio of the time taken for water to cross the cell wall to the time taken for the cell to elongate. If water transport is rapid, which is a conventional assumption, then $\phi\theta^2 h/k \ll 1$, $P \approx \Pi$ and

$$\frac{1}{l}\frac{\mathrm{d}l}{\mathrm{d}t}=\phi\theta[\Pi-Y/\theta]_{+}. \tag{3.10}$$

This demonstrates how the cell acts as an osmotic pump. Provided solutes can be concentrated sufficiently (increasing the osmotic pressure Π), then water is drawn quickly into the cell, increasing its volume. Lockhart's model treats growth as a viscoplastic process, with molecular bonds being broken irreversibly in the cell wall to allow its elongation. The assumption that the wall thickness h remains constant as the cell elongates implicitly accounts for active metabolic processes that deliver new material to the cell wall (which would be expected to thin as it stretches), allowing the wall to maintain its thickness as it expands. More complex treatments examine the coupling between water fluxes and growth.[23]

Finally, in some circumstances it is helpful to incorporate the elastic response of the cell as well, in which case eqn (3.3) leads to

$$\frac{1}{l}\frac{\mathrm{d}l}{\mathrm{d}t}=\phi\theta[\Pi-Y/\theta]_{+}+\theta E^{-1}\frac{\mathrm{d}\Pi}{\mathrm{d}t}. \tag{3.11}$$

> *Technical Comment.* An alternative formulation that turns out to be useful[24] in higher dimensions is as follows. Write $e=e_\mathrm{e}+e_\mathrm{g}$, decomposing the strain in adjacent Maxwell elements into that due to the elastic component $e_\mathrm{e}=\sigma/E$, and that arising from plastic deformation (used to mimic growth). Let $e_\mathrm{Y}=Y/E$ be the yield strain associated with the yield stress Y. Then eqn (3.3) can be re-expressed as
>
> $$e_\mathrm{e}=E^{-1}\sigma,\quad \dot{e}_\mathrm{g}=\phi E[e_\mathrm{e}-e_\mathrm{Y}]_{+},\quad e=e_\mathrm{e}+e_\mathrm{g}, \tag{3.12}$$
>
> so that a load σ induces an (internal) elastic strain which, if large enough, increases the unloaded length of the material. For eqn (3.11), we see that $e_\mathrm{e}=\Pi\theta/E$. If one introduces $\bar{l}$ as the length that the cell would take due to growth alone, a variation of eqn (3.12) has
>
> $$e_\mathrm{e}=\Pi\theta/E,\quad \dot{\bar{l}}/\bar{l}=\phi E[e_\mathrm{e}-e_\mathrm{Y}]_{+},\quad l=e_\mathrm{e}\bar{l}. \tag{3.13}$$
>
> We will see later how this model is formulated in 3D.

3.3 One-dimensional Models for Elongation of Slender Tissues

Roots and shoots can be long and slender, encouraging the use of a spatially 1D description. Following the pioneering work of Silk,[25] let us consider how to model a line of elongating cells, moving from a discrete description of individual cells formulated as sets of ordinary differential equations (ODEs) to a continuous description formulated using PDEs.

3.3.1 A Continuous Model for Primary Root Growth

A primary root has a meristem at its tip, in which cell division and differentiation take place. Immediately behind the meristem is an *elongation zone* (EZ), in which tightly adherent cells lengthen rapidly before maturing. In this simple treatment, we will assume no cell division takes place in the EZ. In the frame of reference of the root tip, there is a flux of cells through the EZ in the shootward direction, leaving the EZ at some speed v_∞. Mature cells are stationary with respect to the plant and its environment, implying that in this frame of reference the elongation zone and meristem propagate downward through soil at speed v_∞.

Let x measure distance along a line of root cells, measured from the meristem. We suppose cells are introduced at $x=0$ through a process of cell division. Let cell j occupy $X_{j+1}(t)<x<X_j(t)$, so that it has length $L_j=X_j-X_{j+1}$. Its speed with respect to the meristem is $V_j=\frac{1}{2}(\dot{X}_j+\dot{X}_{j+1})$. (It is convenient to label cells entering the tissue with increasing values of j.) We assume each cell undergoes elongation according to a Lockhart law before maturing at the downstream end of the elongation zone. If cells enter the domain with fixed period τ, then the tissue will appear stationary when viewed through a stroboscope with the same period, *i.e.* $X_j(t)=X_{j+1}(t+\tau)$, as illustrated in Figure 3.2.

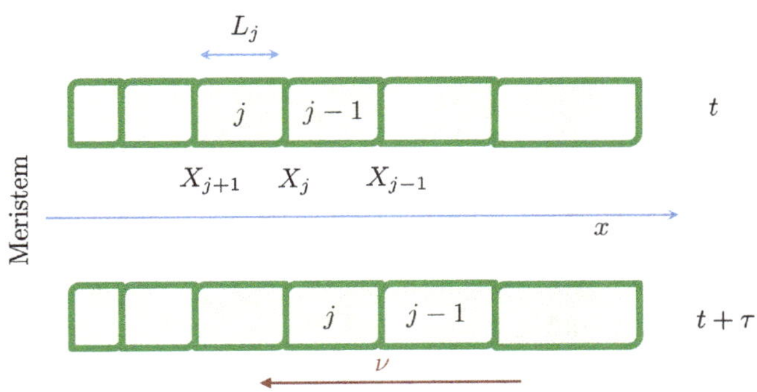

Figure 3.2 A row of cells at time t (top) and at time $t+\tau$ (bottom), where τ is the division time over which a new cell enters the domain, in the frame of reference of the meristem (left) from which the cells emerge. Mature elongated cells exit the domain at the right. x measures distance from the meristem; ν is a variable counting (and anchored to) cells, increasing in the opposite direction.

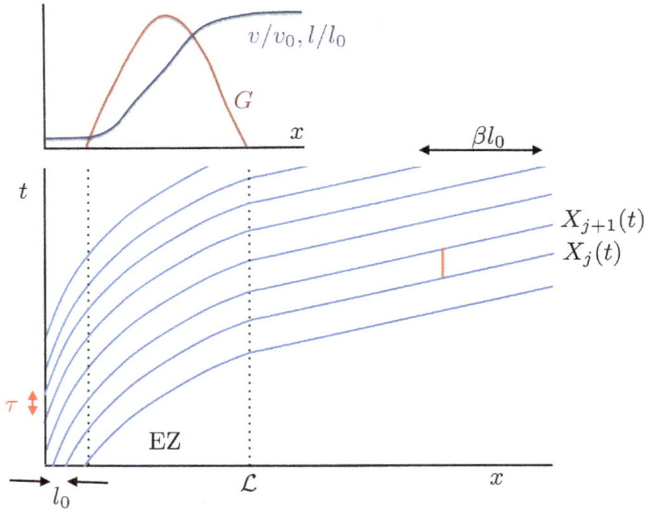

Figure 3.3 The elongation zone (EZ) along a root (between vertical dotted lines) is characterised by a region of positive RER, indicated by the function $G(x)$ (sketched, top), where x measures distance from the meristem. Cells enter the EZ at length l_0, speed v_0 at rate $1/\tau$ and leave with length βl_0 and speed βv_0, where the amplification factor β satisfies (17); the increase of relative length l/l_0 and speed v/v_0 across the EZ is sketched (top). The space–time diagram (bottom) shows cell vertex locations $X_j(t)$. The vertical red line indicates how $X_j(t) = X_{j+1}(t+\tau)$; horizontal distances between curves indicate cell lengths.

The trajectories of cell vertices are shown in a space–time plot in Figure 3.3. Horizontal distances between the curves show increasing cell lengths. The slope of each curve diminishes with respect to x, reflecting increasing speed with respect to the meristem. To understand how spatial observations reveal the time-evolution of individual cells, we approximate the discrete model for individual cells as a continuous model for a tissue.

Technical Comment. The following argument reveals an important relationship between *Eulerian* (laboratory-based) and *Lagrangian* (material-based) quantities. If the cell length and speed vary sufficiently smoothly with respect to j, we can imagine interpolating the discrete functions L_j and V_j with smooth analogues $l(\nu, t)$ and $v(\nu, t)$, where ν is a Lagrangian variable such that $l(j, t) = L_j(t)$ and $v(j, t) = V_j(t)$. The location of cell vertices is $x(j, t) = X_j(t)$. Approximating differences with derivatives, so that

$$\left.\frac{\partial x}{\partial \nu}\right|_t \approx X_{j+1} - X_j, \quad \tau \left.\frac{\partial x}{\partial t}\right|_\nu \approx X_j(t+\tau) - X_j(t), \tag{3.14}$$

we have (to leading order)

$$l=-\frac{\partial x}{\partial \nu}, \quad v=\frac{\partial x}{\partial t}, \quad \tau\frac{\partial x}{\partial t}+\frac{\partial x}{\partial \nu}=0. \tag{3.15}$$

We see immediately that $l=\tau v$, *i.e.* that the length of a cell is directly related to its speed. It also follows *via* careful use of the chain rule that

$$\left.\frac{\partial v}{\partial x}\right|_{t}=\frac{\partial v/\partial \nu}{\partial x/\partial \nu}=\frac{\partial^2 x/\partial \nu\partial t}{-l}=\left.\frac{1}{l}\frac{\partial l}{\partial t}\right|_{\nu}\approx\frac{\dot{L}_j}{L_j}. \tag{3.16}$$

Eqn (3.16) shows that the RER of individual cells ($\dot{L}_j/L_j$) is directly related to the spatial velocity gradient of the tissue ($\partial v/\partial x$). Because cell length and speed are related by $l=\tau v$, the RER can be measured[26,27] from an image *via* $(\partial l/\partial x)|_t/\tau$, which we will call $G(x)$, say. (For a steadily growing root, $G(x)$ will not vary with time.) Eqn (3.16) also shows how the velocity gradient can be written $\dot{F}F^{-1}$, where $F=\partial x/\partial \nu$ is a form of *deformation gradient*.

Noting that spatial hormone distributions (of auxin in particular) regulate cell expansion, we can model growth directly in a continuous framework using position-dependent functions of extensibility and yield, which we encompass in an RER distribution $G(x)$ (Figure 3.3). We can integrate $\partial v/\partial x=G(x)$ to recover $v(x)=v_0+\int_0^x G(x')\mathrm{d}x'$ and $l(x)=\tau v$, where $l_0=\tau v_0$ is the length of cells where they enter the domain. $G(x)$ is a hump-like function,[26] confined (by definition) to an elongation zone ($0<x<\mathcal{L}$, say). Mature cells leave the elongation zone with length βl_0 where

$$\beta=1+\frac{\tau}{l_0}\int_0^{\mathcal{L}} G(x')\mathrm{d}x'. \tag{3.17}$$

In a root, mature cells are stationary with respect to the surrounding soil, implying that the speed with which the root tip penetrates the soil is $\beta v_0=\beta l_0/\tau$. In *Arabidopsis*, β can be as large as 30.[26]

Technical Comment. The evolution of a scalar field $m(x, t)$ (such as a hormone concentration) within individual cells satisfies

$$\left.\frac{\partial m}{\partial t}\right|_{\nu}=\left.\frac{\partial m}{\partial t}\right|_{x}+\left.\frac{\partial m}{\partial x}\right|_{\nu}\frac{\partial x}{\partial \nu}=\frac{\partial m}{\partial t}+v\frac{\partial m}{\partial x}, \tag{3.18}$$

which students of fluid mechanics will recognise as a *material* or *Lagrangian derivative*. This is equivalent to evaluating $\mathrm{d}m/\mathrm{d}t$ along the characteristic $\mathrm{d}x/\mathrm{d}t=v$. The cell vertex locations in Figure 3.3 illustrate the characteristics. A little algebra shows that the cell density, $\rho = 1/l$, satisfies

$$\frac{\partial \rho}{\partial t} + \frac{\partial}{\partial x}(\rho v) = 0, \tag{3.19}$$

which is equivalent to the mass conservation equation for a compressible flow in one dimension. This demonstrates that the *cell flux* in the frame of the root tip is ρv (which is a constant, $1/\tau$ in the present simple example). In the presence of cell division, eqn (3.19) would be supplemented by a source term on its right-hand-side.

3.3.2 Growth Against an External Load

As an illustration, working on the assumption that spatial hormone patterns regulate cell extensibility and yield, making $\phi = \phi(x)$ and $Y = Y(x)$ functions of position in eqn (3.10), the steady cell length distribution can be calculated from

$$\frac{1}{\tau}\frac{\mathrm{d}l}{\mathrm{d}x} = \phi(x)\theta[\Pi - \Sigma - Y(x)/\theta]_{+}. \tag{3.20}$$

Here eqn (3.16) has been used to write the RER as a spatial gradient and the term Σ has been added to represent the compressive stress that runs along a line of cells that are expanding against an axial load imposed at the tip of the organ (*e.g.* a primary root), neglecting other loads on the cells. A crude model of an elongation zone has $Y = Y_0$ in $0 < x < \mathcal{L}$ and $Y = Y_1$ in $x > \mathcal{L}$, where Y_0 is sufficiently small to allow growth but Y_1 is sufficiently large to suppress growth. Then, assuming cells grow from length l_0 at $x = 0$ to their mature length βl_0 at $x = \mathcal{L}$, where they move at speed v_∞ with respect to the meristem, it follows that

$$v_\infty = \frac{\beta l_0}{\tau} = v_0 + \theta\left[\Pi - \Sigma - \frac{Y_0}{\theta}\right]\Phi, \quad \text{where } \Phi = \int_0^{\mathcal{L}} \phi(x)\mathrm{d}x. \tag{3.21}$$

Assuming mature cells are stationary with respect to their environment (soil or agar), then the meristem moves with speed v_∞ with respect to the environment. Eqn (3.21) illustrates how the speed of the root tip is determined by the collective expansion of all the cells in the EZ. If this motion induces a drag $\Sigma = k_d v_\infty$ for some $k_d > 0$, then

$$v_\infty = \frac{(l_0/\tau) + [\Pi\theta - Y_0]\Phi}{1 + \theta k_\mathrm{d}\Phi}. \tag{3.22}$$

This expression illustrates how growth rates may be suppressed by an environmental drag k_d *via* passive mechanical resistance; the drag may also induce actively regulated changes of growth rate *via thigmotropic* or touch-sensitive responses, which have not been modelled here.

3.4 Quasi-1D Models for Bending of Slender Tissues

Previously, we scaled up the stress/strain-rate relation for a cell wall (eqn (3.6)) to a single cell (eqn (3.10)), introducing the geometric parameter θ. Let us now repeat the exercise for multiple cells in a root cross-section (Figure 3.4). We idealise it as a set of thin cell walls under tension containing cytoplasm under turgor pressure $P \equiv \Pi$. Integration of a force balance across the whole root cross-section[28] (assuming no external load on the root), splitting the area integral into contributions from cytoplasm and cell walls, gives

$$P\mathcal{A} = \langle Y \rangle + \left\langle \frac{1}{\phi} \mathrm{RER}_{\mathrm{cell}} \right\rangle, \quad \text{for } P\mathcal{A} > \langle Y \rangle, \tag{3.23}$$

balancing the total force due to turgor with the net tension induced in walls, modelled by eqn (3.6). Here area integrals of stresses over slender cell walls have been reduced to line integrals of stress resultants, so that angle brackets denote line integration along cell walls, accounting for variable wall thickness if necessary; in the single-cell example in eqn (3.7), for example, $\langle Y \rangle$ was expressed as phY. $\mathcal{A}$ is the root cross-sectional area (approximating the area containing cytoplasm) and turgor is assumed uniform across the root. Suppose all cells elongate at the same rate, *i.e.* with the same RER. Then

$$RER = \frac{1}{\langle \phi^{-1} \rangle} (P\mathcal{A} - \langle Y \rangle), \quad \text{for } P\mathcal{A} > \langle Y \rangle, \tag{3.24}$$

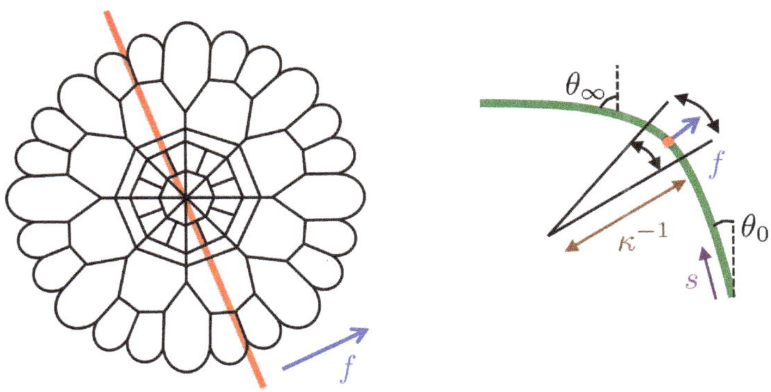

Figure 3.4 Left: An idealised representation of the cross-section of an *Arabidopsis* primary root (adapted from[28]). In a tropic response, material properties on one side of the root are altered with respect to the other, inducing bending of the root centreline by differential expansion. The red line indicates the axis around which the root bends and f measures perpendicular distance from this line. Right: a transverse view of the root shows how the elongation rate is greater on the outer bend ($f > 0$) than the inner bend ($f < 0$); κ^{-1} is the local radius of curvature.

defining the effective extensibility $\langle \phi^{-1} \rangle^{-1}$ and yield $\langle Y \rangle$ of the cross-section. This is the Lockhart equation once more, now upscaled (or homogenised) for a complete root cross-section. An immediate consequence is that the peripheral (epidermal) cells have the greatest contribution to $\langle \phi^{-1} \rangle$ and $\langle Y \rangle$, by virtue of having the largest length of cell walls of all layers.[28]

If material properties vary slightly across the root cross-section, then there may be a slight gradient in RER across the cross-section, inducing bending of the root centreline. This is encoded using the curvature κ of the centreline. Let f measure perpendicular distance from a diameter across the root (Figure 3.4), such that the diameter is parallel to the axis of curvature of the root. Then a geometric argument (in the *Technical Comment* below) tells us that $\mathrm{RER_{cell}} \approx \mathrm{RER} + \mathrm{CGR}f$, where the curvature growth rate $(\mathrm{CGR}) = \mathrm{d}\kappa/\mathrm{d}t$ for a root segment. A *moment balance* across the cross-section, assuming no external couple on the root (multiply eqn (3.23) by f and integrate over the cross-section[28]) reveals that

$$-\left\langle \frac{f^2}{\phi} \right\rangle \mathrm{CGR} = \langle fY \rangle + \left\langle \frac{f}{\phi} \right\rangle \mathrm{RER} - P \int f \mathrm{d}A, \quad \text{for } PA > \langle Y \rangle. \tag{3.25}$$

This shows how asymmetries in yield, extensibility and turgor (measured by first spatial moments as integrals over cell walls) can each generate curvature of the structure.

Technical Comment. To relate $\mathrm{RER_{cell}}$ to CGR, consider a short segment of root, with centreline arclength δ_0 and curvature κ, subtending an angle ψ at its centre of curvature, a distance $1/\kappa$ from the centreline. Thus $\psi = \delta_0/(1/\kappa) = \kappa\delta_0$. The arclength δ through the segment a radial distance f from the centreline subtends the same angle ψ. Thus $\kappa\delta_0 = \delta/[(1/\kappa) + f]$ and so $\delta = \delta_0 (1 + \kappa f)$. Assuming δ_0 and κ are functions of time, and that $\kappa f \ll 1$, then $\dot{\delta}/\delta = \dot{\delta}_0/\delta_0 + \dot{\kappa} f$, *i.e.* $\mathrm{RER_{cell}} = \mathrm{RER} + \mathrm{CGR}\,f$ for a cell a distance f from the centreline of the segment.

Eqn (3.25) shows how transverse gradients in ϕ are amplified by stretching to induce bending. Considering this effect in isolation, we can write $\mathrm{d}\kappa/\mathrm{d}t = A_0 \mathrm{RER}$ for the root segment, where A_0 characterises the material asymmetry in extensibility across the root. If the curvature remains sufficiently weak, we may define distance along the structure with arclength s instead of linear distance x, but still exploit the 1D model derived in Section 3.3. From eqn (3.18), we may write the CGR as a material derivative, so that

$$\frac{\partial \kappa}{\partial t} + v \frac{\partial \kappa}{\partial s} = A_0 \frac{\mathrm{d}v}{\mathrm{d}s}, \tag{3.26}$$

showing how curvature of a root segment is amplified by stretching as it is swept through the elongation zone.

Technical Comment. We can use eqn (3.26) to write down a simple model of root gravitropism. Let θ_0 be the angle of the root tip to the vertical and θ_∞ the angle of root to the vertical at the top of the elongation zone (Figure 3.4). Then $\theta_\infty = \theta_0 + \int_0^{\mathcal{L}} \kappa \mathrm{d}s$. Assuming mature cells are immobile with respect to their surroundings, that there are no external forces constraining lateral motions of the elongation zone, and that the gravitropic signal is communicated rapidly (by auxin transport) to tissues in comparison to the timescale for growth, then a model for the gravitropic response of the root is

$$\frac{\partial \kappa}{\partial t} + v\frac{\partial \kappa}{\partial s} = \chi \frac{\mathrm{d}v}{\mathrm{d}s} \sin\left[\theta_\infty - \int_0^{\mathcal{L}} \kappa \mathrm{d}s\right], \quad \frac{\mathrm{d}\theta_\infty}{\mathrm{d}t} = -\kappa_\infty v_\infty \tag{3.27}$$

where eqn (3.27) must be solved with $\kappa(0, t) = 0$ to determine $\kappa(\mathcal{L}, t) \equiv \kappa_\infty$. Here χ is a parameter that measures the degree of asymmetry induced by the gravitropic signal, arising from gravity sensing at the root tip. The path of the root tip with respect to the surroundings is determined by the evolution of θ_∞ as tissue is steadily extruded from the elongation zone at axial speed v_∞. The constraints of the environment on the root's motion are an important omission.

The simple model (eqn (3.27)) is a variant of the famous 'sine law' model for shoot gravitropism, which takes the form $\partial\kappa/\partial t = -\chi \sin\theta - \gamma\kappa$, where $\kappa = \partial\theta/\partial s$.[29–31] The χ term reflects the local response to the gravity stimulus that takes place in the shoot (the advective derivative has been neglected). The γ term models *proprioception*, an inherent tendency of the organ to straighten; for a comprehensive review of posture control see Moulia *et al.*[32]

3.5 Constitutive Models for 3D Anisotropic Growing Materials

To generalise the simple constitutive models of Section 3.2 from one to three spatial dimensions, at either the cell or tissue level, we must reformulate them using tensor calculus. Let us first briefly review descriptions of materials that have linear constitutive laws. We write stress as the symmetric tensor $\boldsymbol{\sigma} = \boldsymbol{\sigma}^T$ (where T denotes transpose) such that the (vector) force per unit area acting on a surface with unit normal $\mathbf{n}$ is $\boldsymbol{\sigma}\cdot\mathbf{n}$. A small deformation $\mathbf{u}$ is characterised by strain $\mathrm{e} = \frac{1}{2}(\nabla\otimes\mathbf{u} + (\nabla\otimes\mathbf{u})^T)$. An *incompressible* material has $\nabla\cdot\mathbf{u} = \mathrm{tr}(\mathrm{e}) = 0$ (where tr denotes trace), implying that material elements do not change volume.

Technical Comment. To clarify notation, vectors are represented as 3×1 column vectors with respect to Cartesian coordinates, and tensors as 3×3 matrices. ∇ is the spatial gradient operator, and can be written $(\partial/\partial x, \partial/\partial y, \partial/\partial z)^T$ or

$(\partial/\partial x_1,\ \partial/\partial x_2,\ \partial/\partial x_3)^T$ with respect to a right-handed Cartesian coordinate system with unit vectors $\hat{\mathbf{x}}, \hat{\mathbf{y}}$ and $\hat{\mathbf{z}}$. The dot product of two vectors $\mathbf{a}=(a_1, a_2, a_3)^T$ and $\mathbf{b}=(b_1,\ b_2,\ b_3)^T$ is $\mathbf{a}\cdot\mathbf{b}=\mathbf{a}^T\mathbf{b}=a_1b_1+a_2b_2+a_3b_3$. Thus if $\mathbf{u}=(u_1,\ u_2,\ u_3)$, then $\nabla\cdot\mathbf{u}=\partial u_1/\partial x_1+\partial u_2/\partial x_2+\partial u_3/\partial x_3$. The *outer product* $\otimes$ is defined such that $\{\mathbf{a}\otimes\mathbf{b}\}_{ij}\equiv\{\mathbf{ab}^T\}_{ij}=a_ib_j$ for i and j taking values 1, 2 and 3, and $(\mathbf{a}\otimes\mathbf{b})^T=(\mathbf{ab}^T)^T=\mathbf{ba}^T=\mathbf{b}\otimes\mathbf{a}$.

In linear elasticity, we assume a linear relation between stress and strain, of the form $\boldsymbol{\sigma}=\mathrm{C}{:}\mathrm{e}$, where C is a fourth-order *stiffness tensor*. (In terms of components, $\sigma_{ij}=\sum_{k,l} C_{ijkl}e_{kl}$.) For a *homogeneous isotropic* material (a very poor description of plant tissues), C is characterised by just two parameters (despite having 81 components), the elastic *Lamé constants* λ and μ.

Technical Comment. Using index notation, symmetry and related arguments, it can be shown that

$$C_{ijkl}=\lambda\delta_{ij}\delta_{kl}+\mu(\delta_{ik}\delta_{jl}+\delta_{il}\delta_{jk}) \tag{3.28}$$

where $\delta_{ij}=1$ if $i=j$ and 0 otherwise, giving $\boldsymbol{\sigma}=\lambda\mathrm{Itr}(\mathrm{e})+2\mu\mathrm{e}$, where I is the second-order identity tensor. In the incompressible limit, $\lambda\to\infty$ and $\mathrm{tr}(\mathrm{e})\to 0$ with their product remaining finite, in the form of a Lagrange multiplier ($-p$, a pressure). In this case, stress and strain are related in 3D by

$$\boldsymbol{\sigma}=-p\mathrm{I}+2\mu\mathrm{e}, \tag{3.29}$$

with μ sometimes termed the *shear modulus*. To illustrate how this generalises the 1D model presented in Section 3.2.1, consider an incompressible rod undergoing uniaxial extension, with $\mathbf{u}=\alpha\left(z\hat{\mathbf{z}}-\frac{1}{2}x\hat{\mathbf{x}}-\frac{1}{2}y\hat{\mathbf{y}}\right)$, so that $\mathrm{e}=\alpha\left(\hat{\mathbf{z}}\otimes\hat{\mathbf{z}}-\frac{1}{2}\hat{\mathbf{x}}\otimes\hat{\mathbf{x}}-\frac{1}{2}\hat{\mathbf{y}}\otimes\hat{\mathbf{y}}\right)$. The transverse components of stress (in the x and y directions) vanish if $p=-\mu\alpha$, in which case the axial (z) component of stress is $3\mu\alpha$. Thus we recover the 1D description of linear elasticity, with a linear relation between axial stress and axial strain α, with coefficient $E=3\mu$.

An incompressible isotropic viscous material with a linear stress/strain-rate relation (*i.e.* a Newtonian fluid) satisfies a relation analogous to that of an incompressible elastic material, namely $\boldsymbol{\sigma}=-p\mathrm{I}+2\eta\dot{\mathrm{e}}$, where η is a viscosity and $\dot{\mathrm{e}}$ is a rate-of-strain tensor, defined in terms of the velocity field $\mathbf{v}$ as $\dot{\mathrm{e}}=\frac{1}{2}(\nabla\otimes\mathbf{v}+(\nabla\otimes\mathbf{v})^T)$ with $\nabla\cdot\mathbf{v}=0$. Pressure p appears again as an isotropic component of the stress, ensuring incompressibility.

3.5.1 Anisotropy

The plant cell wall can be modelled as a composite *fibre-reinforced* material, having cellulose microfibrils embedded in a pectin matrix. The anisotropic structure of the cell wall, combined with the orientation of cell walls within a tissue, is crucial to the plant's ability to alter its shape in order to adapt to its environment. It is instructive to examine the form of the elastic stiffness tensor for a segment of cell wall, treating it as a thin sheet (so using a 2D approximation). Let **a** be a unit vector in the fibre direction, lying in the plane of the sheet, and assume all fibres are locally parallel (an over-simplification). The material is then considered to be *transversely isotropic*, requiring us to distinguish quantities aligned in the fibre direction ($\|$) from those in the perpendicular direction within the plane of the sheet ($\perp$).

Consider small deformations, for which the stress–strain relation remains linear (a nonlinear theory is also available[33]). We define the *structure tensor* $A = \mathbf{a} \otimes \mathbf{a}$, the strain magnitude in the fibre direction $e_{\|} = \mathbf{a}\cdot\mathrm{e}\cdot\mathbf{a}$, and the strain in the fibre direction $\mathrm{e}_a = (\mathrm{e}\cdot\mathbf{a}) \otimes \mathbf{a} + \mathbf{a} \otimes (\mathrm{e}\cdot\mathbf{a})$. Then, in the incompressible limit, the stiffness tensor can be simplified (using symmetry arguments[34]) to give a stress–strain relation of the form

$$\boldsymbol{\sigma} = -p\mathrm{I} + \mu_2 e_{\|}\mathrm{A} + 2\mu_{\perp}(\mathrm{e}-\mathrm{e}_a) + 2\mu_s e_a, \tag{3.30}$$

where μ_2, $\mu_{\perp}$ and μ_s are constants (elastic moduli). Thus in addition to the familiar stress components $(-p\mathrm{I} + 2\mu_{\perp}\mathrm{e})$ there are additional contributions due to fibres, involving two additional material parameters. We can interpret the constants by examining the stress–strain relation in some special cases.

Technical Comment. In a 2D plane containing the fibre direction, a local basis is provided by orthogonal unit vectors **a** and **b** where $\mathbf{a}\cdot\mathbf{b} = 0$, $\mathbf{a}\cdot\mathbf{a} = 1$ and $\mathbf{b}\cdot\mathbf{b} = 1$. Consider a 2D deformation with strain $\mathrm{e} = \lambda(\mathbf{a} \otimes \mathbf{a} - \mathbf{b} \otimes \mathbf{b})$ for some constant λ, implying extension in the fibre direction and compression orthogonal to this (with no deformation in the direction normal to the wall, making this a *plane strain* approximation). Then $e_{\|} = \lambda$, $\mathrm{e}_a = \lambda\mathbf{a} \otimes \mathbf{a}$ is the tensile strain in the fibre direction, $\mathrm{e} - \mathrm{e}_a = -\lambda\mathbf{b} \otimes \mathbf{b}$ is the compressive strain orthogonal to the fibres, and $\boldsymbol{\sigma} = -p\mathrm{I} + (\mu_2 + 2\mu_s)\mathrm{e}_a + 2\mu_{\perp}(\mathrm{e} - \mathrm{e}_a)$. For this deformation, reading off the stresses in the $\|$ and $\perp$ directions, we see that the effective extensional stiffness in the fibre direction is $\mu_{\|} = \mu_s + \frac{1}{2}\mu_2$, while that perpendicular to the fibres is $\mu_{\perp}$. Likewise, a shear deformation, for which $\mathrm{e} = \gamma\,(\mathbf{a} \otimes \mathbf{b} + \mathbf{b} \otimes \mathbf{a})$ for some constant γ, has $\boldsymbol{\sigma} = -p\mathrm{I} + 2\mu_s\mathrm{e}$, identifying μ_s as an elastic shear modulus.

Typically, fibre-reinforced materials can be very stiff in the fibre direction, potentially restricting any stretching, so that $\mu_{\|} = \mu_s + \frac{1}{2}\mu_2 \gg \mu_{\perp}$. This is a key property exploited by plant cell walls to avoid radial expansion while allowing axial elongation. In the extreme limit, $\mu_2 \to \infty$ and $e_{\|} \to 0$ with their product remaining finite. We can introduce a Lagrange multiplier T to eqn (3.30), representing a tension induced in the fibres, giving

$$\boldsymbol{\sigma} = -p\mathrm{I} + T\mathrm{A} + 2\mu_{\perp}(\mathrm{e}-\mathrm{e}_a) + 2\mu_s\mathrm{e}_a. \tag{3.31}$$

In this case, just two parameters, the matrix stiffness $\mu_\perp$ and shear stiffness μ_S, regulate the restricted motion, and the induced tension T ensures no stretching in the fibre direction.

Technical Comment. This approach has a direct analogue for anisotropic fluid motion, with strains e, e_a, $e_\parallel$ replaced by strain rates $\dot{e}$, $\dot{e}_a$, $\dot{e}_\parallel$ and elastic moduli replaced by viscosities η_2, $\eta_\perp$ and η_S.[35] Fluids have the advantage that a linear constitutive law can be used to describe large deformations, but then the fibres will deform with the fluid. They evolve according to

$$\frac{\partial \mathbf{a}}{\partial t} + (\mathbf{v} \cdot \nabla)\mathbf{a} + \zeta\mathbf{a} = (\mathbf{a} \cdot \nabla)\mathbf{v} \tag{3.32}$$

where $\zeta = \frac{1}{2}\mathbf{a} \cdot (\nabla \otimes \mathbf{v} + (\nabla \otimes \mathbf{v})^T) \cdot \mathbf{a}$ is the component of the rate-of-strain tensor in the fibre direction. We recognise the first two terms as a material derivative; the remaining terms suppress stretching of **a** by the flow, ensuring that $\mathbf{a}{\cdot}\mathbf{a} = 1$ throughout.

So far, we have described the effects of a population of aligned fibres. For a material with fibre orientations $\mathbf{a}(\Omega)$ having a distribution $\rho(\Omega)$ over *solid angle* Ω, the stiffness tensor can have an isotropic component of the form (3.28), plus an anisotropic component of the form

$$\mathrm{C}_f = Y_f \int \mathrm{A} \otimes \mathrm{A}\rho(\Omega)\mathrm{d}\Omega \tag{3.33}$$

for some stiffness parameter Y_f. If the fibres lie in a plane, the integral is over the polar angle θ and six independent components of C_f can be derived.[36,37]

In a 2D computational model of individual cells in an elongated tissue, Fozard *et al.*[38] combined a finite element discretisation of an anisotropic linear constitutive law (eqn (3.30)) for periclinal cell walls (parallel to the tissue axis) with a scalar Lockhart model for anticlinal cell walls (orthogonal to the tissue axis), to mimic root bending.

3.5.2 Growth in Three Dimensions

Large deformations require a nonlinear elastic theory, which brings further technical complications, many of which are bypassed in the short account below. First, it is necessary to distinguish between a reference (Lagrangian) state and a deformed state, with deformations described by a mapping between them. From this one can define a *deformation gradient* F where $\mathrm{d}\mathbf{x} = \mathrm{F}\mathrm{d}\mathbf{X}$ or $F_{ij} = \partial x_i/\partial X_j$. Here **X** labels material points in the undeformed configuration and $\mathbf{x}(\mathbf{X}, t)$ gives their location in the deformed state.

Within this framework, growth is described by considering two mappings: from the original state to an intermediate (grown) state; and from the grown state to the physical state, accounting for elastic deformations. This decomposition,[15,16] which appears also in the theory of nonlinear plasticity, has deformation gradient $\mathrm{F} = \mathrm{F_e F_g}$. $\mathrm{F_g}(t)$ represents time-dependent growth at some time t and $\mathrm{F_e}$ the instantaneous elastic deformation that is required to accommodate an imposed load.

The velocity of a material particle is $\mathbf{v} = \dot{\mathbf{x}}$ (for fixed $\mathbf{X}$), so that the velocity gradient with respect to Lagrangian variables is $\partial\mathbf{v}/\partial\mathbf{X} = (\partial\mathbf{v}/\partial\mathbf{x})\mathrm{F}$ but also $\partial\mathbf{v}/\partial\mathbf{X} = \dot{\mathrm{F}}$. Thus the velocity gradient with respect to Eulerian variables is $\dot{\mathrm{F}}\mathrm{F}^{-1} = \partial\mathbf{v}/\partial\mathbf{x}$, generalising eqn (3.16). In terms of the composite deformation, $\dot{\mathrm{F}}\mathrm{F}^{-1} = \mathrm{F_e}\dot{\mathrm{F}}_\mathrm{g}\mathrm{F_g}^{-1}\,\mathrm{F_e}^{-1}$. Here the $\mathrm{F_e}$ tensors are (in the jargon) *pushing forward* the velocity gradient in the intermediate state to the current state, so we can identify $\dot{\mathrm{F}}_\mathrm{g}\mathrm{F_g}^{-1}$ as the velocity gradient of the unstressed grown state. This is taken as the analogue of $\dot{e}_\mathrm{g}$ in eqn (3.12) and (3.13).

Technical Comment. We can now see how the Lockhart model (eqn (3.3) and (3.12)) can be formulated for 3D anisotropic materials,[24,37] although we will only touch on a few of the technical details, staying as close as possible to the simpler models we have seen so far. Considering first purely elastic deformations, the *second Piola–Kirchoff stress* $\mathrm{S_e}$ and the *Lagrangian strain* $\mathrm{E_e} = \frac{1}{2}(\mathrm{F_e^T F_e} - \mathrm{I})$ are related *via* a *strain energy function* $\mathcal{W}$ as $\mathrm{S_e} = \partial\mathcal{W}/\partial\mathrm{E_e}$. In general, $\mathcal{W}$ is a function of *strain invariants*, which may account for anisotropy. A simple approach uses the *hyperelastic* Hookean model $\mathcal{W} = \frac{1}{2}\mathrm{E_e}:\mathrm{C}:\mathrm{E_e}$ where C is a fourth-order stiffness tensor that accommodates anisotropy, so that $\mathrm{S_e} = \mathrm{C}:\mathrm{E_e}$ and $\mathrm{C} = \partial^2\mathcal{W}/\partial^2\mathrm{E_e}$. This is generalised to incorporate growth using

$$\mathrm{E_e} = \mathrm{C}^{-1}:\mathrm{S_e}, \qquad \dot{\mathrm{F}}_\mathrm{g}\mathrm{F_g}^{-1} = \gamma[\mathrm{E_e} - \mathrm{E_Y}]_+ \tag{3.34}$$

for some constant γ. This can be written in terms of the driving stress $\mathrm{S_e}$, which depends on turgor, and yield stress $\mathrm{S_0}$ as

$$\dot{\mathrm{F}}_\mathrm{g}\mathrm{F_g}^{-1} = \gamma[\mathrm{C}^{-1}:(\mathrm{S_e} - \mathrm{S_0})]_+. \tag{3.35}$$

Here the threshold term $[\cdot]_+$(a tensor ramp function[37]) is a simplified version of more complex plasticity approaches,[24] such that for a tensor T decomposed into its eigenvalues λ_n and eigenvectors $\mathbf{t}_n$ we have

$$\mathrm{T} = \sum_{n=1}^{d} \lambda_n \mathbf{t}_n \otimes \mathbf{t}_n \quad \text{and} \quad [\mathrm{T}]_+ = \sum_{n=1}^{d} \max(\lambda_n, 0)\mathbf{t}_n \otimes \mathbf{t}_n. \tag{3.36}$$

Note that in this model, a single stiffness tensor C captures anisotropy both in the viscous creep (3.35) and in the elastic response (3.34).

Boudon *et al.*[24] have implemented this model by resolving individual cell walls in a tissue using a 3D finite-element discretisation. A similar framework was adopted by Bozorg *et al.*[39] Models incoporating mechanical feedback on growth are reviewed by Ali *et al.*[3]

3.6 Discrete Modelling Approaches for Multicellular Tissues

A popular class of models resolves the structure of individual cells in an inherently discrete formulation. This approach can be computationally intensive, but it avoids the requirement to average over cells in order to describe a tissue, thereby resolving features that may be lost in spatial averaging. Furthermore, there is the advantage that biochemical processes can be resolved in individual cells and coupled to mechanical processes.[40–42] The continuum models of Section 3.5 are of sufficient complexity to require solution by (typically) a finite-element method. In this top-down approach, bulk (averaged) descriptions of tissue mechanics are projected down onto individual elements. In contrast, in a bottom-up approach the mechanical properties of each element are defined directly. The two approaches may not always be consistent, as the top-down description requires fields to vary smoothly from cell to cell and may rest on *ad hoc* approximations that are not immediately related to properties at the cell scale. There is therefore a logic in identifying individual cells as natural 'elements' in the spatial discretisation.

Cell geometry can be represented in various ways. A popular grid-based model is the *Cellular Potts* scheme, where an individual cell is represented using sets of neighbouring elements on a fixed spatial grid. Using a sufficiently fine grid, this method will resolve elaborate cell shapes, albeit at high computational cost. At a much coarser level, *cell-centre* models assign cells to a single spatial location and model cell–cell interactions. We focus here on intermediate *vertex-based* descriptions, where a 2D (3D) cell is treated as a polygon (polyhedron), allowing its shape to be defined economically by the location of its vertices. In 2D, a monolayer is represented as a tiling of the plane by polygons; in 3D, a tissue is defined by space-filling polyhedra. The polygons need not be regular or periodic. Growth and mechanics are simulated by deriving appropriate rules for the motion of the vertices. For plant cells, which typically have a cytoplasm dominated by a large vacuole at fixed turgor pressure, this is a natural approach as attention can be focused on the mechanically influential cell walls.

Vertex-based models differ in their choice of constitutive assumption. One popular model assigns to each cell a mechanical energy,[43–46] which can account for strain of individual cells and cell–cell adhesion. The arrangement of cells in space is then determined by allowing cell configurations to adjust until the global mechanical energy is minimised. Direct minimisation identifies one or more equilibria (the system can be 'glassy' with a rough energy landscape, with different minima accessed from different initial conditions). Alternatively, a model for viscous dynamics is employed to track the unsteady evolution to an equilibrium. Growth is modelled through cell expansion (for example *via* a Lockhart-based model[38]) or through cell division (requiring a model for cell cycling and division orientation[47]).

3.6.1 The Mechanical Energy of a Cell

Cell-based models typically rely on simple geometric invariants (the perimeter, surface area or volume of a cell), defined with respect to 'target' values of each at which the corresponding energy contribution is minimal. We illustrate this for a 2D layer of cells; generalisations to 3D follow naturally. (Mechanical quantities below are therefore defined per unit length in the direction orthogonal to the plane of interest.) The energy of a single cell (labelled by i) is often written[43,45,48]

$$U_i = \frac{1}{2}K_a(A_i - A_0)^2 + \frac{1}{2}K_l(L_i - L_0)^2, \tag{3.37}$$

where K_a and K_l are positive parameters. The first term describes the energy associated with the cell's area A_i (so that large K_a constrains A_i to remain close to a reference area A_0), and so is a measure of the stiffness of the cytoplasm. The second term characterises the energy associated with the cell's perimeter, and its deviation from a reference perimeter L_0. Separately, the contributions define a pressure and a tension for the cell defined by $\partial U_i/\partial A_i$ and $\partial U_i/\partial L_i$, respectively, as

$$P_i = K_a(A_i - A_0), \quad T_i = K_l(L_i - L_0). \tag{3.38}$$

For a single cell, U_i might be minimised by a configuration in which $A_i < A_0$ and $L_i > L_0$, with the competition between bulk compression ($P_i < 0$) and peripheral tension ($T_i > 0$) giving the cell an intrinsic rigidity (*via* so-called *tensegrity*). The energy of a tissue formed of confluent cells can then be written $U = \sum_i U_i$; additional terms might measure the strength of adhesion between particular cell types.

Growth can be accommodated by allowing for cell expansion (varying A_0 or L_0) or by explicitly modelling cell division. Conversely, the process of cell death and extrusion of an individual cell from the monolayer can also be described. Both division and extrusion change the topology of the monolayer, by introducing or removing cells. Cells may also undergo neighbour exchange (through a so-called *T1 transition*). This reorganisation of the tissue is a microscopic manifestation of plasticity. It is a common feature of developing animal tissues but is less common in plants, where cells typically adhere tightly to their neighbours.

Technical Comment. The strain energy of an elastic continuum is normally defined in terms of *strain invariants*, measuring the deformation from a reference state, expressed in terms of eigenvalues of (say) the right Cauchy–Green tensor $\mathbf{F}^{\mathrm{T}}\mathbf{F}$. For a growing elastic material, the elastic strains are measured relative to an evolving intermediate state. In contrast, the vertex-based model uses a cell-based formulation that exploits natural geometric invariants (cell area and cell perimeter) but which lacks a well-defined reference state, leading to analogies with granular materials.[49]

Plant-specific variants of eqn (3.37) include the following model[47] for the superficial cells of the shoot apical meristem:

$$U=\sum_{j\in\text{walls}}\frac{1}{2}K_{\mathrm{w}}\left(l_j-l_j^0\right)^2-\sum_{i\in\text{cells}}\Pi_i A_i-\sum_{i\in\text{cells}}P_{i,\text{int}}V_{i,\text{int}}. \tag{3.39}$$

The first term describes tensile forces in anticlinal walls (with reference length l_j^0); the second describes forces normal to anticlinal walls due to cell turgor pressure Π_i (assumed uniform in all cells); the third represents pressure from the underlying tissues, producing a force acting in the direction normal to the cell layer. Anisotropy is introduced by making anticlinal wall stiffness K_{w} a function of orientation. Minimisation of U, by movement of cell vertices, brings the cell layer to equilibrium; slower cell growth was implemented in this model[47] using a Lockhart law

$$\frac{\mathrm{d}l_j^0}{\mathrm{d}t}=k_{\mathrm{g}}\left[\frac{l_j-l_j^0}{l_j^0}-T_{\mathrm{g}}\right]_+ \tag{3.40}$$

for some extensibility-like parameter k_{g} and a yield strain T_{g}. This framework is readily extended to account for fluid transport between adjacent cells.[50] For now, we restrict attention to the simpler model (eqn (3.37)). Related approaches address turgor[50,51] and cell shape.[52]

3.6.2 Cell Topology and Geometry

For large-scale discrete models, careful book-keeping is needed in order to keep track of relationships between vertices, cell edges, cell faces, and so on as a tissue evolves. Here we briefly outline one approach to address this, illustrating the approach in 2D.[49]

An array of cells is defined in terms of a set of vertices (position vectors) $\mathbf{r}_k$, $k=1,\ldots,N_{\mathrm{v}}$, a set of *oriented edges* $\mathbf{t}_j$ (of length $l_j=|\mathbf{t}_j|$), $j=1,\ldots,N_{\mathrm{e}}$ and a set of *oriented cell faces* $\mathbf{a}_i$ (of area A_i), $i=1,\ldots,N_{\mathrm{c}}$ (Figures 3.5 and 3.6). Here $\mathbf{r}_k$ and $\mathbf{t}_j$ are vectors in a 2D (x, y) plane, while $\mathbf{a}_i=A_i\boldsymbol{\varepsilon}_i$, where $\boldsymbol{\varepsilon}_i$ represents a rotation

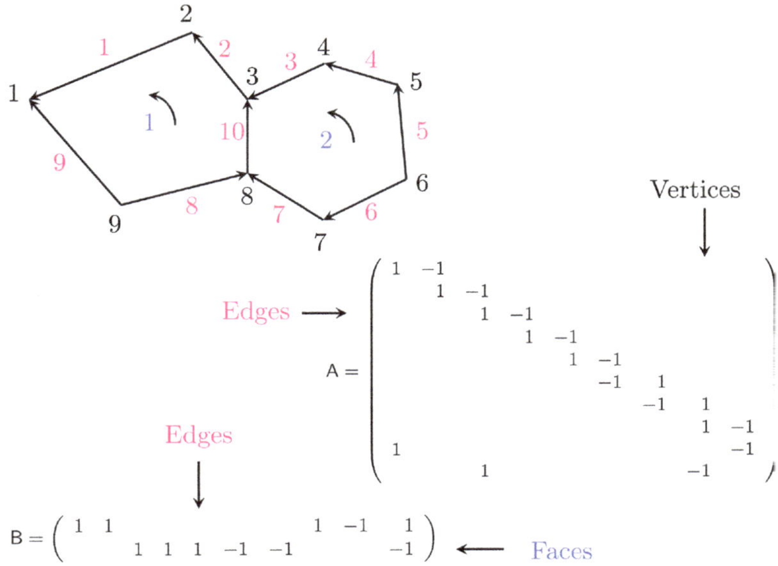

Figure 3.5 An illustration of the representation of cell topology through incidence matrices. Two cells, joined by a single edge, are defined by 9 vertices, 10 edges and 2 faces. The orientations of edges and faces are prescribed arbitrarily. The 10×9 matrix A shows which edges and vertices are connected (empty spaces are zeros), with 1 (-1) indicating an edge pointing into (out of) a vertex. The 2×10 matrix B shows which edges neighbour which face, and which are coherent with (+1) or not coherent with (-1) the face.

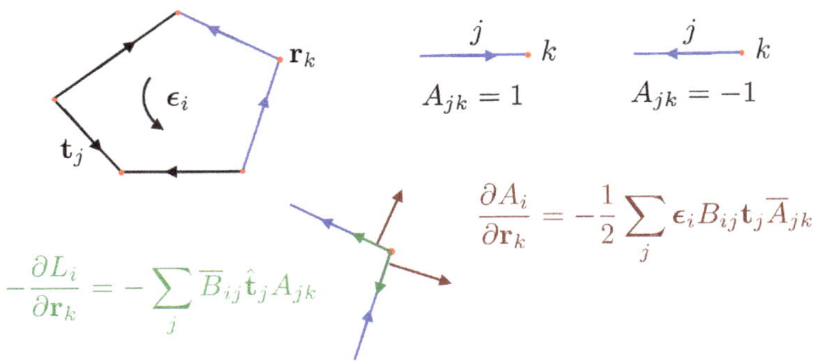

Figure 3.6 A schematic summarising the forces associated with area and perimeter changes in a vertex-based model. Unsigned incidence matrices indicate if edge j neighbours vertex k ($\bar{A}_{jk} = 1$) or if edge j neighbours cell i ($\bar{B}_{ij} = 1$). ε_i is a $\pi/2$ rotation indicating the prescribed orientation of cell i (top left); edges $B_{ij}\mathbf{t}_j$ of cell i are coherent with ε_i so that $-B_{ij}\varepsilon_i\mathbf{t}_j$ rotates edges by $\pi/2$ to construct outward normals to cell i, contributing to the area change of cell i if vertex k is displaced (brown, bottom right). The incidence matrix $A_{jk} = 1$ or -1 if oriented edge j points toward or out of vertex k (top right), and is zero otherwise, so that $-A_{jk}\hat{\mathbf{t}}_j$ denotes unit vectors along edges pointing away from vertex k, contributing to the perimeter change of cell i if vertex k is displaced (green, bottom left).

by $\pm\pi/2$, *i.e.* $\begin{pmatrix} 0 & \mp 1 \\ \pm 1 & 0 \end{pmatrix}$ in terms of Cartesian coordinates. Orientations of edges and faces are prescribed but arbitrary.

The *topology* of the monolayer (what connects with what) can be defined in terms of two *incidence matrices*[53] (Figure 3.5). A, with elements A_{jk}, is an $N_e \times N_v$ matrix with elements 1 (or -1) when edge j is oriented into (or out of) vertex k, and zero otherwise. B, with elements B_{ij}, is an $N_c \times N_e$ matrix with elements ± 1 when edge j is on the boundary of cell i, taking values $+1$ if the edge is coherent with the orientation of the cell face and -1 if not; otherwise, all elements of B are zero. Replacing -1 with 1 in each matrix produces *unsigned incidence matrices* $\overline{\text{A}}$ and $\overline{\text{B}}$, which identify neighbours but do not indicate orientation. The $N_c \times N_v$ matrix $\overline{\text{C}} = \frac{1}{2}\overline{\text{B}}\overline{\text{A}}$ takes the value $\overline{C}_{ik} = 1$ if vertex k neighbours cell i, and is zero otherwise.

Technical Comment. A and B have interpretations as *discrete derivatives*, while A^T and B^T are *boundary operators*, identifying vertices bounding an edge or edges bounding a face.[53] Since the edges of cells form closed loops, they have no boundary and so $\text{A}^T\text{B}^T = 0$ and $\text{BA} = 0$.

The *geometry* of the monolayer (measuring its size) is defined by magnitudes of lengths and areas. Edges are defined by $\mathbf{t}_j = \sum_k A_{jk}\mathbf{r}_k$ (summing over all vertices), from which $l_j = \sqrt{\mathbf{t}_j \cdot \mathbf{t}_j}$ can be evaluated, along with the unit vector $\hat{\mathbf{t}}_j = \mathbf{t}_j / l_j$. The perimeter of cell i is then $L_i = \sum_j \overline{B}_{ij} l_i$. (summing over all edges). It follows that

$$\frac{\partial l_j}{\partial \mathbf{r}_k} = \hat{\mathbf{t}}_j A_{jk} \quad \text{and} \quad \frac{\partial L_i}{\partial \mathbf{r}_k} = \sum_j \overline{B}_{ij} \hat{\mathbf{t}}_j A_{jk}. \tag{3.41}$$

$-\partial L_i / \partial \mathbf{r}_k$ identifies two unit vectors aligned with edges of cell i where it meets vertex k, pointing out of the vertex (Figure 3.6). Some tedious algebra[49,54] reveals that

$$\frac{\partial A_i}{\partial \mathbf{r}_k} = -\frac{1}{2}\sum_j \varepsilon_i B_{ij} \mathbf{t}_j \overline{A}_{jk} \equiv \frac{1}{2}\sum_j \mathbf{n}_{ij} \overline{A}_{jk}. \tag{3.42}$$

$\partial A_i / \partial \mathbf{r}_k$ identifies two outward normal vectors associated with the edges of vertex k bordering cell i (Figure 3.6).

3.6.3 Vertex Dynamics

Armed with eqn (3.41) and (3.42), showing how the length and perimeter of cell i change when vertex k moves, we can now evaluate $\mathbf{f}_{ik} = \delta U_i / \delta \mathbf{r}_k$, the first variation of the energy of cell i with respect to a small displacement of vertex $\mathbf{r}_k$. This determines the elastic restoring force at $\mathbf{r}_k$ acting on cell i as

$$\mathbf{f}_{ik} = \sum_j \left[\frac{1}{2} P_i \varepsilon_i B_{ij} \mathbf{t}_j \overline{A}_{jk} + T_i \overline{B}_{ij} \hat{\mathbf{t}}_j A_{jk} \right]. \tag{3.43}$$

As Figure 3.6 illustrates, the force at vertex k due to cell i has contributions from P_i acting along normals at the vertex, and from T_i acting along tangents. In equilibrium, the net force at vertex k and the net force on cell i must both vanish, *i.e.*

$$\sum_i \mathbf{f}_{ik} = \mathbf{0}, \quad \sum_k \mathbf{f}_{ik} = \mathbf{0}. \tag{3.44}$$

A common model applies a drag η_d to each vertex, so that the equilibrium is reached by timestepping N_v coupled ODEs for $\mathbf{r}_k(t)$ of the form

$$\eta_d \frac{d\mathbf{r}_k}{dt} = -\sum_i \mathbf{f}_{ik}. \tag{3.45}$$

This is the evolution equation (or rather, N_v ordinary differential equations) for vertex locations $\mathbf{r}_k$, coupled through the dependence of $\mathbf{f}_{ik}$ on $\mathbf{r}_k$ *via* A_i and L_i. An *initial condition* for cell vertices may come from an image, or may be constructed using a *Voronoi tessellation* of the plane. Suitable *boundary conditions* for the array of cells must also be chosen (a periodic box is a popular choice). Eqn (3.45) is integrated until it reaches an equilibrium, satisfying (44a).

3.6.4 Cell and Tissue Stress

For a monolayer in equilibrium, evaluating $\sum_k \mathbf{r}_k \otimes \mathbf{f}_{ik} \equiv A_i \sigma_i$, where $\boldsymbol{\sigma}_i$ is the stress (force moment) associated with cell i, one finds after some algebra[54–57] that

$$\boldsymbol{\sigma}_i = P_i \mathrm{I} + \frac{T_i L_i}{A_i} \mathrm{Q}_i, \quad \text{where} \quad \mathrm{Q}_i \equiv \frac{1}{L_i} \sum_j \overline{B}_{ij} l_j \hat{\mathbf{t}}_j \otimes \hat{\mathbf{t}}_j. \tag{3.46}$$

A simple way of interpreting these terms is to consider that under an imposed uniform strain E, A_i changes by $A_i \mathrm{I}{:}\mathrm{E} \equiv A_i \mathrm{tr}(\mathrm{E})$ and L_i changes by $L_i \mathrm{Q}_i{:}\mathrm{E}$.[57] It turns out[54] that the principal axes of the cell stress tensor (more specifcally, of Q_i) align with those of the cell's shape tensor $\sum_k \overline{C}_{ik} (\mathbf{r}_k - \mathbf{R}_i) \otimes (\mathbf{r}_k - \mathbf{R}_i)$, where $\mathbf{R}_i$ is the centroid of each cell, showing that cell shape and cell stress are intimately coupled. The isotropic component of the stress in each cell $\frac{1}{2}\mathrm{tr}(\sigma_i)$ reveals the *effective pressure* in each cell

$$P_{\text{eff}} = P_i + \frac{T_i L_i}{2A_i} \tag{3.47}$$

with contributions from both the bulk and the periphery. For an isolated cell in equilibrium, with $P_{\text{eff}}=0$, we expect $P_i<0$ (because $A_i<A_0$) and $T_i>0$ (because $L_i>L_0$), the typical state of a turgid plant cell. For example a cell at fixed turgor, as in (1.39), has $P_i=-\Pi_i$.

The stress of the monolayer as a whole is

$$A\boldsymbol{\sigma}=\sum_i A_i\boldsymbol{\sigma}_i=\sum_i (P_iA_i\mathrm{I}+T_iL_i\mathrm{Q}_i). \tag{3.48}$$

These expressions for cell and tissue stress (eqn (3.46) and (3.48)) can be compared with, for example, eqn (3.31), showing how the structure tensor for each cell wall $\hat{\mathbf{t}}_j \otimes \hat{\mathbf{t}}_j$ contributes additively (*via* Q_i) to the total stress, resembling the structure tensor of fibres A.

Patterns of P_{eff} can show inherently discrete features (such as *force chains*) across cell monolayers.[54] The cell stress (eqn (3.46)) and tissue stress (eqn (3.48)) have been derived from an underlying energy (eqn (3.37)), and sit outside a traditional continuum framework. However, the tissue as a whole exhibits linearly elastic properties when perturbed from an equilibrium, and plastic properties if cell rearrangements take place.

Technical Comment. Perturbing the monolayer about an equilibrium reveals its stiffness tensor as

$$C=\frac{1}{A}\sum_i\left[K_\mathrm{a}A_i^2\mathrm{I}\otimes\mathrm{I}+K_\mathrm{l}L_i^2\mathrm{Q}_i\otimes\mathrm{Q}_i+L_iT_i(\mathrm{B}_i-\mathrm{Q}_i\otimes\mathrm{I})\right] \tag{3.49}$$

where B_i is a fourth-order tensor[57] defined in terms of $\mathbf{t}_j$ such that under strain E, the change in $L_i\mathrm{Q}_i$ is $L_i\mathrm{B}$: E. This is the discrete analogue of eqn (3.28) or eqn (3.33). This expression shows explicitly how cell orientations can induce anisotropy in the tissue. The effective tissue bulk and shear moduli can be derived from C. If L_0 becomes so large that T_i becomes negative for some cells, then the cells lose resistance to shear (*via* an unjamming *phase transition*[58]). Tissue rigidity is promoted by inducing large tensions in cell walls. Notice that the tissue properties are recovered as sums, without requiring (for example) a periodicity assumption, commonly made when homogenising cellular materials.[59]

Further adaptions of the vertex model include hybridisation with a centreline model of the kind described in Section 3.4.[60]

3.7 Plant Cell Wall Mechanics and the Origins of the Lockhart Model

The primary plant cell wall is formed from an array of cellulose microfibrils, crosslinked by hemicellulose, embedded in a pectin matrix. Enzymes target distinct components: pectin methylesterase (PME) targets the matrix;

xyloglucan endotransglucosylase/hydrolase (XTH) and expansin disrupt crosslinks; and so on. Microstructural models of the cell wall can be used to understand how enzyme action influences mechanical attributes such as yield stress or extensibility, and how the wall's architecture influences cell and tissue dynamics.[19] Cellulose self-assembles in a synthase complex that migrates along cortical microtubules; thus the microtubules influence the wall's structure. The mechanosensitivity of microtubules and details of the wall's microstructure are active areas of research.[5,11,47,61]

3.7.1 The Matrix

We start with a very simple model of a growing cell wall as a thin viscous sheet, deriving the relation between the axial stress resultant T in terms of its strain rate, thickness h and viscosity η.

Technical Comment. Consider a sheet of initial length L_0, thickness h_0 that is subject to an extensional stretching flow of velocity magnitude U_0. For an incompressible Newtonian fluid, satisfying $\nabla\cdot\mathbf{v}=0$ and $\nabla\cdot\boldsymbol{\sigma}=0$, where $\boldsymbol{\sigma}=-p\mathbf{I}+\eta(\nabla\otimes\mathbf{v}+(\nabla\otimes\mathbf{v})^T)$ (see Section 3.5), the 2D Stokes' equations are

$$\partial_x u+\partial_y v=0,\quad 0=-\partial_x p+\eta(\partial_x^2 u+\partial_y^2 u),\quad 0=-\partial_y p+\eta(\partial_x^2 v+\partial_y^2 v). \tag{3.50}$$

where the velocity field $\mathbf{v}=(u,\ v)$ is defined with respect to coordinates (x, y) and ∂_x is shorthand for ∂/∂_x. The components of the stress tensor are

$$\sigma_{xx}=-p+2\eta\partial_x u,\quad \sigma_{yy}=-p+2\eta\partial_y v,\quad \sigma_{xy}=\eta(\partial_y u+\partial_x v). \tag{3.51}$$

We assume the sheet is stress-free at its surfaces $y=0$ and $y=h$ (*i.e.* $\sigma_{xy}=\sigma_{yy}=0$ there).

Eqn (3.50) and (3.51) can be simplified by exploiting the sheet's slender geometry, using the small parameter $\varepsilon\equiv h_0/L_0\ll 1$. Write $(x,\ y,\ h)\ =L_0$ $(x^*,\ \varepsilon y^*,\ \varepsilon h^*)$, $(u,\ v)=U_0(u^*,\ \varepsilon v^*)$, $(p,\ \boldsymbol{\sigma})=\eta(U_0/L_0)(p^*,\ \boldsymbol{\sigma}^*)$. Then in $0<y^*<h^*$, we re-write (1.50, 1.51) in terms of dimensionless (starred) variables as

$$\begin{aligned}\partial_{x^*}u^*+\partial_{y^*}v^*=0,\quad 0=-\partial_{x^*}p^*+\partial_{x^*}^2u^*+\varepsilon^{-2}\partial_{y^*}^2u^*,\\ 0=-\partial_{y^*}p^*+\varepsilon^2\partial_{x^*}^2v^*+\partial_{y^*}^2v^*,\end{aligned} \tag{3.52}$$

with

$$\begin{aligned}\sigma_{xx}^*=-p^*+2\partial_{x^*}u^*,\quad \sigma_{yy}^*=-p^*+2\partial_{y^*}v^*,\\ \sigma_{xy}^*=\varepsilon^{-1}\partial_{y^*}u^*+\varepsilon\partial_{x^*}v^*.\end{aligned} \tag{3.53}$$

Now expand all variables using $u^*=u_0^*+\varepsilon^2u_1^*+\ldots$, *etc.* At leading order, $\partial_{y^*}^2u_0^*=0$; furthermore $\sigma_{xy}^*=0$ on $y^*=0$ and $y^*=h^*$ implies $\partial_{y^*}u_0^*=0$

there. Thus $u_0^* = u_0^*(x^*,t^*)$. $\partial_{y^*} v_0^* = -\partial_{x^*} u_0^*$ implies $v_0^* = -y^* \partial_{x^*} u_0^*$ (assuming $v_0^* = 0$ on $y^* = 0$). Thus $\partial_{y^*}^2 v_0^* = 0$, implying $\partial_{y^*} p_0^* = 0$. $\sigma_{0yy}^* = 0$ on $y^* = 0$ and h^* implies $p_0^* = 2\partial_{y^*} v_0^* = -2\partial_{x^*} u_0^*$, giving finally that $\sigma_{0xx}^* = 4\partial_{x^*} u_0^*$. Restoring this expression to dimensional units we obtain the stress/strain-rate relation $\sigma_{xx} = 4\eta \partial_x u$, implying that the sheet has *extensional viscosity* 4η.

The stress resultant $T = \sigma_{xx} h$ therefore satisfies $T = 4\eta h \partial_x u$, so that the extensibility in eqn (3.6) (at least, that part of it arising from stretching of the matrix) can be expressed in terms of wall thickness and viscosity as

$$\phi = 1/(4\eta). \tag{3.54}$$

For a wall of spatially uniform thickness, the extensional flow of material in the wall (relative to a point on its outer surface) is $\mathbf{v} = (\alpha x, -\alpha y)$, where α is the RER. In the absence of sources of wall material, the wall thickness satisfies $\mathrm{d}h/\mathrm{d}t = -\alpha h$ (stretching induces thinning). In practice, we expect metabolic processes to be depositing material on the wall's inner surface at $y = h$ in order to maintain the wall's thickness during growth (Figure 3.7).

3.7.2 Fibres, Crosslinks and the Origins of the Lockhart Model

Now consider a cylindrical cell, elongating along its axis, with fibres oriented in the transverse (azimuthal) direction within the thin cell wall. Suppose the fibres resist stretching, so that the strain-rate in the wall is $\dot{\mathrm{e}} = \alpha \hat{\mathbf{z}} \otimes \hat{\mathbf{z}} - \alpha \hat{\mathbf{r}} \otimes \hat{\mathbf{r}}$. Then the viscous analogue of eqn (3.31) gives the stress field as $\boldsymbol{\sigma} = -p\mathrm{I} + T\hat{\boldsymbol{\theta}} \otimes \hat{\boldsymbol{\theta}} + 2\eta_\perp \dot{\mathrm{e}}$. In this simple example, the hoop stress $\sigma_{\theta\theta} = T$ that inhibits swelling decouples from the axial extension. In practice, fibres that are not exactly transverse will be rotated as the wall extends (*via* eqn (1.32)), complicating the story.[62] For example, fibres deposited at a fixed angle at the inner wall will rotate toward the cell's axis as the wall stretches, while moving toward the outer wall; as they rotate, they suppress cell elongation.

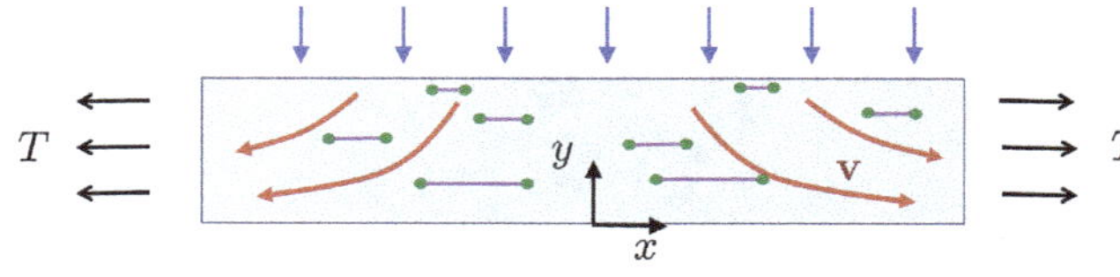

Figure 3.7 A schematic of an element of wall undergoing stretching. At the inner surface, matrix, fibres and crosslinks are assembled and deposited (blue arrows). Fibres (green dots) are assumed to be normal to the page. Crosslinks (purple) are extended by the stretching of the wall, breaking before they reach the outer wall at $y = 0$. The wall elongates with velocity field $\mathbf{v} = \alpha(x, -y)$, where α is the RER.

The cell's effective extensibility then depends on the fibre orientation integrated across the wall, which depends on the manner in which fibres were deposited at earlier times.

To examine the role of crosslinks, consider the simplest possible scenario, when fibrils are deposited transversely on the inner wall (perpendicular to the axis of the cell), with hemicellulose crosslinks formed between them. As the cell elongates, fibres migrate toward the outer wall of the cell, while being separated by the stretching of the wall (Figure 3.7). The distance between fibres satisfies $\mathrm{d}L/\mathrm{d}t = \alpha L$ on $\mathrm{d}y/\mathrm{d}t = -\alpha y$, while the density of crosslinks between fibres, $n(y, t)$, satisifes $\partial_t n - \alpha y \partial_y n = -k_{\text{off}} n$, for some breakage rate k_{off}. We can model each crosslink as a spring with stiffness κ_s and unstressed length L_0. Taking an off-rate (breakage rate) of the form

$$k_{\text{off}} = k_0 \exp\left(\beta^2 \frac{\kappa_s (L - L_0)^2}{2k_b T}\right), \tag{3.55}$$

where $k_b T$ is a unit of thermal energy and $\beta \ll 1$ is a constant, then bonds break close to the outer wall where crosslinks are sufficiently extended. (The kinetics depend here on a ratio of mechanical to thermal energy; other models[63] estimate the mechanical energy as a force times a distance, which is linear rather than quadratic in strain.) We are here assuming that broken crosslinks cannot re-form within the wall. The stress resultant in the wall can be modified to incorporate the additional effect of crosslinks as

$$T = \int_0^h n(y, t) k_s (L - L_0) \mathrm{d}y + 4\eta h \alpha. \tag{3.56}$$

This simple model can be used to determine the dependence of stress on the RER α. If the RER is sufficiently small, crosslinks break close to the inner surface of the wall (they break before migrating very far). Increasing the RER carries crosslinks into the wall and extends them, increasing n and thus T. If the RER is sufficiently large, crosslinks migrate close to the outer surface of the wall before breaking once very extended. This is a nearly saturated state, becoming insensitive to further increases in the RER. The net effect is a nonlinear response between T and α, with characteristics of the plastic response in eqn (3.2). For a strain increasing linearly in time at rate α, the model shows an initial elastic response (with stiffness $\mathrm{E} = n_0 \kappa_s L_0 h$), a long-time viscous response characteristic of eqn (3.3).[64] The yield stress has magnitude $\mathrm{E}\log(1/\beta)$. Although crude, this simple model provides insights into the microstructural components of the cell wall that regulate its rheological properties.

3.8 Outlook

Significant work needs to be done to further develop microscale models[65] of cell wall structure that incorporate relevant biochemistry and thermodynamics,[66] and then to scale these up to mechanical models of whole tissues, to understand how cell wall architecture influences plant morphology, building on recent progress in connecting microscale anisotropy[67] or differential growth[68] in slender structures to macroscopic curving, twisting and torsion. As demonstrated in a recent integrative model[69] of shoot gravitropism, there is much to be gained by using careful mathematical approximations that capture the core features of key processes in an economic way, enabling tractable simulations of whole-organ behaviour. While many challenges remain, particularly in incorporating feedback processes driven by environmental stimuli, this review hopefully demonstrates the variety and value of relatively simple theoretical models in understanding the mechanisms and mechanics of plant growth.

Acknowledgements

This chapter is developed from lectures given at the 24th CISM-IUTAM International Summer School on Plant Biomechanics in 2019.

References

1. O. E. Jensen and J. A. Fozard, Multiscale models in the biomechanics of plant growth, *Physiology*, 2015, **30**, 159–166.
2. A. Goriely, *The Mathematics and Mechanics of Biological Growth*, vol. 45, Springer, 2017.
3. O. Ali, V. Mirabet, C. Godin and J. Traas, Physical models of plant development, *Annu. Rev. Cell Dev. Biol.*, 2014, **30**, 59–78.
4. D. Kierzkowski and A. L. Routier-Kierzkowska, Cellular basis of growth in plants: geometry matters, *Curr. Opin. Plant Biol.*, 2019, **47**, 56–63.
5. V. Mirabet, P. Das, A. Boudaoud and O. Hamant, The role of mechanical forces in plant morphogenesis, *Annu. Rev. Plant Biol.*, 2011, **62**, 365–385.
6. E. Smithers, J. Luo and R. Dyson, Mathematical principles and models of plant growth mechanics: from cell wall dynamics to tissue morphogenesis, *J. Exp. Bot.*, 2019, **70**, 3587–3600.
7. B. Moulia and M. Fournier, The power and control of gravitropic movements in plants: a biomechanical and systems biology view, *J. Exp. Bot.*, 2009, **60**, 461–486.
8. M. R. Nelson, L. R. Band, R. J. Dyson, T. Lessinnes, D. Wells and C. Yang, *et al.*, A biomechanical model of anther opening reveals the roles of dehydration and secondary thickening, *New Phytol.*, 2012, **196**, 1030–1037.
9. E. De Langre, Effects of wind on plants, *Annu. Rev. Fluid Mech.*, 2008, **40**, 141–168.
10. T. I. Baskin and O. E. Jensen, On the role of stress anisotropy in the growth of stems, *J. Exp. Bot.*, 2013, **64**, 4697–4707.
11. D. J. Cosgrove, Plant cell wall extensibility: connecting plant cell growth with cell wall structure, mechanics, and the action of wall-modifying enzymes, *J. Expt. Bot.*, 2015, **67**, 463–476.

12. R. Vandiver and A. Goriely, Tissue tension and axial growth of cylindrical structures in plants and elastic tissues, *Europhys Lett.*, 2008, **84**, 58004.
13. L. Beauzamy, M. Louveaux, O. Hamant and A. Boudaoud, Mechanically, the shoot apical meristem of Arabidopsis behaves like a shell inflated by a pressure of about 1 MPa, *Front. Plant Sci.*, 2015, **6**, 1038.
14. D. Kierzkowski, N. Nakayama, A. L. Routier-Kierzkowska, A. Weber, E. Bayer and M. Schorderet, *et al.*, Elastic domains regulate growth and organogenesis in the plant shoot apical meristem, *Science*, 2012, **335**, 1096–1099.
15. E. K. Rodriguez, A. Hoger and A. D. McCulloch, Stress-dependent finite growth in soft elastic tissues, *J. Biomech.*, 1994, **27**, 455–467.
16. R. Skalak, G. Dasgupta, M. Moss, E. Otten, P. Dullemeijer and H. Vilmann, Analytical description of growth, *J. Theor. Biol.*, 1982, **94**, 555–577.
17. A. Piatnitski and M. Ptashnyk, Homogenization of biomechanical models for plant tissues, *Multiscale Model. Simul.*, 2017, **15**, 339–387.
18. M. Fakih, J. Y. Delenne, F. Radjai and T. Fourcaud, Root growth and force chains in a granular soil, *Phys. Rev. E*, 2019, **99**, 042903.
19. D. J. Cosgrove, Diffuse growth of plant cell walls, *Plant Physiol.*, 2018, **176**, 16–27.
20. D. Bonn, M. M. Denn, L. Berthier, T. Divoux and S. Manneville, Yield stress materials in soft condensed matter, *Rev. Mod. Phys.*, 2017, **89**, 035005.
21. C. Prior, J. Moussou, B. Chakrabarti, O. E. Jensen and A. Juel, Ribbon curling via stress relaxation in thin polymer films, *Proc. Natl. Acad. Sci.*, 2016, **113**, 1719–1724.
22. J. A. Lockhart, An analysis of irreversible plant cell elongation, *J. Theor. Biol.*, 1965, **8**, 264–275.
23. I. Cheddadi, M. Genard, N. Bertin and C. Godin, Coupling water fluxes with cell wall mechanics in a multicellular model of plant development, *PLoS Comput. Biol.*, 2019, **15**, e1007121.
24. F. Boudon, J. Chopard, O. Ali, B. Gilles, O. Hamant and A. Boudaoud, *et al.*, A computational framework for 3D mechanical modeling of plant morphogenesis with cellular resolution, *PLoS Comput. Biol.*, 2015, **11**, e1003950.
25. W. K. Silk and R. O. Erickson, Kinematics of plant growth, *J. Theor. Biol.*, 1979, **76**, 481–501.
26. G. T. Beemster and T. I. Baskin, Analysis of cell division and elongation underlying the developmental acceleration of root growth in Arabidopsis thaliana, *Plant Physiol.*, 1998, **116**, 1515–1526.
27. L. R. Band, S. Úbeda-Tomás, R. J. Dyson, A. M. Middleton, T. C. Hodgman and M. R. Owen, *et al.*, Growth-induced hormone dilution can explain the dynamics of plant root cell elongation, *Proc. Natl. Acad. Sci.*, 2012, **109**, 7577–7582.
28. R. J. Dyson, G. Vizcay-Barrena, L. R. Band, A. N. Fernandes, A. P. French and J. A. Fozard, *et al.*, Mechanical modelling quantifies the functional

importance of outer tissue layers during root elongation and bending, *New Phytol.*, 2014, **202**, 1212–1222.
29. R. Bastien, T. Bohr, B. Moulia and S. Douady, Unifying model of shoot gravitropism reveals proprioception as a central feature of posture control in plants, *Proc. Natl. Acad. Sci.*, 2013, **110**, 755–760.
30. R. Bastien, S. Douady and B. Moulia, A unified model of shoot tropism in plants: photo-, gravi-and propio-ception, *PLoS Comput. Biol.*, 2015, **11**, e1004037.
31. Y. Meroz, R. Bastien and L. Mahadevan, Spatio-temporal integration in plant tropisms, *J. R. Soc., Interface*, 2019, **16**, 20190038.
32. B. Moulia, R. Bastien, H. Chauvet-Thiry and N. Leblanc-Fournier, Posture control in land plants: growth, position sensing, proprioception, balance, and elasticity, *J. Exp. Bot.*, 2019, **70**, 3467–3494.
33. G. A. Holzapfel, R. W. Ogden and S. Sherifova, On fibre dispersion modelling of soft biological tissues: a review, *Proc. R. Soc. A*, 2019, **475**, 20180736.
34. A. J. M. Spencer, *Deformations of Fibre-reinforced Materials*, Oxford University Press, 1972.
35. J. Ericksen, Transversely isotropic fluids, *Colloid Polym. Sci.*, 1960, **173**, 117–122.
36. H. Cox, The elasticity and strength of paper and other fibrous materials, *Br. J. Appl. Phys.*, 1952, **3**, 72.
37. H. Oliveri, J. Traas, C. Godin and O. Ali, Regulation of plant cell wall stiffness by mechanical stress: a mesoscale physical model, *J. Math. Biol.*, 2019, **78**, 625–653.
38. J. A. Fozard, M. Lucas, J. R. King and O. E. Jensen, Vertex-element models for anisotropic growth of elongated plant organs, *Front. Plant Sci.*, 2013, **4**, 233.
39. B. Bozorg, P. Krupinski and S. H. Jönsson, A continuous growth model for plant tissue, *Phys. Biol.*, 2016, **13**, 065002.
40. L. Dupuy, J. Mackenzie, T. Rudge and J. Haseloff, A system for modelling cell–cell interactions during plant morphogenesis, *Ann. Bot.*, 2007, **101**, 1255–1265.
41. H. Jönsson, M. G. Heisler, B. E. Shapiro, E. M. Meyerowitz and E. Mjolsness, An auxin-driven polarized transport model for phyllotaxis, *Proc. Natl. Acad. Sci.*, 2006, **103**, 1633–1638.
42. R. S. Smith, S. Guyomarc'h, T. Mandel, D. Reinhardt, C. Kuhlemeier and P. Prusinkiewicz, A plausible model of phyllotaxis, *Proc. Natl. Acad. Sci.*, 2006, **103**, 1301–1306.
43. R. Farhadifar, J. C. Röper, B. Aigouy, S. Eaton and F. Jülicher, The influence of cell mechanics, cell-cell interactions, and proliferation on epithelial packing, *Curr. Biol.*, 2007, **17**, 2095–2104.
44. A. G. Fletcher, M. Osterfield, R. E. Baker and S. Y. Shvartsman, Vertex models of epithelial morphogenesis, *Biophys. J.*, 2014, **106**, 2291–2304.
45. T. Nagai and H. Honda, A dynamic cell model for the formation of epithelial tissues, *Philos. Mag. B*, 2001, **81**, 699–719.

46. M. Weliky and G. Oster, The mechanical basis of cell rearrangement. I. Epithelial morphogenesis during Fundulus epiboly, *Development*, 1990, **109**, 373–386.
47. O. Hamant, M. G. Heisler, H. Jönsson, P. Krupinski, M. Uyttewaal and P. Bokov, *et al.*, Developmental patterning by mechanical signals in Arabidopsis, *Science*, 2008, **322**, 1650–1655.
48. D. B. Staple, R. Farhadifar, J. C. Roeper, B. Aigouy, S. Eaton and F. Jülicher, Mechanics and remodelling of cell packings in epithelia, *Eur. Phys. J. E*, 2010, **33**, 117–127.
49. O. E. Jensen, E. Johns and S. Woolner, Force networks, torque balance and Airy stress in the planar vertex model of a confluent epithelium, *Proc. R. Soc. A*, 2020, **476**, 20190716.
50. Y. Long, I. Cheddadi, G. Mosca, V. Mirabet, M. Dumond and A. Kiss, *et al.*, Cellular heterogeneity in pressure and growth emerges from tissue topology and geometry, *Curr. Biol.*, 2020, **30**, 1504–1516.
51. F. Corson, O. Hamant, S. Bohn, J. Traas, A. Boudaoud and Y. Couder, Turning a plant tissue into a living cell froth through isotropic growth, *Proc. Natl. Acad. Sci.*, 2009, **106**, 8453–8458.
52. R. M. Merks, M. Guravage, D. Inzé and G. T. Beemster, VirtualLeaf: an open-source framework for cell-based modeling of plant tissue growth and development, *Plant Physiol.*, 2011, **155**, 656–666.
53. L. J. Grady and J. R. Polimeni, *Discrete Calculus: Applied Analysis on Graphs for Computational Science*, Springer Science & Business Media, 2010.
54. A. Nestor-Bergmann, G. Goddard, S. Woolner and O. E. Jensen, A vertex-based model relating cell shape and mechanical stress in an epithelium, *Math. Med. Biol.*, 2018, **35**, 1–27.
55. S. Ishihara and K. Sugimura, Bayesian inference of force dynamics during morphogenesis, *J. Theor. Biol.*, 2012, **313**, 201–211.
56. B. Guirao, S. U. Rigaud, F. Bosveld, A. Bailles, J. Lopez-Gay and S. Ishihara, *et al.*, Unified quantitative characterization of epithelial tissue development, *eLife*, 2015, **4**, e08519.
57. A. Nestor-Bergmann, E. Johns, S. Woolner and O. E. Jensen, Mechanical characterization of disordered and anisotropic cellular monolayers, *Phys. Rev. E*, 2018 May, **97**, 052409.
58. D. Bi, J. H. Lopez, J. M. Schwarz and M. L. Manning, A density-independent rigidity transition in biological tissues, *Nat. Phys.*, 2015, 1074–1079.
59. N. Murisic, V. Hakim, I. G. Kevrekidis, S. Y. Shvartsman and B. Audoly, From Discrete to Continuum Models of Three-Dimensional Deformations in Epithelial Sheets, *Biophys. J.*, 2015, **109**, 154–163.
60. J. A. Fozard, M. J. Bennett, J. R. King and O. E. Jensen, Hybrid vertex-midline modelling of elongated plant organs, *Interface Focus*, 2016, **6**, 20160043.
61. S. A. Braybrook and H. Jönsson, Shifting foundations: the mechanical cell wall and development, *Curr. Opin. Plant Biol.*, 2016, **29**, 115–120.

62. R. J. Dyson and O. E. Jensen, A fibre-reinforced fluid model of anisotropic plant cell growth, *J. Fluid Mech.*, 2010, **655**, 472–503.
63. O. Ali and J. Traas, Force-driven polymerization and turgor-induced wall expansion, *Trends Plant Sci.*, 2016, **21**, 398–409.
64. R. J. Dyson, L. R. Band and O. E. Jensen, A model of crosslink kinetics in the expanding plant cell wall: yield stress and enzyme action, *J. Theor. Biol.*, 2012, **307**, 125–136.
65. H. Yi and V. M. Puri, Architecture-based multiscale computational modeling of plant cell wall mechanics to examine the hydrogen-bonding hypothesis of the cell wall network structure model, *Plant Physiol.*, 2012, **160**, 1281–1292.
66. A. Barbacci, M. Lahaye and V. Magnenet, Another brick in the cell wall: biosynthesis dependent growth model, *PLoS One*, 2013, **8**, e74400.
67. H. Wada, Hierarchical helical order in the twisted growth of plant organs, *Phys. Rev. Lett.*, 2012, **109**, 128104.
68. D. E. Moulton, T. Lessinnes and A. Goriely, Morphoelastic Rods III: Differential Growth and Curvature Generation in Elastic Filaments, *J. Mech. Phys. Solids*, 2020, 104022.
69. D. E. Moulton, H. Oliveri and A. Goriely, Multiscale integration of environmental stimuli in plant tropism produces complex behaviors, *Proc. Natl. Acad. Sci.*, 2020, **117**(51), 32226–32237.

CHAPTER 4

Negative Pressure and Cavitation Dynamics in Plant-like Structures

OLIVIER VINCENT

CNRS, Univ. Lyon, Univ. Claude Bernard Lyon 1, Institut Lumière matière, F-69622, Villeurbanne, France
Email: olivier.vincent@cnrs.fr

4.1 Introduction

4.1.1 Negative Pressure

Liquids, similarly to solids, have internal cohesion. The individual molecules have attractive interaction and a restoring force brings them back together if one tries to pull them apart. Due to this internal cohesion, liquids can sustain tensile stress, and can thus be put in a state where their pressure is absolutely negative. Going back to the definition of pressure as a force per unit area, a state of *negative pressure* means that the force exerted by the fluid on a surface is directed towards the liquid and not towards the surface as it is for $P>0$ (see Figure 4.1a–b). Note that because pressure in a liquid is isotropic, the tensile stress applies in all directions, while a solid can develop anisotropic stresses, *e.g.* with tensile stress only in one direction and compressive stresses in the other directions.

A useful analogy is that of a chain of springs: if one stretches the chain, each individual spring gets stretched and starts pulling on its neighbors; a person holding the chain would feel a force directed towards the chain. Similarly, a liquid at negative pressure is stretched: intermolecular distances

Soft Matter Series No. 15
Soft Matter in Plants: From Biophysics to Biomimetics
Edited by Kaare H. Jensen and Yoël Forterre

Published by the Royal Society of Chemistry, www.rsc.org

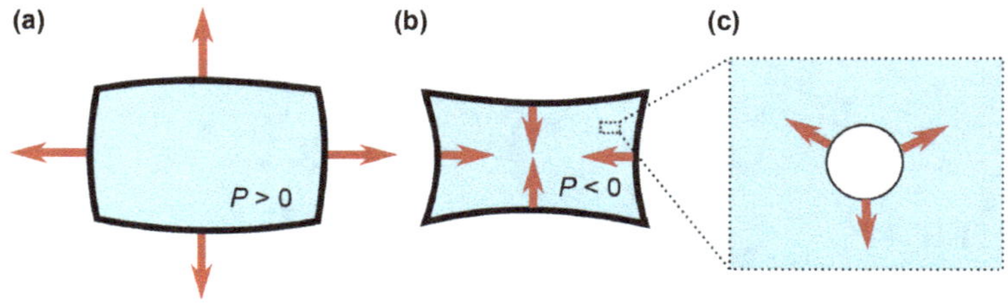

Figure 4.1 (a) A liquid at positive pressure exerts an outward force on the walls of a container, while (b) a liquid at negative pressure is in a tensile state and exerts an inward force on the walls. The walls are represented as being deformable, to illustrate the force applied by the fluid. (c) The tensile state as in (b) not only "pulls" on the walls but also on other interfaces; in particular, this tends to make sufficiently large bubbles grow, resulting in cavitation.

are larger than in ambient conditions, which results in a global attractive force between molecules, but also between the molecules and the container they are in. Thus, while a liquid at positive pressure pushes on the walls of a container, a liquid at negative pressure pulls on them (see Figure 4.1a–b). Obviously, the state of negative pressure can be maintained only if adhesion between the walls and the liquid is good. In the case of water, this means that the walls of the container have to be somewhat hydrophilic. Because of this stretched, tensile state, a liquid at negative pressure is also called *under tension*, *under stress*, or simply *stretched*. This stretched state is metastable and can relax by cavitation (see below).

It is useful to note here already that it is not only the sign of the pressure that matters in cavitation phenomena, but its value compared to the *saturation vapor pressure*, P_{sat}. In particular, a liquid at positive pressure but with $P < P_{sat}$ is also metastable and potentially subject to cavitation, and crossing the value $P = 0$ for the liquid does not have particular consequences. Thus, the term "negative pressure" in the following should be broadly interpreted as $P - P_{sat} < 0$, and "stretched" as meaning "more stretched than at saturation". For the phenomena we discuss in this chapter, this distinction remains anecdotal, because P_{sat} is several orders of magnitude below the typical negative pressures of interest ($-P \gg P_{sat}$), so that considering $P_{sat} \simeq 0$ does not make a significant difference.

4.1.2 Cavitation

In a state of negative pressure, the liquid tends to "pull" on any interface it is in contact with, and in particular it will tend to make a bubble grow (see Figure 4.1c), if the bubble is sufficiently large.† Bubble growth only stops when the negative pressure vanishes, which happens when bubble

†Following the remark about $P < P_{sat}$, bubble growth also happens if $0 < P < P_{sat}$, because the bubble fills with water vapor at P_{sat}, and the pressure in the liquid is below P_{sat}, so that there is a net outwards driving force on the bubble.

expansion has allowed the stretched liquid to contract back into a non-stretched state. This process of bubble expansion associated with the disappearance of negative pressure and the relaxation of metastability is what we define as *cavitation*.

Following this definition, cavitation requires a *germ*, or *nucleus*, *i.e.* a microscopic bubble in the liquid that gets stretched by the tensile forces in the liquid into a macroscopic bubble. While such germs can pre-exist (*e.g.* bubbles trapped in walls, or stabilized free-floating bubbles), they can also form spontaneously by activation from thermal fluctuations.

It may seem at first sight that liquid at negative pressure should be unstable, because cavitation would always occur from existing or activated germs. However, another driving force counteracts the stretching effect of the negative pressure on bubbles: surface tension, which accounts for the energetic cost of having a liquid–vapor interface and tends to make it as small as possible, *i.e.* make the bubble collapse. Because the surface-to-volume ratio increases when bubble size decreases, surface tension dominates for small bubbles. In other words, germs have to be above a *critical size* to lead to cavitation. Since large germs are less likely than small germs to pre-exist or to form spontaneously, a liquid at negative pressure is in fact not unstable but metastable, and can be long-lived. In practice, cavitation occurs only if the magnitude of the negative pressure exceeds some threshold that greatly depends on the microscopic mechanism at the origin of cavitation.

Cavitation is a particular type of *nucleation* due to pressure changes. Boiling (nucleation of vapor due to temperature changes) is another very similar nucleation phenomenon. We note that we use a broad definition of cavitation (destruction of a state of negative pressure by the formation of a macroscopic bubble) that thus naturally includes the case of *air seeding* frequently discussed in plants, where the initial origin of the cavitation bubble is the aspiration of a meniscus through a membrane or porous structure. In fact, the cavitation dynamics phenomena that we describe in this chapter are mostly independent on the microscopic mechanism that leads to nucleation.

4.1.3 Negative Pressure and Cavitation in Plants

In the context of plants, water at negative pressure is ubiquitous. Vascular plants like trees actually use that negative pressure as a suction force to drive sap flow from the root to the leaves in the *xylem* tissue. This process is driven by transpiration, *i.e.* natural evaporation from the xylem into the atmosphere in the leaves.[6–8] Most trees routinely experience tens of atmospheres of negative pressure in their xylem during the day[7–9] and thus contain large amounts of water under stress, which is metastable and susceptible to cavitation (see Figure 4.2a). Here, cavitation is detrimental, because it eventually results in air-filling (*embolism*) of xylem conduits, which disrupts the upwards flow of sap (see Figure 4.2b). Spreading of embolism through the whole vascular structure would be catastrophic, but xylem presents a

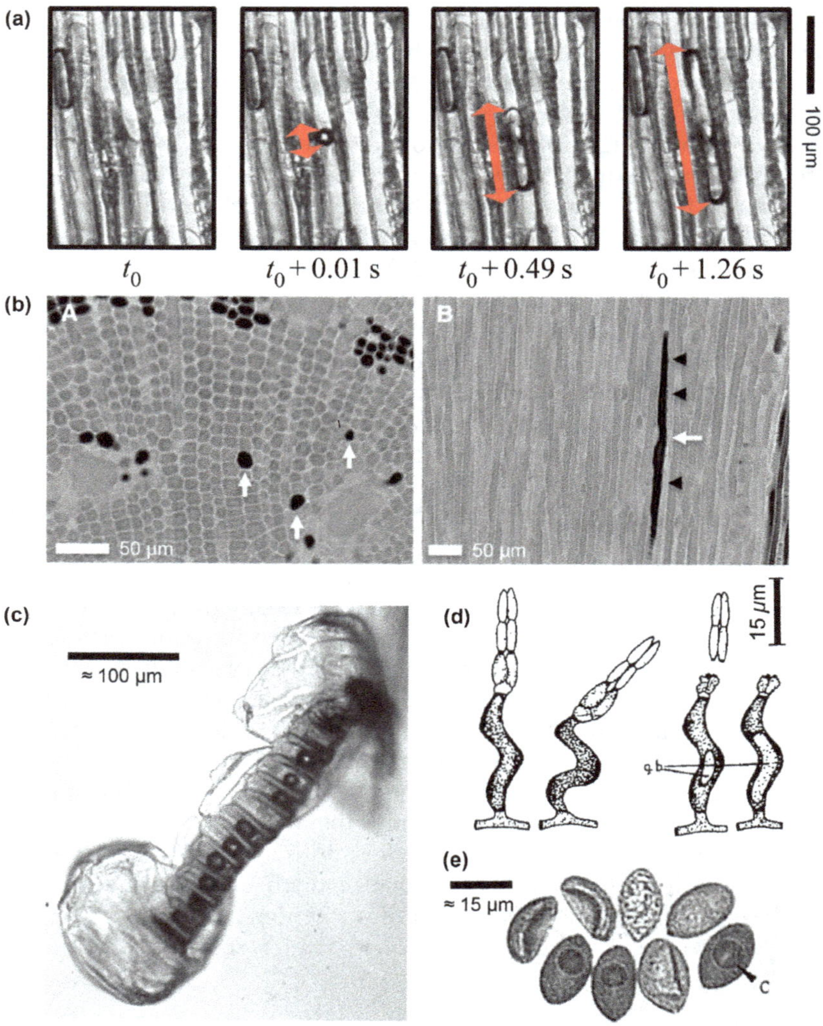

Figure 4.2 (a) Birth and growth of a cavitation bubble in a dehydrating slice of vascular tissue (xylem) of a pine tree. (b) In-vivo observation of the resulting embolism using X-ray microtomography (transverse and longitudinal cross-sections on left and right, respectively); black areas are gas-filled. (c) Simultaneous cavitation in neighboring cells of a fern sporangium, at the basis of a catapult-like spore ejection mechanism. (d) Another cavitation-based spore ejection mechanism in a fungus. (e) Cavitation induced by osmotic dehydration in fungal spores. Panels (a) and (c) reproduced from ref. 1 and 2 with permission from the Royal Society, Copyright 2014 and 2016 respectively; permission conveyed through Copyright Clearance Center, Inc. Panels (b) and (d) reproduced from ref. 3 and 4 with permission from Oxford University Press, Copyright 2016 and 1963, respectively. Panel (e) reproduced from ref. 5 with permission from John Wiley and Sons, Copyright © 1970 New Phytologist Trust.

segmented structure made of a large number of conducting elements (*vessels*, *tracheids*) interconnected by nanoporous membranes (*pits*); if one of the elements fails by cavitation and gets embolized, spreading to the adjacent elements is prevented by the pits (at least temporarily), and embolism is contained (see *e.g.* Figure 4.2b, right panel where a single embolized tracheid is surrounded by intact water-filled tracheids).

In other situations, however, plants use cavitation to their advantage. For example, some species of ferns use simultaneous cavitation in an annular structure as a trigger for a catapult-like ejection mechanism for their spores (ref. 2 and 10 see Figure 4.2c). There, the slow build-up of negative pressure prior to cavitation also originates from natural dehydration of the cells in the annulus due to evaporation into the surrounding air when conditions get dry. Not far from the plant kingdom, negative pressure and cavitation are also observed in fungi (see Figure 4.2d–e), where they can be at the basis of other spore dispersion strategies (ref. 5 and 11 see Figure 4.2d).

We briefly note that negative pressure and cavitation can also be found in the animal kingdom. In particular, several species of shrimp use cavitation bubbles as a hunting tool (pistol shrimp,[12] mantis shrimp[13]); octopuses and squids can also generate a few atmospheres of negative pressure, and cavitation is thought to limit their adhesion strength.[14]

The situations described above for plants and fungi have common features. First, water under negative pressure is contained in closed, cellular structures that are rigid enough to sustain the inner tension force without collapsing. Second, these structures allow for exchange of water with the surrounding environment (other cells or air) through the porous walls and/or through specialized porous membranes (pits). Last, negative pressure is generated by dehydration (transpiration at the leaf level for trees, evaporation through cell walls for the other structures) and sustains typical values in the range −10 to −100 bars for large amounts of time.

4.1.4 Chapter Contents

In this chapter, we discuss the general physics of negative pressure and cavitation in water for situations similar to those found in plants and fungi: large, static negative pressures produced by dehydration in a liquid confined in porous, cellular structures.

First, we introduce important concepts related to the mechanics and thermodynamics of water (Section 4.2). Then, we explain how negative pressures are generated, focusing on the effect of dehydration from cells (Section 4.3). Next, we discuss various microscopic mechanisms that may lead to cavitation in liquids at negative pressure (Section 4.4). In the next section, we present an extension of classical nucleation theory that describes cavitation in closed, elastic cells (Section 4.5). The last two sections illustrate the various theoretical concepts established previously with experimental results on artificial systems reproducing some features found in plants: we first describe the rich dynamics of a cavitation in a single cell, that spans several orders of magnitude of timescales (Section 4.6), before discussing the spatio-temporal patterns of cavitation (and its propagation) in extended systems containing many, interconnected cells (Section 4.7).

This chapter presents a physicist's perspective on the basics of negative pressure and cavitation in plant-like, cellular structures and thus focuses on a local level. There are many other aspects of water relations and flow in plants that are not discussed here (*e.g.* global sap flow, stomatal control in the leaves, exchange between xylem and phloem, *etc.*). For more details on the physiology and water transport mechanisms in plants and fungi, we refer the reader to classic books[6,11,15] and recent reviews on various aspect of water stress and transport in plants.[7–9,16] We also recommend exhaustive books[17,18] and reviews[19,20] on the physics of metastable water and cavitation.

4.2 Water Properties

In this section, we introduce and define several properties of water that are relevant in the context of negative pressure and for the phenomenon of cavitation.

4.2.1 Cohesion and Surface Tension

Liquid water has large internal cohesion, which is mostly due to the presence of hydrogen bonds. As a result, since molecules strongly bond together, exposing water molecules at an interface is highly unfavorable, and the surface tension of liquid water is particularly high: $\gamma = 72 \times 10^{-3}$ N m^{-1} at 25 °C. As a comparison, the surface tension of organic liquids such as ethanol or acetone is less than a third of that value ($\sim 22 \times 10^{-3}$ N m^{-1}).

From the value of the surface tension, one can illustrate liquid water's cohesion with the following thought experiment. Imagine "pulling" on water sufficiently strongly to separate the molecules further than their typical interaction distance, δ, thus "breaking open" a liquid column. The energy associated with the newly created interfaces, $E = 2\gamma S$, must be equal to the work $W \simeq F_{\text{pull}} \times \delta$ of the force pulling the liquid apart. One can thus estimate that in order to "break open" a water column it takes a stress $\Delta P_{\text{pull}} = F_{\text{pull}}/S = 2\gamma/\delta \sim 300$ MPa, assuming $\delta \sim 0.5$ nm. From this rough estimate, we can already see that liquid water should be able to withstand very large tensile stresses, *i.e.* negative pressure. We refine this estimate in Section 4.2.2.

Note that contrary to liquids, gases do not have internal cohesion. As a result, they cannot be brought to a state of negative pressure: "pulling" on a gas will just dilute it towards the limit of zero pressure for infinite expansion. A direct consequence of this remark is that a liquid at negative pressure cannot coexist with a gas, except in situations where something (the surface tension of an interface, a membrane, *etc.*) allows for a pressure mismatch between the two phases.

4.2.2 Compressibility and Spinodal

Liquid water under negative pressure is stretched, and the amount by which it stretches is related to the compressibility of the liquid. Thus, compressibility plays a natural role in cavitation phenomena and will be involved in several of the sections of this chapter. Also, as we discuss briefly at the end of this section, the evolution of compressibility with pressure provides a way to estimate the tensile strength of the liquid through its *spinodal*.

The value of the isothermal compressibility, $\chi_\ell = (-1/V_\ell)(\partial V_\ell/\partial P)_T$, for liquid water is $\chi_\ell = 0.45\ \mathrm{GPa}^{-1}$ *i.e.* its bulk modulus is $K_\ell = 1/\chi_\ell = 2.2$ GPa; in other words, a pressure of 22 MPa will contract water's volume, V_ℓ, by 1%.

Generally, as long as the variations of pressure are not too large ($|\chi_\ell \Delta P| \ll 1$), it is a good approximation to relate the variation of volume of the liquid to variations of pressure through the linear approximation

$$(V_\ell - V_{\ell,\mathrm{ref}})/V_{\ell,\mathrm{ref}} = -\chi_\ell(P - P_{\mathrm{ref}}) \tag{4.1}$$

where we consider deviations from a reference state at volume and pressure ($V_{\ell,\mathrm{ref}}$,P_{ref}). In other words, eqn (4.1) assumes χ_ℓ constant in the range of pressure of interest. This linear approximation expressed by eqn (4.1) is largely sufficient to describe the situations we consider in this chapter, where the negative pressures have a magnitude of at most $\simeq 20$ MPa, *i.e.* two orders of magnitude below $K_\ell = 1/\chi_\ell$.

For negative pressures of larger magnitude, it would be necessary to take into account the variations of χ_ℓ with P. In fact, it is interesting to note that the variation of χ_ℓ with pressure offers another way to estimate the cohesive tensile strength of liquid water. Indeed, going back to our thought experiment from above (Section 4.2.1), compressibility is the reason why molecules get further apart when we try to "pull" on a liquid. However, because the range of interaction of the molecules is not infinite, there must be a point where the interaction starts getting weaker and cannot resist the pulling force anymore. At this point, tensile stress as a function of volume reaches a maximum, in other words the bulk modulus becomes zero and the compressibility χ_ℓ diverges. A way to estimate the tensile strength of the liquid is thus to extrapolate the equation of state of the liquid to estimate where $K_\ell = 1/\chi_\ell$ is zero. Following this approach, values of tensile strength in the vicinity of 200 MPa (*i.e.* a negative pressure of −200 MPa) can be found at ambient temperature.[20,21] This validates the rough estimate made above from the value of the liquid's surface tension.

The value of $P \simeq -200$ MPa actually corresponds to the *spinodal* of the liquid, where it becomes mechanically unstable and spontaneously breaks apart. In practice, this point is never reached because cavitation appears beforehand by various mechanisms (see Section 4.4).

4.2.3 Saturation Pressure, Phase Diagram

So far we have discussed mechanical properties of water. Now, we move to thermodynamic considerations. Generally, the most stable phase of water

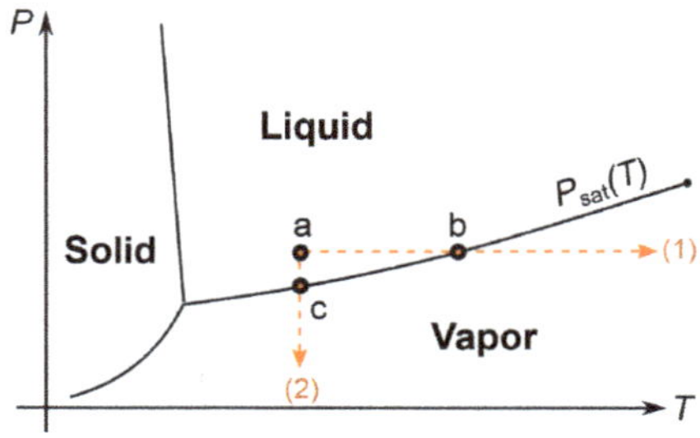

Figure 4.3 Schematic pressure–temperature phase diagram of water. $P_{sat}(T)$ is the saturation vapor pressure, which represents the coexistence line between the zones of stability of the liquid and vapor phases. When a liquid (point a) is heated up (path 1), vapor can become the stable phase (beyond point b): vapor bubbles can spontaneously nucleate, corresponding to *boiling*. When pressure is decreased from point a (path 2), vapor can also become the stable phase (below point c) and the nucleation of vapor bubbles is called *cavitation* in this case.

depends on conditions of pressure and temperature. It is useful to represent these conditions as a point in the pressure–temperature (P–T) phase diagram (see Figure 4.3). The diagram defines three zones corresponding to the zones where ice, liquid water and water vapor are the most stable, respectively.‡ The line separating the zones of stability of the vapor and liquid phases between the triple point and the critical point defines the *saturation vapor pressure*, $P_{sat}(T)$; in ambient conditions (*e.g.* $T = 25$ °C), $P_{sat} \simeq 3$ kPa, a value much smaller than the atmospheric pressure, $P_{atm} \simeq 100$ kPa.

Most of us are familiar (*e.g.* when cooking pasta) with the fact that when heated up, water boils at 100 °C because one crosses the liquid–vapor coexistence line and one enters the domain where the vapor is most stable (see Figure 4.3, path 1). Boiling occurs because for $T > 100$ °C, $P_{sat}(T)$ is larger than atmospheric pressure P_{atm} The coexistence line can also be crossed by decreasing the pressure (see Figure 4.3, path 2). In this case, we do not call the nucleation of vapor bubbles *boiling*, but *cavitation*. Cavitation occurs because the liquid pressure falls below the value of P_{sat} at ambient temperature. In practice, neither boiling nor cavitation necessarily happen exactly when the coexistence line $P_{sat}(T)$ is crossed, because of the existence of metastable states.

4.2.4 Metastable States

The liquid-to-vapor phase transition requires the formation of a bubble, *i.e.* the creation of a liquid–vapor interface. Due to the large surface tension of water, this process is particularly unfavorable and an energy barrier

‡In reality, more zones are present in the ice domain because there exist different types of ice.

(and a critical bubble size, see Introduction) must be overcome to form a bubble. As a result, the liquid state can persist even when the coexistence line is crossed (*i.e.* for $P<P_{sat}$). This situation is metastable: small fluctuations relax back to the liquid state, but a large, less likely fluctuation can make the system cross the energy barrier and create a stable bubble, *i.e.* induce cavitation or boiling. The further away from the coexistence line one penetrates into the zone of vapor stability in the *P–T* phase diagram, the smaller the energy barrier is and the more likely nucleation is; see the section on nucleation theory for an illustration of these aspects (Section 4.5).

Similar metastable states exist for the vapor to liquid phase transition (supersaturated vapor), or the liquid to ice transition (supercooled liquid).[§] Metastable liquid water below P_{sat} is often called *superheated* when following the boiling route, and *stretched* when following the cavitation route. Indeed, lowering the pressure results in an expansion of the liquid due to water's compressibility (see Section 4.2.2).

4.2.5 Chemical Potential, Water Potential

Thermodynamic driving forces and equilibria are conveniently described by the *chemical potential* of water, μ (J mol^{-1}), which combines energetic and entropic contributions and enables comparison of the relative stability of the same substance in different phases: liquid water (μ_ℓ), water vapor (μ_v). Molecules spontaneously move from areas of high μ to areas of low μ and equilibrium is attained when the chemical potential is identical in all parts of a system. In particular, at the coexistence line ($P=P_{sat}(T)$), $\mu_\ell(P_{sat})=\mu_v(P_{sat})=\mu_{sat}(T)$.

The concept of *water potential* (Ψ, in Pa) is more widely used in the plant science community instead of chemical potential:

$$\Psi=\frac{\mu-\mu_{ref}}{v_m} \tag{4.2}$$

where $\mu_{ref}(T)$ is the chemical potential of a reference state, usually chosen to be pure liquid water at atmospheric pressure ($P=P_{atm}$), at the temperature of interest, T; v_m is the molar volume of liquid water in the reference state ($v_m=1.807\times10^{-5}$ m^3 mol^{-1} at $T=25$ °C). Thus, chemical potential and water potential only differ by a multiplicative factor and an additive constant and fundamentally represent the same quantity. Water potential is convenient to use, because it has units of pressure and is zero for bulk water at ambient conditions.

The reason of using $P=P_{atm}$ instead of $P=P_{sat}$ as a reference state is not often discussed, but this choice has physical meaning. Indeed, P_{sat} corresponds to the equilibrium between liquid water and pure water vapor; in

[§]For example, supercooled water can be brought down to $\simeq-40$ °C before nucleation of ice occurs. Supercooled water plays a large role in the atmosphere and for the survival of some living organisms in the winter.[17]

this situation, both phases are at the same pressure. Now, let us consider liquid water situated in air at atmospheric pressure (*e.g.* in an open bottle) and evaluate equilibrium with water vapor in the air. The liquid phase, due to mechanical equilibrium with air, is at $P = P_{atm}$, while the vapor phase is characterized by its *partial vapor pressure*, $p \neq P$. Due to this pressure mismatch, equilibrium between the two phases occurs at a different vapor pressure, $p_{sat} \neq P_{sat}$ in the presence of air: the condition for equilibrium is $(p = p_{sat}, P = P_{atm})$, with $p_{sat} > P_{sat}$ (see Appendix A). The difference between p_{sat} and P_{sat} is small and difficult to measure in practice (*e.g.* at 25 °C, $P_{sat} = 3170$ Pa, $p_{sat} = 3172$ Pa, *i.e.* a difference of 0.07%), so that both saturation pressures are usually assumed to be the same. However, it is useful to keep the distinction between the two definitions in mind, in order to avoid thermodynamic inconsistencies in calculations.

The conclusion from the previous paragraph is that while P_{sat} corresponds to a liquid–vapor equilibrium of the pure water substance (*i.e.* in vacuum), the equilibrium state with liquid water at total pressure, $P = P_{atm}$ and with water vapor at partial pressure, $p = p_{sat} > P_{sat}$ is a true saturation state of the system when considering water in air at atmospheric pressure. It is thus natural to use this state as a reference when dealing with processes occurring in air. Using this reference, eqn (4.2) yields

$$\Psi_\ell = \Delta P - \Pi \tag{4.3}$$

for liquid water, where $\Delta P = P - P_{atm}$ is the pressure difference with respect to atmospheric pressure, and Π is the *osmotic pressure*; $\Pi = 0$ for pure water, and $\Pi \simeq RTC > 0$ when solutes are present at a concentration C (mol m^{-3}).[8,15,22] ΔP is a purely mechanical contribution, while Π is an entropic contribution associated with the colligative decrease of chemical potential that occurs when mixing water with a solute. Here, we neglect the contribution of gravity ($\Delta\Psi_g = \rho g z$, with ρ the density of liquid water, g the acceleration of gravity and z the elevation) that will not be needed for the discussions in this chapter, and is only significant to compare water at large height differences, *e.g.* in tall trees ($\rho g \simeq 0.01$ MPa m^{-1}). We have also neglected the contribution due to the compressibility of water. Indeed, for consistency with eqn (4.1), an additional term $\Delta\Psi_\chi = \chi_\ell \Delta P^2/2$ should be included. However, this contribution is also negligible for the conditions considered in this chapter ($|\Delta\Psi_\chi/\Psi| < 0.5\%$ for $\Psi > -20$ MPa); we thus neglect this term for simplicity.

For water vapor in air,

$$\Psi_v = \frac{RT}{v_m} \ln\left(\frac{p}{p_{sat}}\right) \tag{4.4}$$

where p_{sat} is the saturation vapor pressure *in air* defined above; $a = p/p_{sat}$ is the *activity* of water vapor or its *relative humidity*: for example $a = 0.85$ corresponds to a relative humidity of 85%RH. Although this is rarely mentioned, Ψ_v is impacted by gravity in the exact same way as Ψ_ℓ (*i.e.*, $\Delta\Psi = \rho g z$, where ρ is still the density of liquid water).[¶] Again, we neglect this effect here.

Similarly to chemical potential (μ), water potential (Ψ) is the natural driving force for transport and equilibria within a single phase or between phases. Transport occurs towards areas of lower Ψ and equilibrium occurs when the driving force vanishes, *i.e.* equal Ψ in different parts of the system.

4.2.6 Evaporation *vs.* Cavitation

Eqn (4.4) implies that the water potential of water vapor in air is always negative when relative humidity is less than 100%RH ($a<1$). This means that an open container with pure, liquid water ($\Psi_\ell=0$, see eqn (4.3)) always tends to evaporate, except in rare cases of 100%RH humidity, because water molecules lower their water potential by moving from the liquid phase to the vapor phase. This corresponds to the fact that when $a<1$, there are more molecules per unit time leaving the liquid towards the vapor than molecules coming back from the vapor into the liquid, so that over time the liquid loses mass. Evaporation thus occurs at the interface between liquid and air due to a transport imbalance between the liquid and vapor phases. In the example above, the two phases (liquid at P_{atm}, vapor at $p<=p_{sat}$) are both in their stability zone of the phase diagram, but evaporation can also occur from a metastable liquid into a sub-saturated vapor through a membrane (see Section 4.3.2). In contrast to evaporation, cavitation and boiling correspond to the nucleation of the stable vapor phase *within* the metastable liquid, and do not happen in a stable liquid.

4.3 Origins of Negative Pressure

As explained in the Introduction, negative pressure in plants mainly originates from dehydration (*e.g.* evaporation, osmosis). Below, we first explain the connection between negative pressure and dehydration from a mechanical perspective. Then, we describe the driving forces and equilibrium states involved in dehydration from a thermodynamic perspective. Finally, we briefly mention other ways negative pressures can be obtained.

4.3.1 Dehydration: Mechanics

Let's consider liquid water that is enclosed in a cell of volume, V_c, at ambient pressure P_{atm}. Now, we examine a situation where the cell dehydrates, *i.e.* loses water molecules (by an amount of substance Δn), *e.g.* by transport through the

[¶]The fact that the density of the liquid, ρ, is involved in the equation for water vapor is simply due to the way water potential is defined, *i.e.* by dividing water potential differences $\Delta\mu$ by the molar volume of the *liquid*; for water vapor, $\Delta\mu=RT\ln(a)+Mgz$, where M is the molar mass of water. Not taking into account the effect of gravity on the water potential of water vapor can result in the erroneous prediction that one can build a perpetual-motion pumping machine based on liquid/vapor local equilibrium, *i.e.* a tall column of liquid water could never be in equilibrium with the surrounding vapor.

porous walls. We assume that liquid adhesion to the cell walls is good so that no cavitation occurs and the liquid remains homogeneous in the process.

If the cell is infinitely rigid (see Figure 4.4a), V_c does not change in the process, and the cell now contains fewer molecules occupying the same volume. Necessarily, the average distance between molecules in the liquid increases; in other words, the liquid gets stretched and its pressure drops: $P=P_{atm}+\Delta P$, with $\Delta P<0$. Another way to look at the situation is that the molecules initially present in the cell (volume, V_c) now occupy a larger volume, $V_c+\Delta V$, where $\Delta V=v_m\Delta n$ is the liquid volume occupied by the molecules removed from the cell. We can evaluate the pressure required to stretch the liquid by an amount ΔV using eqn (4.1):

$$\Delta P=-\frac{1}{\chi_\ell}\frac{\Delta V}{V}. \tag{4.5}$$

Because of the low compressibility of water ($\chi_\ell=0.45\ \text{GPa}^{-1}$; see Section 4.2.2), removing only a small fraction of the water molecules within the cell results in massive pressure changes. For example, withdrawing only 0.1% of the molecules ($\Delta V/V=0.001$) is sufficient to make the pressure drop from ambient to more than 20 bars of negative pressure, $\Delta P=-2.2$ MPa.

In real systems, the cell enclosing the liquid is not infinitely rigid and deforms under liquid pressure changes. In particular, if P decreases, the cell tends to contract so that V_c also decreases (see Figure 4.4b). We can quantify this effect using an effective cell compressibility $\chi_c=(1/V_c)(\partial V_c/\partial P)_T$; note the opposite sign in this definition compared to the liquid's compressibility (see Section 4.2.2) that allows us to keep all coefficients positive. $K_c=1/\chi_c$ is the effective bulk modulus of the cell. Now, removing a volume $\Delta V=v_m\Delta n$ of liquid water from within the cell results in both contraction of the cell and

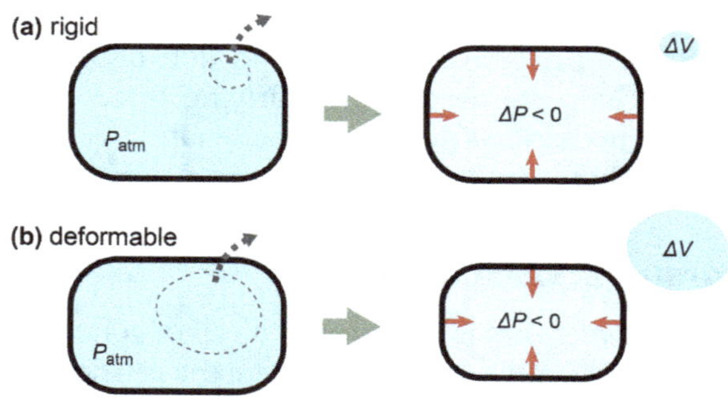

Figure 4.4 Mechanics of negative pressure generation by dehydration of a cell containing liquid water. (a) If the confining cell is infinitely rigid, withdrawing a volume ΔV of the liquid stretches the liquid by the same amount, which makes the pressure drop (see eqn (4.5)). (b) If the cell is deformable, the same process both stretches the liquid and makes the cell shrink. More liquid volume needs to be removed to produce a similar drop in pressure (see eqn (4.6)). Red arrows represent the force exerted by the liquid on the walls; lighter blue color indicates a less dense liquid.

stretching of the liquid. Using the definitions of χ_ℓ and χ_c, it is straightforward to show that the corresponding drop in pressure is given by

$$\Delta P = -\frac{1}{\chi_\ell + \chi_c}\frac{\Delta V}{V} \tag{4.6}$$

as long as the pressure variations are small with respect to both bulk moduli ($|\chi_\ell \Delta P| \ll 1, |\chi_c \Delta P| \ll 1$). We recall that ΔV is the liquid volume withdrawn from the cell, not the change in the cell volume, ΔV_c; ΔV and ΔV_c are different in general because of the compressibility of the liquid. Because χ_c is positive, it requires more dehydration of the cell (larger ΔV) to achieve a similar drop in pressure compared to the case of an infinitely rigid cell discussed previously, as can be seen by comparing eqn (4.5) and (4.6).

Eqn (4.6) shows the interplay of water compressibility (χ_ℓ) and cell deformability (χ_c). When dehydration occurs, water expansion and cell contraction occur simultaneously. One can have different views on the generation of negative pressure by dehydration. Some people would prefer to consider that water leaving the cell induces an elastic strain in the cell wall, which "pulls" on the liquid and generates negative pressure. Other people would think of the same process as water leaving the cell that stretches the inner liquid, resulting in negative pressure that pulls on the cell walls and makes the cell shrink. However, these are just two point of views on a unique, coupled phenomenon. Note that even with a very stiff cell, a tiny deformation of the walls has to occur to maintain equilibrium with the pressure force exerted by the liquid.

We have assumed a linear response between the pressure and the cell volume, which is an excellent approximation as long as $|\Delta P| \ll K_c$ (small deformations). For larger deformations, one would need to know in more detail the stress–strain relationship of the cell to quantify deviations to eqn (4.6). Also, instabilities such as creasing or buckling can occur (some are visible in Figure 4.2e) and may result in collapse of the structure. To avoid too large deformations (that reduce the available volume) or collapse, structures conducting water at negative pressure should be stiff ($K_c \gg |\Delta P|$). In trees (see Figure 4.2a–b), for example, typical negative pressures in xylem are -1 to -10 MPa, while K_c is in the range 0.1–1 GPa; deformations of wood under the effect of water negative pressure are thus small but measurable.[23,24] For cavitation-based spore ejection mechanisms (see Figure 4.2c–d), however, the structure is much more compliant, which allows the system to deform significantly and store a large amount of elastic energy when negative pressure builds up; this stored energy is suddenly released when cavitation occurs. In these situations, the cell elastic moduli and the negative pressures are of the same order of magnitude (~10 MPa).[‖]

‖For the fern sporangium,[2] typical negative pressures of -10 MPa develop, and the cell elastic modulus is $K_c = 2B/h$ where $B \simeq 300$ N/m is the bending modulus of the annulus, and $h \simeq 40$ µm is the height of the cells. Thus, $K_c \simeq 20$ MPa.

4.3.2 Dehydration: Thermodynamics

In the previous discussion, we have explained how dehydration can generate negative pressure through the coupled stretching of the liquid and strain of the enclosing structure that it implies. Now, we discuss the thermodynamic driving forces for dehydration and the potential associated equilibria.

As discussed in Section 4.2.5, driving forces can be described using *water potential*, Ψ. In the same way as a drop of water ($\Psi_\ell = 0$) that spontaneously evaporates in a subsaturated vapor ($a = p/p_{sat} < 1$, hence $\Psi_v < 0$) as discussed in Section 4.2.6, a cell structure containing water, initially at ambient pressure ($\Psi_\ell = 0$ if there are no solutes), tends to dehydrate spontaneously if the membrane permits exchange with air that is subsaturated in water vapor (see Figure 4.5a). From the mechanical mechanisms described in the previous section (see Figure 4.4), this dehydration produces a drop in pressure, which decreases the water potential of the liquid within the cell (see eqn (4.3)). As dehydration progresses, Ψ_ℓ within the cell becomes more and more negative (as does the liquid pressure) and gets closer to Ψ_v of the vapor, so that the driving force for dehydration decreases. An equilibrium can be reached when $\Psi_\ell = \Psi_v$, or, from eqn (4.3) and (4.4),

$$\Delta P = \frac{RT}{v_m} \ln\left(\frac{p}{p_{sat}}\right) + \Pi \qquad (4.7)$$

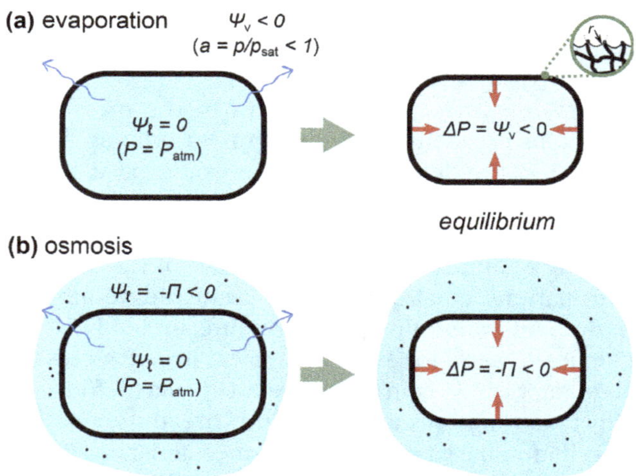

Figure 4.5 Thermodynamics of negative pressure originating from dehydration. (a) Evaporation in a subsaturated vapor dehydrates the cell until the negative pressure in the cell balances the negative water potential of the external water vapor (see eqn (4.7)). Equilibrium between the liquid and vapor at different pressures can be mediated through the formation of curved liquid–vapor menisci (inset). (b) Dehydration in an osmotic solution produces the same effect; equilibrium is reached when the negative pressure in the cell balances the external osmotic pressure (see text).

where $\Delta P = P - P_{atm}$, and Π is the osmotic pressure associated with potential solutes present in liquid water. In trees, xylem contains solutes at low concentrations, and the associated osmotic pressure is usually neglected.[6,15] However, other structures such as a fern leptosporangium can contain solutes with Π in the MPa range;[2] in such a situation, as can be seen in eqn (4.7), the presence of solutes makes the equilibrium pressure at given external humidity less negative than for pure water.

For pure water, the negative pressure implied by eqn (4.7) is rapidly increasing in magnitude as the relative humidity of the air is decreased below 100%RH (see Table 4.1) so that equilibrium with a subsaturated vapor is a very efficient way to generate massive negative pressures. With solutes, the values of liquid pressure indicated in Table 4.1 have to be shifted up by the amount Π (*e.g.*, the equilibrium pressure of a cell containing solutes at $\Pi = 2$ MPa at 92.9%RH would be -8 MPa).

We note that dehydration can also be driven by osmosis instead of evaporation. Taking the example of a cell containing pure liquid water immersed in a solution at ambient pressure (see Figure 4.5b), if the cell membrane is able to maintain the concentration difference (*i.e.* excludes the solute), equilibrium is obtained when a the negative pressure in the inner liquid balances the osmotic pressure, *i.e.* $\Delta P = -\Pi$. This strategy is sometimes used to artificially induce negative pressure and cavitation in plant cells or artificial structures (see *e.g.* ref. 2, 5, 22 and 25–27). More generally, with solutes both inside and outside of the cell, $\Delta P = \Pi_{in} - \Pi_{out}$ at equilibrium. The dehydration effect still occurs if the membrane only hinders solute transport instead of being fully permeable, but with a reduced efficiency: $\Delta P = \sigma(\Pi_{in} - \Pi_{out})$, with the *reflection coefficient* $\sigma < 1$.[15,22,28] If not actively maintained, the concentration difference, however, tends to vanish over time due to solute diffusion, and the osmotic pressure difference becomes zero. In some sense, osmosis can be seen as less efficient than evaporation as a driving force to induce negative pressures, because large concentrations of solutes are required to generate significant osmotic pressures (*e.g.* 1 mol L^{-1} of sucrose for $\Pi \simeq 3$ MPa[15]), while small changes in humidity generate large negative pressure differences (see Table 4.1).

Of course, the negative pressure described by eqn (4.7) is only reached if equilibrium can be attained. This means that the membrane separating the liquid at negative pressure ($P_0 + \Delta P$, $\Delta P < 0$) and the air (pressure, P_0) must resist collapse and/or mechanical failure under the pressure difference ΔP, but also allow for the coexistence of liquid and air at very different pressures.

Table 4.1 Equilibrium liquid pressure, $P = P_{atm} + \Delta P$, and radius of curvature, r, of the liquid/vapor interface as a function of water vapor humidity in the air (%RH $= p/p_{sat} \times 100$), calculated from eqn (4.7)–(4.8) for a temperature of $T = 298.15$ K.

Air humidity (%RH)	100	99.93	99.2	92.9	83.3	69.4	48.2
Liquid pressure (MPa)	0.1 (=P_{atm})	0	−1	−10	−25	−50	−100
Radius of curvature (nm)	∞	1498	131	14	5.7	2.9	1.4

One way to achieve the latter is if the membrane is wetted by the liquid and allows curved liquid–vapor menisci to develop (see Figure 4.5a, inset). Assuming for simplicity that the menisci are hemispherical caps with the same curvature $1/r$ in both directions, they maintain a pressure difference

$$\Delta P = -2\gamma/r \tag{4.8}$$

from the Laplace law. The lower the humidity, the larger the pressure difference and the larger the curvature of the menisci. Note that the combination of eqn (4.7) and (4.8) is known as the Kelvin–Laplace equation. Table 4.1 indicates radii of curvature predicted from the Kelvin–Laplace equation; in order for the curvature to develop, the constrictions within the membrane (pore radius r_p for a porous medium, typical half mesh size of the polymer network for a hydrogel, *etc.*) must be smaller than r, assuming the medium to be well wetted by water. For less hydrophilic surfaces, the constrictions must be even smaller (*e.g.* for a pore, $r_\mathrm{p} < r \times \cos\theta$, where θ is the receding water contact angle on the pore wall).

Of course, meniscus invasion in a complex nanoporous structure cannot be simply represented by a single constriction size (and the concept of a meniscus is somewhat debatable in polymer structures), so these estimates must be taken as order of magnitudes only. However, the values in Table 4.1 indicate that only membranes containing networks with typical dimensions in the nanometer range are suitable to be able to develop large negative pressures. Static negative pressures described by eqn (4.7) down to −20 to −25 MPa could in fact be obtained in artificial systems consisting in large voids (∼10–100 μm in size) separated from air by membranes based on hydrophilic hydrogels[25,29,30] or hydrophilic mesoporous silicon,[22,31,32] both with estimated constrictions below 5 nm.

In fact, in the experiments mentioned in the previous paragraph based on hydrogels or mesoporous silicon, the mechanism limiting the achievable negative pressure to −25 MPa was not capillary failure in the membrane but cavitation of the containing liquid. More generally, the equilibrium described by eqn (4.7) is metastable when the pressure is negative, because of the possibility of bubble nucleation by various mechanisms (see Section 4.4). It is remarkable that as long as the system allows for equilibrium (no mechanical or capillary failure of the membrane, no cavitation), the equilibrium predicted by eqn (4.7) does not depend on the details of the membrane itself.

4.3.3 Dehydration: Discussion

The values of negative pressure listed in Table 4.1 could make one think that trees should routinely endure massive water stresses and cavitation. Indeed, it is not rare for humidity to fall below 50%RH during the day, which corresponds to ∼ −100 MPa of pressure at equilibrium (see Table 4.1), close to the homogeneous cavitation limit (see Section 4.4). In reality, water within the xylem is not at equilibrium with the atmosphere, first because the water-conducting cells are continuously fed with water coming from the roots, and also because trees have protection mechanisms that isolate water in the xylem from the outside when pressure becomes too negative (*e.g.* closure of stomata[6,15]). As a result, the actual negative pressure in the xylem arises from a balance between water potential in the soil, water potential in the atmosphere, and the different resistances to water transport separating the soil and the atmosphere through the xylem tissue.[6,15,16,20] However, the general idea that dehydration (in this case by transpiration in the leaves) is the driving force for negative pressure generation remains true; simply the local water potential (*i.e* dehydration state) in the xylem is not as severe as one could expect, due to continuous flow or isolation. Other natural structures that rely on cavitation for spore ejection (see Figure 4.2) are in more direct contact with the atmosphere and potentially quickly reach equilibrium with the local humidity surrounding them. As a result, they require the humidity to drop below a critical value for their mechanisms to trigger.

4.3.4 Other Origins of Negative Pressure

Although negative pressure in plants originates from dehydration-related processes in natural conditions, there exist other methods to induce negative pressure and cavitation. We only give a brief summary here; more details can be found in other references.[19,20,25]

a. *Traction*. This is perhaps the most intuitive way to stretch a liquid, *i.e.* by literally pulling on it, for example using the piston in a syringe. Any air pocket must be carefully removed because the pulling force would just make the gas expand without pulling on the liquid. As noted by Huygens in the 17th century, the liquid's own weight in a vertical tube can also lead to a tensile state, and two centuries later Reynolds obtained negative pressures of a few atmospheres using this method.[19]
b. *Hydrodynamics.* Accelerating a fluid can result in a drop in pressure (the Venturi effect) sufficiently large to make the pressure negative. This effect is well known by mechanical engineers, because cavitation around fast-spinning boat propellers results in damage from the violent collapse of the bubbles on the blades.[33] The order of magnitude of negative pressure achievable with this technique is typically of a few bars at maximum, probably due to remaining air bubbles in the fluid that act as cavitation nuclei.[34]
c. *Centrifugation*. Spinning a tube containing a liquid, around an axis perpendicular to the length of the tube, makes the liquid "pull on itself" due to the centrifugal effect, resulting in a gradient of pressure with the minimum value at the rotation axis. Briggs obtained negative pressures down to −28 MPa with this technique, which has also been adapted to study cavitation in xylem by spinning branches.[35,36,37]
d. *Isochoric cooling*. Known as the Berthelot method,[38] this technique consists in cooling down a liquid trapped in a pocket within a solid: the liquid would naturally contract, but if there is good adhesion with the

walls, the liquid is forced to remain at a volume larger than at equilibrium and gets stretched. After cavitation, the pocket of liquid can be "reset" by heating up the system sufficiently so that the liquid expands again and occupies the whole volume. The rare experiments that reached what seems to be cavitation by homogeneous nucleation at −120 to −140 MPa used the Berthelot technique with micrometric inclusions in quartz.[39,40]

e. *Acoustic waves.* Acoustic waves are an oscillation of pressure around ambient pressure. If the amplitude of the wave is large enough (*e.g.* due to strong focalization in a liquid), the low-pressure part of the oscillation cycle can be below $P=0$. Despite the fact that the probed volume and timescales are small and that the wave can be focalized away from any surface, even careful experiments do not seem to be able to reach pressures below −30 MPa.[19,41] One experiment using reflected shock waves reports pressures down to −60 MPa, albeit with a more indirect way of estimating the pressure, based on comparison with simulations of shockwave propagation.[42]

4.4 Cavitation Mechanisms

Although water is theoretically able to withstand ∼ −200 MPa of pressure before spontaneously breaking apart (at the spinodal limit; see Section 4.2.2), cavitation always occurs at more moderate values of negative pressure, so that the spinodal is never reached in practice. The pressure at which cavitation happens, P_{cav}, strongly depends on the microscopic mechanism leading to the formation of a bubble. Below, we briefly introduce some commonly discussed cavitation mechanisms.

4.4.1 Homogeneous Nucleation

Cavitation by homogeneous nucleation (see Figure 4.6a) corresponds to the spontaneous nucleation of a vapor bubble in the liquid from thermal fluctuations. Although this process is obviously stochastic, the probability of cavitation is very nonlinear as a function of pressure and varies very quickly around a typical value of P_{cav}. Theoretical estimates for P_{cav} are usually based on classical nucleation theory (CNT)[17,43] and are typically in the range −120 to −150 MPa at ambient temperatures, depending on considered volumes, timescales and refinements of CNT.[19,40,44] In section IB, we present an extended version of CNT that takes into account finite volume and the effect of compressibility (of the liquid and of the containing cell).

The limit $P_{cav} \sim -120$ to −150 MPa is at more moderate negative pressures than the spinodal (∼ −150 to −200 MPa[20,21]) and can be seen as the ultimate negative pressure that one can hope to reach in water, as even thermal fluctuations are sufficient to produce cavitation at that point. Reaching this limit experimentally is difficult, as one must eliminate all other mechanisms that can trigger nucleation at less negative pressures (see examples below).

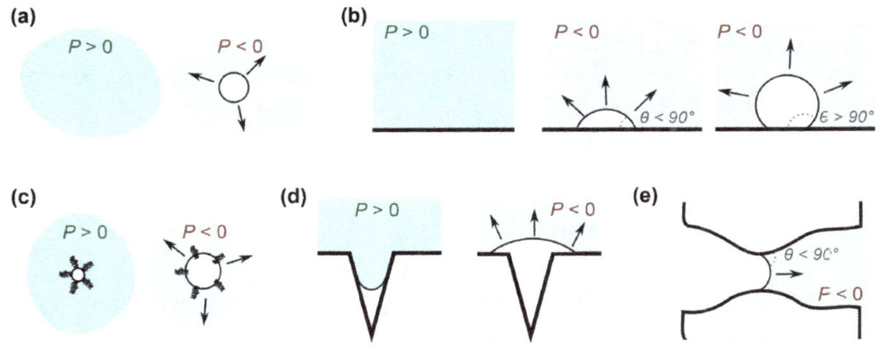

Figure 4.6 Examples of cavitation mechanisms. (a) Homogeneous nucleation. (b) Surface-aided nucleation (left: hydrophilic surface; right: hydrophobic surface). (c) Cavitation from a floating gas bubble (black elements represent potential organic molecules stabilizing the surface of the bubble). (d) Cavitation from air trapped in a crevice. (e) Cavitation from meniscus aspiration through a pore. Black arrows represent motion, dashed areas represent solids. Bubbles contain gas (water vapor and/or air).

The rare experiments that have reached such extreme negative pressures have used thermal cycling in water pockets trapped in quartz inclusions.[39,40] Many other methods (including acoustic waves focused in a pure liquid far from any walls) seem to find a cavitation threshold at −20 to −30 MPa, even with careful purification and preparation; there is no clear understanding on the reason of such a difference between these experiments and those with quartz inclusions.[20]

The other mechanisms discussed below can be all considered as *heterogeneous* mechanisms, by opposition to the homogeneous nucleation mechanism described here.

4.4.2 Surface-aided Nucleation

Homogeneous nucleation is difficult because creating a liquid–vapor interface costs energy. Any solid surface that the liquid does not completely wet (*i.e.* contact angle, $\theta > 0$) reduces this energetic cost by enabling pathways of nucleation where the bubble is a truncated sphere (see Figure 4.6b). This effect can easily be incorporated into CNT;[19,43] the predicted P_{cav} depends on θ but only deviates significantly from that of homogeneous nucleation when the surface is hydrophobic ($\theta > 90^\circ$). For example, a contact angle greater than 150° would be required to predict $P_{cav} = -20$ MPa.[25] Globally, any cavitation pressure can be obtained between the homogeneous nucleation value (−120 to −150 MPa) and $P = P_{sat}$ by varying the contact angle between 0 and 180°. Non-flat surfaces (*e.g.* crevices or tips) can also modify the cavitation pressure.[45]

Fundamentally, the surface-aided mechanism is not very different from that of homogeneous nucleation in the sense that it still involves a vapor bubble that is spontaneously nucleated in the liquid, but with a reduction in energy cost due to a surface. Cavitation through this mechanism can be seen as a *loss of adhesion* between the liquids and the surface. The other mechanisms discussed below, however, are fundamentally different because they involve a pre-existing germ; the liquid–vapor interface already exists and is simply expanded when cavitation occurs.

4.4.3 Seeded Cavitation

Here we discuss several cavitation mechanisms that involve pre-existing germs, *i.e.* microscopic bubbles that can contain gas (typically, air) or water vapor, which are stable at moderate negative pressures but become unstable at some critical pressure value P_{cav}, where they expand into a macroscopic bubble, *i.e.* cavitation.

For a gas bubble in a liquid, $P_{cav} = P_{sat} - 4\gamma/3R$, where R is the radius of the bubble,** which is known as the *Blake threshold*. The cavitation pressure of a liquid containing bubbles is thus dictated by the largest bubble it contains.

It is worth noting, however, that a bubble containing gas that is soluble in the surrounding liquid is never stable and spontaneously dissolves due to the combined effect of surface tension and diffusion.[18,47] Air bubbles dissolve quickly (*e.g.* ~1 minute for an initial radius of ~10 μm) so in the context of plants where negative pressures last for hours, it is difficult to imagine situations where air bubbles would trigger cavitation before having completely dissolved. Gas germs thus have to be stabilized against dissolution in some way.

One way is for the bubble to have an organic coating (see Figure 4.6c) that blocks diffusion and/or reduce surface tension. Blake's threshold is supposed to still be a good estimate of P_{cav} in this situation.[34,48] Another stabilizing mechanism is that of air trapped in a crevice (see Figure 4.6d): with some conditions on geometry and surface wettability, such a bubble can be stable and gets "extracted" from the crevice if the pressure gets too low in the liquid, resulting in cavitation.[30,49–51] Even with a conical crevice, there are many cases to consider, but the result is that, again, any cavitation pressure between P_{sat} and the homogeneous nucleation threshold can be obtained by varying the crevice size, angle, gas content and wettability.

A related mechanism is that of a pore separating the liquid at negative pressure from an air-filled (or vapor-filled) space (see Figure 4.6e). The situation becomes unstable if the pressure difference $\Delta P = P - P_{gas}$ falls below a critical value ΔP_{cav}, which is set by the maximum magnitude of the Laplace pressure that the geometry can allow: for a cylindrical pore of radius r_p, $\Delta P_{cav} = -2\gamma\cos\theta/r_p$.†† This simple estimate can be adapted for more

---

**Obviously, the radius of the bubble depends on the pressure in the surrounding liquid. Here, the radius to consider is the one when $P = P_{cav}$; solving for P_{cav} requires using the equilibrium condition given by the Laplace law.[18,46]

††For a surface that exhibits contact angle hysteresis, θ would represent the receding contact angle in this situation.

realistic pore geometries, *e.g.* for pit membranes in xylem, with the interesting conclusion that ΔP_{cav} not only depends on the pore size distribution but also strongly on the thickness of the membrane.[52] Nevertheless, the simple equation for a cylindrical pore allows us to illustrate the general idea that cavitation occurs at more negative pressures for smaller and more wettable pores. This mechanism is often referred to as *air seeding* in the plant science literature.

Because ΔP can also be lowered by increasing P_{gas} without changing the liquid pressure P, air seeding can be triggered by pressurizing the gas phase instead of putting the liquid under tension. In fact, external air pressurization is one of the popular methods used by plant scientists to estimate vulnerability of xylem to cavitation, although it can lead to various artifacts.[37] One could argue that air seeding is not a *true* cavitation mechanism, because it does not require the liquid to be thermodynamically metastable (*e.g.* in the situation of air pressurization described previously). However, in natural situations the gas pressure is not artificially changed, and the pore sizes (*e.g.* in xylem pits) are sufficiently small (sub-micron) to allow for liquid pressures lower than P_{sat} (*i.e.* thermodynamically metastable) to develop in the liquid. In such situations, the formation of a bubble by air seeding falls under our definition of cavitation, as the expansion of the bubble makes the liquid relax from a metastable state to a stable state.

4.4.4 Discussion

From the description of cavitation mechanisms above, we can extract the following conclusions:

- The most negative pressure achievable in water at ambient temperature is set by homogeneous nucleation and is larger than 1000 atmospheres ($P_{cav}^{(hom)} = -120$ to -150 MPa) in magnitude.
- -20 to -30 MPa seems to be a more practical lower limit, while plants typically develop only -1 to -10 MPa (see ref. 7–9).
- Heterogeneous mechanisms can set the cavitation pressure of a system to any value between P_{sat} and $P_{cav}^{(hom)}$.
- Resistance to cavitation is better if defects in the system (pores or crevices in the walls, hydrophobic patches, floating bubbles, *etc.*) are smaller (and/or more wettable when applicable).

This last point is illustrated by the fact that cavitation pressure scales as γ/L, where L is a typical size (*e.g.* bubble radius for Blake's threshold, pore size for air seeding, critical bubble radius for homogeneous nucleation, see Section 4.5.1).

It is usually difficult to pinpoint a specific cavitation mechanism, but some mechanisms can be ruled out by various tests. For example, cavitation seeded by air trapped in crevices can be modified or even suppressed by prepressurization of the system. Examples can be found of such tests in artificial, xylem-like systems based on hydrogels[30] or porous silicon,[32] with a careful discussion of potential cavitation mechanisms.

Other cavitation mechanisms may exist. For example, recent results show that a negative pressure of −10 MPa is sufficient to tear apart lipid bilayers, which could thus be responsible for cavitation by loss of adhesion within the bilayer due to the tensile force of water acting on it.[53] More generally, there has been a growing interest in understanding the role of surfactants present in sap or in cell walls on cavitation in plants.[54,55]

Generally, the size separation between the microscopic germ responsible for cavitation and the final cavitation bubble is so large that the exact cavitation mechanism has little impact on the dynamics of the cavitation bubble after nucleation. As a result, the dynamical phenomena (bubble dynamics, cavitation propagation, *etc.*) presented in the end of this chapter are mostly insensitive to the microscopic mechanism at the origin of the cavitation bubble.

4.5 Confined Cavitation Theory

Here, we present a single theoretical framework that will allow us to discuss both thermodynamic aspects and dynamical aspects of negative pressure and cavitation in closed environments. Mainly, it consists in an extension of CNT[17,18,19,43] to take into account the compressibility of the liquid and the elasticity of the containing cell.[25,46] As we have shown in Section 4.3, these two effects are important in the generation of negative pressure. We will show below that they also play a significant role for the equilibrium states and dynamics of cavitation bubbles.

The idea is to evaluate the minimum work of formation of a bubble in a liquid at negative pressure in a closed, elastic container (see Figure 4.7). The minimum work identifies with the *Helmholtz free energy* difference, ΔF between an initial state without bubble and with the liquid at negative pressure, $P_0 < 0$ (see Figure 4.7, state A) and a state with a bubble of volume V and surface area A (states B − D). ΔF also corresponds to the potential energy of the system and dictates the driving forces that shape the dynamics.

We consider a bubble forming within a liquid, *i.e.* homogeneous nucleation (see Section 4.4 for a discussion of other mechanisms). We also assume that the liquid does not contain dissolved air, or that if it does, transport of air by diffusion into the bubble is negligible during the timescales of cavitation. As a result, the saturation vapor pressure of interest is P_{sat} (but using p_{sat} would result in negligible differences anyway; see Section 4.2.5). It can be shown (see Appendix B for a detailed derivation) that in this situation,

$$\Delta F = \gamma A + (P_0 - P_{sat})V + \frac{1}{2}\frac{1}{V_0(\chi_\ell + \chi_c)}V^2 \tag{4.9}$$

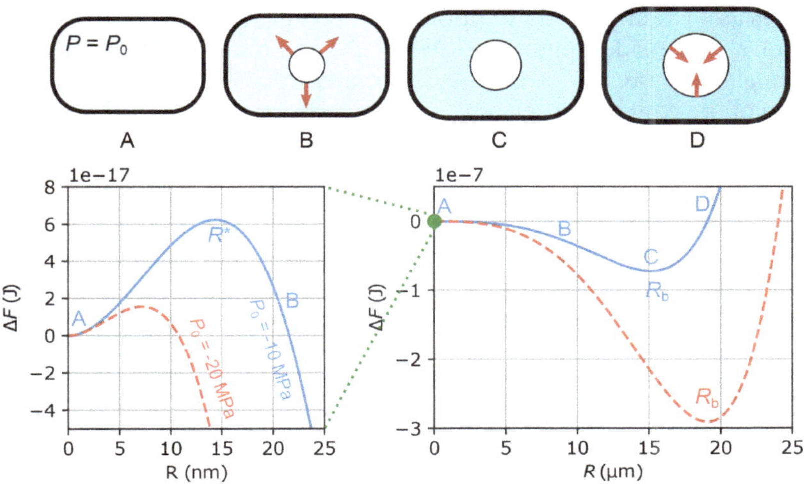

Figure 4.7 Cavitation in a closed container (cell), assuming a spherical bubble and an effective container compressibility of $\chi_c = 1\ \text{GPa}^{-1}$. State A corresponds to the homogeneous liquid at negative pressure ($P_0 < 0$). In state B, a bubble has overcome the critical nucleation radius R^*: the effective pressure P_{eff}, representing the net driving force applied on the bubble (red arrows), is negative and tends to make the bubble grow. State C corresponds to equilibrium ($P_{eff} = 0$, $R = R_b$). In state C, the bubble is larger than its equilibrium size and a net inwards force (P_{eff}) tends to make it go back to its equilibrium size. Oscillations between states B and D around equilibrium are possible due to inertia in the system. In the schematics, darker shades of blue indicate larger pressure and larger density in the liquid. In the graphs, the continuous, blue line corresponds to an initial negative pressure of $P_0 = -10$ MPa while the dashed, red curve corresponds to $P_0 = -20$ MPa. The two graphs represent the same $\Delta F(R)$ function (eqn (4.9)) on different ranges of bubble size. Notice the large scale separation between them.

where V_0 is the initial container volume (equivalently, initial liquid volume), χ_ℓ is the isothermal compressibility of liquid water introduced previously, and χ_c the effective compressibility of the container (see Section 4.3.1).

In eqn (4.9), the first term represents the energetic cost of creating a liquid – vapor interface (surface tension, γ, see Section 4.2.1), the second term is associated with the initial driving force associated with the pressure in the liquid, and the third term describes how this driving force changes as a function of bubble size due to compressibility effects. Indeed, eqn (4.9) assumes that during the timescales of interest (~μs, see below), no significant transport happens through the walls of the containing cell, so that the number of water molecules in the cell remains constant. As a result, the only way to make the bubble grow is by compressing the surrounding liquid and/or inflating the containing cell, hence the $\chi_\ell + \chi_c$ factor in the expression. The compression of the liquid as the bubble grows results in an increase in pressure in the liquid (see eqn 4.32): the negative pressure progressively vanishes as a result. The assumption of short timescales also means that the value of χ_c to consider might be different than in quasi-static situations, *e.g.* for poroelastic systems.[56] Generally, χ_c represents here the pressure/volume relation of the containing cell at the current conditions of dehydration in the medium, and in a no-flow situation (constant water content in the cell).

Eqn (4.9) is applicable for any bubble shape. In the following, we assume the bubble to be spherical, so that $A = 4\pi R^2$ and $V = 4\pi R^3/3$, where R is the bubble radius. The quantity

$$P_{\text{eff}} = \frac{\text{d}\Delta F}{\text{d}V} = \frac{2\gamma}{R} + (P_0 - P_{\text{sat}}) + \frac{1}{\chi_\ell + \chi_c}\frac{V}{V_0} \tag{4.10}$$

is an effective pressure that represents the driving force for the bubble dynamics. If $P_{\text{eff}} < 0$, the system is in tension and the bubble tends to expand (Figure 4.7, state B). Equilibrium occurs when $P_{\text{eff}} = 0$ (Figure 4.7, state C). If the bubble is larger than equilibrium (Figure 4.7, state D), the driving force reverses and tends to make the bubble shrink back to equilibrium.

Figure 4.7 presents the shape of ΔF for two different initial negative pressures, using typical values $\chi_c = 1\ \text{GPa}^{-1}$ and $V_0 = (100\ \text{mm}^3)$ for the containing cell, and tabulated values for the properties of water at 25 °C ($\gamma = 0.072\ \text{N m}^{-1}$, $P_{\text{sat}} = 3170$ Pa, $\chi_\ell = 0.45\ \text{GPa}^{-1}$). In the next sub-section, we discuss the main features of the ΔF potential energy landscape and their consequences.

4.5.1 Critical Radius, Energy Barrier

For small radii in the nanometer range, ΔF presents a maximum at the *critical radius*, R^* (see Figure 4.7, left). Below R^*, $P_{\text{eff}} > 0$ and the bubble tends to collapse. Only when $R > R^*$ does the negative pressure "win" over surface tension ($P_{\text{eff}} < 0$), making the bubble expand and resulting in cavitation. Thus, $\Delta F^* = \Delta F(R^*)$ corresponds to the *energy barrier* for nucleation. This energy barrier decreases when pressure becomes more negative, making nucleation more likely.

With typical container sizes (dimensions 0.01–1 mm and above) for plant cells, the compressibility contribution (third terms in eqn (4.9)–(4.10)) can be shown to be negligible around the critical radius, so that ΔF is well approximated by the first two terms, which correspond to those usually considered in CNT for homogeneous nucleation.[19] In this situation, the critical radius and energy barrier are easily calculated from eqn (4.9)–(4.10), neglecting the last term and requiring $P_{\text{eff}} = 0$; this calculation yields $R^* = 2\gamma/(P_{\text{sat}} - P_0)$ and $\Delta F^* = 4\pi R^{*2}\gamma/3$.

In sub-micron container sizes, however, the surface tension and compressibility terms can become of the same order of magnitude around R^*, with the surprising consequence that a liquid at negative pressure confined at these dimensions can become absolutely stable instead of metastable.[25,46]

The classical approach of CNT is to compare the energy barrier ΔF^* to thermal fluctuations (k_bT) and compute the probability of passing the energy barrier given some attempt frequency. Doing so, one estimates that the probability of cavitation is virtually zero unless pressure reaches values in the -120 to -150 MPa range.[19,40,44] These values depend only weakly on total volume and observation timescales.

The expression (eqn (4.9)) of ΔF can be modified to describe cavitation seeded by gas germs, by adding a term $-3Nk_bT\ln V$; in this case, nucleation occurs from a gas bubble containing N insoluble gas molecules and a confined version of Blake's threshold (see Section 4.4) can be obtained.[46] More generally, the shape of ΔF, the values of the critical radius and energy barrier and the associated typical cavitation pressure depend greatly on the microscopic mechanism of cavitation (see Section 4.4). The shape of the potential $\Delta F(R)$ for radii much larger than the critical radius, however, does not depend significantly on the cavitation mechanism. In the rest of this chapter, we focus on this large-bubble regime, which describes the formation of a macroscopic cavitation bubble once the energy barrier is overcome.

4.5.2 Equilibrium Bubble

At large bubble sizes, ΔF presents a minimum at a point corresponding to the equilibrium bubble (see Figure 4.7, state C). We define the equilibrium bubble volume as V_b and corresponding radius as R_b. Once the energy barrier of nucleation is overcome, the expanding cavitation bubble will tend towards V_b. The equilibrium size of the bubble is found by solving $\mathrm{d}\Delta F/\mathrm{dV}=0$, *i.e.* $P_{\mathrm{eff}}=0$. Using eqn (4.10),

$$V_b = V_0(\chi_\ell + \chi_c)\left(P_{\mathrm{sat}} - P_0 - \frac{2\gamma}{R_b}\right). \tag{4.11}$$

In most situations relevant to this chapter (real and artificial plant-like systems), P_{sat} and the capillary pressure $2\gamma/R_b$ are negligible in magnitude compared to P_0, so that

$$V_b \simeq V_0(\chi_\ell + \chi_c)|P_0| \tag{4.12}$$

to excellent approximation. Eqn (4.11)–(4.12) express mathematically the fact that the volume occupied by the bubble comes from the space liberated by the combined contraction of the liquid and expansion of the cell, due to a change of pressure from $P_0<0$ to $P\simeq 0$. Larger bubble sizes are obtained if the initial negative pressure is larger in magnitude (P_0 more negative), if the cell is more deformable (higher χ_c), or if the cell's volume, V_0, is larger.

We note that with the typical parameters that we have used for the confining cell's volume and elasticity, the size of the equilibrium bubble (R_b) is several orders of magnitude larger than the size of the critical bubble (R^*), as can be seen in Figure 4.7. Consequently, the critical radius and energy barrier are not visible on a graph of ΔF at the length scales and energy scales of the equilibrium bubble (see Figure 4.7, right). This illustrates the fact that the microscopic mechanism for nucleation is irrelevant for the dynamics of formation of the macroscopic bubble, as discussed earlier.

4.5.3 Inertial Oscillations

Around the equilibrium bubble, ΔF has the shape of a potential well. When a cavitation bubble appears, inertia can make the system oscillate around the equilibrium position at the bottom of the well (see Figure 4.7, right, arrows). This results in radial oscillations of the bubble.

a. Effective stiffness. In order to evaluate the oscillation dynamics, it is useful to approximate the system as a harmonic oscillator around its equilibrium point, *i.e.* use a quadratic expansion of $\Delta F(R)$around $R = R_b$: $\Delta F(R) \simeq \Delta F(R_b) + k(R - R_b)^2/2$, where $k = (d^2\Delta F/dR^2)_{R=R_b}$ is the *effective stiffness* of the system. Using the definitions $P_{eff} = d\Delta F/dV$ and $A = dV/dR$, combined with the equilibrium condition $P_{eff}(R_b) = 0$, k can be expressed as

$$k = A_b \left(\frac{dP_{eff}}{dR}\right)_{R=R_b} \tag{4.13}$$

where A_b is the bubble surface area at equilibrium. Using eqn (4.10),

$$k = \frac{A_b^2}{V_0(\chi_\ell + \chi_c)} - 8\pi\gamma \tag{4.14}$$

b. Effective mass. The oscillation dynamics is set by the interplay of the system's stiffness and inertia. In general, inertia (*i.e.*, the kinetic energy, E_k) is complicated to evaluate because it depends on the details of the velocity field in the liquid and in the walls of the containing cell. However, the velocity field typically decreases rapidly away from the bubble surface, so that the dominant contribution to the kinetic energy usually comes from the fluid displaced close to the bubble. As a first approximation, one can thus assume that the kinetic energy is that of a bubble oscillating in an unbounded fluid: $E_k = (1/2)m(dR/dt)^2$, where $m = 4\pi R_b^3 \rho$ is the effective mass of the system.[18] We recall that ρ is the density of the liquid. During the oscillation, the liquid is successively compressed and stretched, so that its density changes, but these variations are small and can be ignored as a first approximation. The main correction to the effective mass comes from the effect of the actual velocity field in the liquid compared to an unbounded situation, and to the potential contribution of the other parts of the system (*e.g.* cell walls and outside medium) to the kinetic energy. These different effects can be bundled into a corrective factor ϕ so that

$$m = 4\pi R_b^3 \rho \phi \tag{4.15}$$

where ϕ is a dimensionless factor describing deviations to the unbounded case ($\phi = 1$ for a bubble in an infinite liquid). In the ideal situation where the cell containing the liquid is a spherical void in a solid with an effective compressibility, χ_c and density, ρ_s that are comparable to those of the liquid (χ_ℓ,ρ) and the bubble is at the center of the cell, $\phi \simeq 1$. However, for an infinitely stiff cell that imposes a strict zero-velocity boundary condition at the cell wall, the effective mass is significantly reduced compared to eqn (4.15), *e.g.* $\phi \simeq 0.65$ for a bubble of radius only one-fifth of the container's radius.[46]

c. Oscillation frequency. With the effective stiffness and effective mass established above, the system is analogous to a mass/spring system, and its natural angular oscillation frequency is given by $\omega = 2\pi f = \sqrt{k/m}$. For simplicity, we assume in the following that the surface tension contribution to the effective stiffness ($-8\pi\gamma$) is small compared to that of the system's compressibility; this is usually a good approximation, but it has to be evaluated depending on the system's parameters.[31,46] Using eqn (4.14)–(4.15),

$$\omega \simeq \sqrt{\frac{4\pi R_b}{(\chi_\ell + \chi_c)\rho V_0}}. \tag{4.16}$$

From eqn (4.11), we also know the bubble size $V_b = 4\pi R_b^3/3$ as a function of the initial negative pressure prior to cavitation and we can rewrite (still neglecting surface tension):

$$\omega \simeq \left(\frac{3}{\rho\phi}\right)^{1/2} \left(\frac{4\pi}{3V_0(\chi_\ell + \chi_c)}\right)^{1/3} (P_{sat} - P_0)^{1/6}, \tag{4.17}$$

Interestingly, eqn (4.17) indicates that the natural oscillation frequency of a cavitation bubble depends very weakly on the initial pressure prior to cavitation (power 1/6). The main parameters governing the oscillation frequency are the cell's volume (V_0), and the system's effective compressibility ($\chi = \chi_\ell + \chi_c$). For the quite different geometry of a cylindrical bubble at the end of a long cylindrical tube (length L), it can also be shown with a similar approach that $\omega \simeq L^{-1}\sqrt{3/(\rho\chi)}$.[25]

Predictions similar to eqn (4.16)–(4.17) can be obtained from estimations of the dispersion relation of acoustic waves related to the oscillation of the bubble within the containing cell,[25,27,46,57] and agree with experimental results[27,31,46,58] and simulations.[59] While these acoustic methods are restricted to small oscillations around equilibrium, the full nonlinear dynamics of the bubble (Rayleigh–Plesset-like equation) can also be directly calculated from the expression of $\Delta F(R)$ and of the kinetic energy, *e.g.* using Euler–Lagrange equations.[25,46] From this Rayleigh–Plesset-like equation it is possible to show that the initial expansion velocity of the bubble when it appears is

$$\left(\frac{dR}{dt}\right)_0 = \sqrt{\frac{2(P_{sat} - P_0)}{3\rho}} \tag{4.18}$$

which is in fact a general result for the rate of expansion of a bubble in a liquid at pressure, P_0.[18] Eqn (4.18) can be used to estimate the negative pressure in the liquid prior to cavitation by measuring the initial rate of growth of the cavitation bubble.[25,46,60]

The radial vibration of the cavitation bubble emits acoustic waves in the medium, and analysis of the frequency of the acoustic signals in relation to the theoretical predictions (*e.g.* eqn (4.16)–(4.17)) potentially allows observers to obtain information about the local properties (cell volume, elasticity) of the system. While this idea works well in artificial structures,[25,27,31] it is less straightforward to implement in a tree, because of the complexity of wave propagation in wood.[61]

In fact, the acoustic emission related to the formation of a cavitation bubble in a liquid enclosed in a stiff container is so strong that it accounts for the majority of the dissipation of the energy initially contained in the stretched liquid (and elastically deformed container) prior to cavitation.[25,46] This effect contrasts with the case of unconfined bubbles, for which damping usually comes from viscous or thermal effects.[18,62] This strong dissipation results in a very quick damping of the radial vibrations, which stop after just a few oscillations; see Section 4.6.

4.6 Cavitation Bubble Dynamics

We now illustrate the theoretical concepts introduced in the previous sections with experimental results on the dynamics of cavitation bubbles appearing in a liquid at negative pressure within an enclosed space. As we will show, this dynamics is rich and spans many orders of magnitude of timescales, from sub-microseconds to minutes.

The discussion in this section is based on a series of experiments using artificial systems that mimic some essential features of cavitation in plants.[25,31,58] The samples (see Figure 4.8, *experimental system*) can be seen as an artificial cells, which consists of spherical voids filled with water, enclosed in a hydrogel (pHEMA). Dehydration of the cells by diffusion of water through the hydrogel results in negative pressure (see Section 4.3.1), and is driven by evaporation (see Section 4.3.2). Cavitation occurs spontaneously in the liquid when $P \simeq -20$ MPa; this value is estimated by equilibrating the system with air at different relative humidities and by using eqn (4.7) to estimate the corresponding pressure in the liquid.[25,29,30] Cavitation can also be triggered at slightly higher pressures by using a laser pulse (for strobe photography measurements, see Section 4.6.1). The hydrogel is stiff enough to resist collapse upon the development of negative pressure in the liquid: its effective compressibility is $\chi_c = 1$ GPa^{-1}, comparable to that of xylem (see discussion in Section 4.3.1). The initial development of such *synthetic xylem* systems was made by Wheeler and Stroock;[29,30] in particular, we refer the reader to their optimization of the mechanical properties of the gel to avoid collapse[29] and to their thorough discussion of possible cavitation mechanisms with respect to those described in Section 4.4 here.[30]

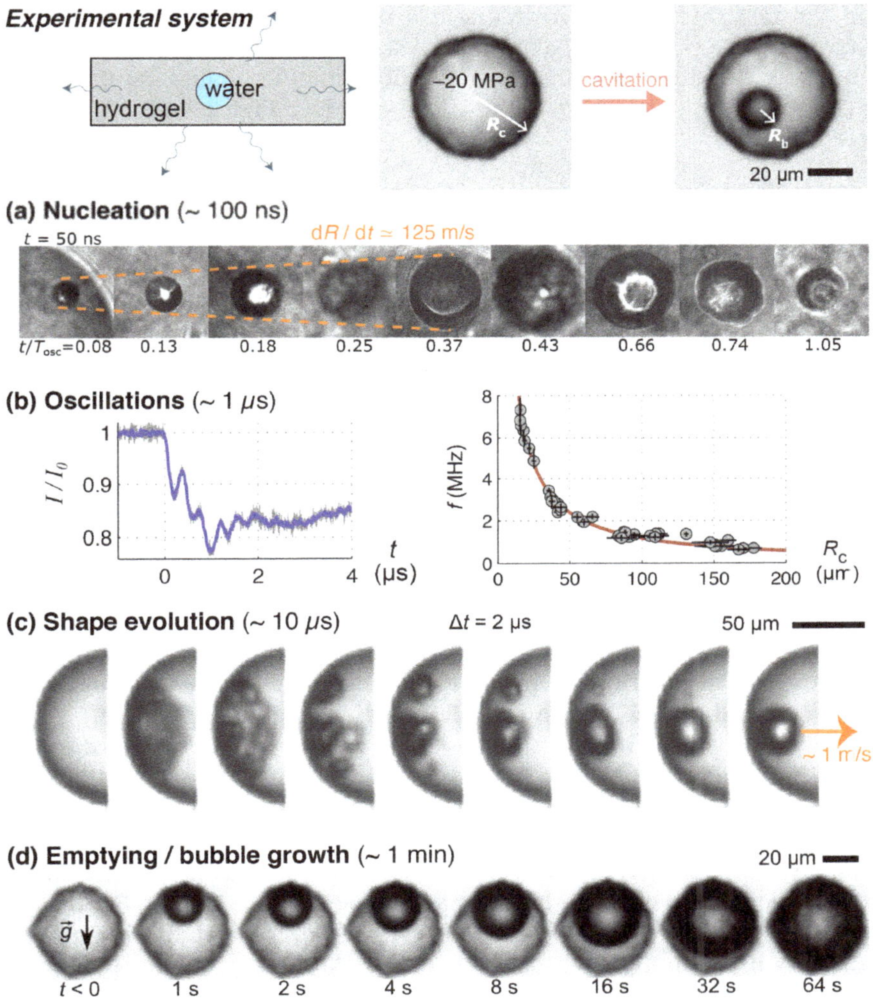

Figure 4.8 Cavitation bubble dynamics in an artificial cell (spherical void in a polymer hydrogel). (a) Nucleation dynamics reconstructed from multiple laser strobe photography experiments[31] (time is normalized by the oscillation period; see (b)). (b) Inertial oscillations measured by light scattering (extinction of transmitted light), and corresponding frequency as a function of void size.[31] (c) High-speed camera recordings of the evolution of the bubble shape.[31] (d) Time-lapse camera recording of the final growth of the bubble and emptying of the void, from a side view.[25]

Other experiments similar to those presented here have been reported in the literature, for example with cubic cells and osmosis-driven dehydration,[27] or with spherical cells in a much softer material, PDMS ($\chi_c \sim 1$ MPa^{-1}).[60] In this latter case, the cell shrinks dramatically during the development of negative pressure, however the nonlinear elasticity of PDMS and the fact that cavitation occurs at a negative pressure of much lower magnitude ($P_{cav} \sim -1$ MPa) allows the system to avoid complete collapse.

Below, we discuss the different stages of formation of a cavitation bubble in the spherical, hydrogel-based artificial cells (see Figure 4.8).

4.6.1 Nucleation

Laser strobe photography experiments consist of triggering nucleation in the metastable liquid by a first laser pulse, then illuminating the system with a second pulse separated from the first one by 50 to 500 ns. Repeating the process several times enables an image reconstruction of the birth of the cavitation bubble (see Figure 4.8a), revealing an ultra-fast expansion. Using eqn (4.18), one can show that the rate of expansion of the bubble, $dR/dt = 125 \pm 20$ m s^{-125} is consistent with the cavitation pressure of -20 MPa estimated from humidity equilibration experiments (see above).

4.6.2 Oscillations

After the initial expansion, the bubble starts to shrink again (see Figure 4.8a, last three frames), because the bubble has overshot its equilibrium position due to inertia (see Section 4.5): this corresponds to the onset of oscillations. Light extinction experiments reveal oscillations with frequencies $f = \omega/(2\pi)$ in the MHz range (see Figure 4.8b, left). This signal corresponds to radial oscillations of the bubble, which are quickly damped due to acoustic radiation (see Section 4.5.3). The frequency decreases as the size of the cell is increased (see Figure 4.8b, right), in excellent agreement with theoretical predictions (eqn (4.17), red line in Figure 4.8b, right). This frequency also depends on the compressibility of water and the elasticity of the cell, but only weakly on the cavitation pressure (see Section 4.5.3). Because of acoustic radiation, these oscillations can also be detected with a high-frequency hydrophone.[1,25,27]

4.6.3 Shape Evolution

A high-speed camera at half a million frames per second cannot resolve the oscillations, which result in image blur (see Figure 4.8c, second frame), but reveals a rich, slower shape-related dynamics that follows the oscillations. The images suggest the evolution of the bubble into a toroidal shape (see Figure 4.8c, third frame) that destabilizes into several bubbles, which collapse back into a single bubble eventually in a form of quick ripening process (see Figure 8c, frames 4 to 9). This peculiar dynamics is thought to arise from a combination of high-speed jets (creating the toroidal bubble[63]) and a Rayleigh-Plateau instability (fragmenting the toroidal bubble[31]), but is not fully understood. Eventually, the bubble returns to a spherical shape, while quickly moving towards the center of the cell (orange arrow); the magnitude and scaling (with bubble size) of this translation dynamics is consistent with a capillary effect driven by surface tension.[25,31]

4.6.4 Temporary Equilibrium

After the fast dynamics described above, the bubble relaxes to its equilibrium shape and size (see Figure 4.8, top right micrograph). The equilibrium bubble volume is proportional to the cell volume (see eqn (4.11)), and is set by the elastic relaxation of both the liquid and the container (see Section 4.5.2).

It is worth noting that the whole process of formation of the equilibrium bubble has a duration on the order of 10 μs and thus requires techniques with very large frame rates to capture it. A regular camera (or the naked eye) would just observe a spherical bubble suddenly appearing, floating in the bulk of the liquid, making the observer think of a homogeneous nucleation process. However, high-speed recordings such as those of Figure 4.8c strongly suggest a heterogeneous process starting on the solid walls.[25,31]

The equilibrium described by the theory in Section 4.5 assumes that the number of water molecules in the containing cell remains constant. However, the walls (here, the hydrogel) are permeable to water, and eventually an outwards flow occurs, because the driving force for dehydration (evaporation or osmosis) is still present. In other words, there is a water potential imbalance (see Section 4.2.5) between water inside the cell, which has relaxed back to $\Psi \simeq 0$ after cavitation, and the surrounding medium ($\Psi < 0$). The initial equilibrium at volume V_b is thus only temporary.

4.6.5 Emptying, Bubble Growth

After cavitation, the surrounding, dehydrated medium tends to "suck out" the liquid remaining in the cell (see previous paragraph), causing bubble growth and eventually resulting in complete emptying of the cell (see Figure 4.8c); during this process the bubble tends to stay at the top of the cell due to buoyancy.

The timescales of emptying strongly depend on the volume of the cell, and of the transport properties and geometry of the surrounding medium. For experiments with hydrogels or synthetic xylem based on porous silicon (see Section 4.7), the emptying time, τ_{empt}, is on the order of minutes, and can be predicted based on the permeability and poroelastic constants of the material.[26,58] In real xylem, values between tens of milliseconds and tens of minutes are predicted, depending on the type of tree.[64] All these estimates are many orders of magnitude slower than the initial dynamics of the cavitation bubble leading to temporary equilibrium (see Section 4.6.4), so that the assumption used in Section 4.5 (*i.e.*, considering that the number of water molecules stays constant in the cell during nucleation and bubble oscillations), is justified.

4.6.6 Discussion

Cavitation bubble dynamics in closed, cell-like environments displays rich features with timescales spanning from sub-microsecond to minutes for a single bubble (see Figure 4.8). In the context of plants, some of these features are potentially interesting. For example, the clear oscillatory signature could help analyze acoustic signals measured on dehydrating plants. Also, cavitation close to surfaces (*e.g.* boat propellers) is usually associated with damage, due to strong, focused jets, shockwaves, *etc.* when the bubble collapses on itself in the vicinity of the surface.[18] We may hypothesize that such damaging phenomena are not present in plants, because cavitation bubbles, rather than collapsing, tend to grow and oscillate around a finite size. One way to explain this difference is that bubbles in plants appear in a tissue that is dehydrated and under static negative pressure, while bubbles close to propellers appear due to a very transient metastable state due to flow, and tend to collapse back completely due to the positive pressure of the stagnant, surrounding fluid. Another noteworthy remark is that the complex bubble dynamics makes it challenging to know the exact location where cavitation was initiated. Indeed, because the bubble oscillates quickly and is then ejected from the walls within microseconds, erroneous conclusions on the initial bubble position can be easily drawn when not recording the dynamics with very large frame rates.

It has to be noted that the conclusions above are based on experiments in model systems with spherical geometry. It is not well known yet how bubble dynamics changes when the shape differs from this ideal situation. Cubic cells have shown a quasi-identical behavior to spherical cells in terms of the oscillatory dynamics,[27] but it would be useful to investigate much more elongated structures such as those found in xylem. From experiments in microfluidics with open channels,[65] one could expect rich shape dynamics driven by the geometry of the confining cell, potentially very different from the spherical case. The last stage of bubble dynamics (slow growth due to cell emptying, see Figure 4.8), however, probably depends less on the exact shape of the cells and more on the global transport properties in the whole tissue. This expansion stage has important consequences for the general dynamics of propagation in multi-cell structures, as we will explain in the next section.

4.7 Propagation of Cavitation

So far we have explored the birth and growth of a cavitation bubble in a single cell. Plant tissues contain many cells connected to one another and the way bubbles from one cell propagate (or not) to other cells has important consequences. In the fern leptosporangium (see Figure 4.2c), the ejection mechanism requires cavitation bubbles to appear simultaneously in a maximum number of cells in the structure; it is thus favorable in such a situation to have a mechanism that allows fast propagation of cavitation through the tissue. On the contrary, cavitation is detrimental in xylem because it results in embolism; there, propagation is unfavorable, and indeed observations in xylem show that cavitation occurs through discrete events (see Figure 4.2a–b). In Sections 4.7.1 and 4.7.2 we will discuss phenomena that either trigger (positive interaction) or suppress (negative interaction) cavitation in neighboring cells. Surprisingly, both effects can result in bursts of cavitation events, but for different reasons and on different timescales.

4.7.1 Triggered Cavitation: Positive Interactions

Simultaneous cavitation in neighboring cells has been observed in the fern leptosporangium (see Figure 4.2c) and also in artificial, hydrogel-based biomimetic systems, either due to spontaneous, neighbor-to-neighbor propagation[66] or due to an externally induced shockwave that travels through all cells.[25] In this last situation, it is straightforward to understand the process: all cells initially contain liquid at negative pressure, and the shockwave passing through the liquid provides a short burst of energy that is sufficient to overcome the energy barrier of nucleation in the metastable liquid. In the first case where simulataneous cavitation is spontaneous, there is no external signal triggering cavitation, but the disturbance associated with the stochastic nucleation of a cavitation bubble in any of the cells can be sufficient to trigger nucleation in neighboring cells (see Figure 4.9a). Cavitation can then propagate from cell to cell in a chain-reaction fashion. While not much is known about the conditions (cell-to-cell separation, wall stiffness, *etc.*) required for cavitation to propagate in this manner, theoretical investigations have confirmed that cell-to-cell propagation is possible due to acoustic coupling between neighboring cells.[67,68] Interestingly, the timescales of propagation are predicted to be strongly linked to the oscillation dynamics of the cavitation bubble: the strongest probability of triggering cavitation in a neighboring cell occurs around the collapse phase after 1 cycle of oscillation. Since oscillation frequencies are typically in the MHz range (see Section 4.6), neighbor-to-neighbor triggering should be almost instantaneous, which is consistent with experimental observations.[10,66] More experimental and modeling investigations are necessary to characterize these propagation phenomena.

4.7.2 Hindered Cavitation: Negative Interactions

If the acoustic coupling between neighboring cells is not sufficiently strong, the positive triggering described in the previous section does not happen. However the cells can communicate hydraulically if their walls (or pits in the case of xylem) allow for water flow. Such flow is reponsible for the last stage of bubble dynamics, *i.e.* the slow growth of the bubble (cell emptying) over long timescales (∼minutes) (see Section 4.6); its origin is the pressure imbalance between the relaxed liquid contained in the cell after cavitation ($P \simeq P_{sat} > 0$) and the liquid in the surrounding cells, which is still at negative

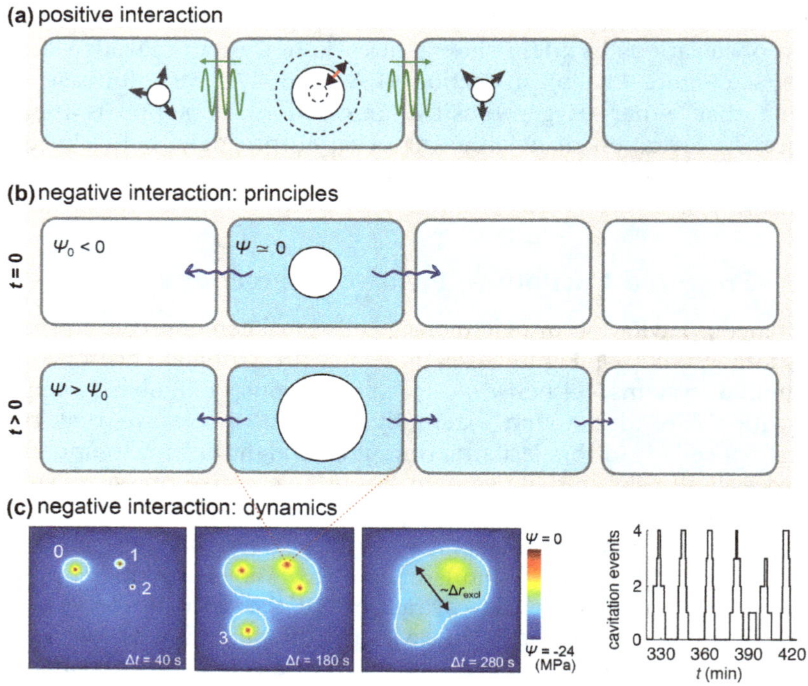

Figure 4.9 Propagation of cavitation in multi-cell systems, either due to positive interactions between cells (a) or negative interactions (b–c), see text for details. (a) The oscillation of a bubble in one cell creates a pressure disturbance sufficient to trigger cavitation in the neighboring cells. (b) Cavitation in one cell can also temporarily prevent cavitation in neighboring cells, because the cavitated cell acts as a source of water that rehydrates its surroundings while the cell is emptying (*i.e.* while the bubble is growing); $t = 0$ refers to the appearance of the cavitation bubble, and colors indicate the evolution of pressure in each cell (darker corresponds to higher pressure). (c) As a result, cavitation events develop an *exclusion zone* of larger pressure around them, where cavitation is hindered. The sequence in the left panel shows a simulation of the evolution of the pressure field due to four distinct cavitation events (labeled 0 to 3 in chronological order) in a multicellular structure; the white lines represent iso-pressure contours at −17 MPa, inside which cavitation probability is essentially zero. The panel on the right is an experimental recording of cavitation "bursts" resulting from the complex relaxation dynamics of the pressure field. Panel (c) reproduced from ref. 26 with permission from American Physical Society, Copyright 2014.

pressure. This outward flow *rehydrates* the neighboring cells, resulting in an increase of pressure in these cells, by a mechanism symmetrical to that described in Section 4.3.1 (see Figure 4.9b). This increase in pressure makes the neighboring cells less likely to cavitate (see Section 4.5), and thus cavitation in one cell tends to hinder cavitation in the immediate vicinity (negative interaction).

The pressure increase progressively spreads to other cells further away, forming an *exclusion zone* around the initial cavitation event, where cavitation is suppressed (see Figure 4.9c, left). The extent, Δr of this zone grows over time, t, and its growth is dictated by the poroelastic transport properties in the medium: $\Delta r \sim \sqrt{Ct}$ where C [m^2 s^{-1}] is an effective diffusivity sometimes called a consolidation coefficient. C scales as the ratio of permeability of the structure to its effective compressibility, $\chi = \chi_\ell + \chi_c$,[26,56] showing once again the importance of compressibility effects in cavitation dynamics. The growth of the exclusion zone is fed by the emptying of the cell where cavitation initially occurred, and this feeding stops when the cell is empty. As a result, the maximum extent of the exclusion zone is $\Delta r_{excl} \sim \sqrt{C\tau_{empt}}$, where τ_{empt} is the emptying time (see Section 4.6); Δr_{excl} can be seen as the radius of influence of a cavitation event in the medium. After the cavitated cell has completely emptied, the associated exclusion zone progressively vanishes. Consequently, the exclusion zone effect is both limited in space (radius of influence Δr_{excl}) and time (duration on the order of τ_{empt}).

The dynamics described above of rehydration with the formation of exclusion zones around cavitation events was demonstrated experimentally in artificial systems based on porous silicon.[26] An interesting consequence is that despite having *negative* interactions between neighboring cells, the propagation of cavitation can occur by bursts of seemingly simultaneous cavitation events (see Figure 4.9c, right). The reason for these bursts is the complex evolution of the pressure field in the system, associated with the regular formation and disappearance of multiple exclusion zones due to stochastic cavitation events. This complex, nonlinear coupling between poroelastic transport and nucleation results in self-organized spatio-temporal patterns that create burst-like behavior, *i.e.* periods with several cavitation events followed by "quiet" periods with no cavitation events. Compared to bursts originating from positive interactions (see previous section), the timescales involved here are much slower: in the artificial, xylem-like systems,[26] individual events within a burst could be separated by tens of seconds, while the bursts themselves happened every 15 minutes, approximately.

Potentially, in real xylem these timescales could be shorter or longer (~hours) depending on the geometry of the cells and physical properties of the tissue (elasticity, permeability). In a living tree, these cavitation/rehydration cycles could provide temporary relief in dry conditions by "sacrificing" some water-conducting cells to rehydrate the surrounding tissues. These cycles could also be related to the puzzling observation that sometimes the sap flow rate and xylem water potential in real plant leaves can oscillate with periods on the order of the hour.[69,70]

4.7.3 Discussion

From the discussion above, it is obvious that the dynamics of cavitation in multi-cell systems is rich and complex, and poorly explored. While a few experiments artificial systems have allowed us to unveil unexpected features of cavitation propagation, probing the dynamics of formation of bubbles in real systems is challenging, in particular for xylem in trees. Non-invasive techniques such as magnetic resonance imaging or X-ray microtomography are powerful but limited in sample size and/or time resolution.[37] Much simpler optical techniques have recently enabled direct visualization of embolism development in xylem slices,[1] living leaves[71] or freshly cut branches;[72] they are limited to samples sufficiently thin to allow for light propagation but are promising as a tool to investigate cavitation and embolism dynamics without the need of heavy experimental equipment. Combined approaches using modeling, experiments with artificial systems, and experiments on real plant tissues are needed to make progress on the understanding of cavitation dynamics and its implications.

4.8 Conclusion

Plants live in an out-of equilibrium world where the atmosphere is often sub-saturated in water vapor (humidity lower than 100%RH). As a result, they naturally dehydrate, putting the liquid they contain into a state of negative pressure, *i.e.* mechanical tension. Plants exploit this metastable state in various ways: in xylem, negative pressure is used as a pulling force to lift water from the roots to the leaves; in some ferns and fungi, cavitation in the metastable liquid provides a sudden release of energy for spore ejection strategies. However, cavitation in xylem also results in embolism, which is detrimental to the plant. Optimized compromises between efficiency and safety have thus evolved since the first plants appeared on land.

In this chapter, we have discussed the basic physics of formation and propagation of cavitation bubbles in cellular structures such as those found in plants and fungi. As we have shown, this physics is rich, from the fluid mechanics of ultra-fast bubble oscillations to the complex spatio-temporal patterns of nucleation in systems with multiple cells in interaction. Most of these results are supported by experimental investigations on artificial systems mimicking some aspects of plants; however, the dynamics of bubbles in real plants is far from being well understood.

The development of synthetic structures has two goals. First, these structures can help in understanding what is happening inside a real plant by studying samples that are much simpler, controlled and accessible than, *e.g.* a living tree. On this aspect, current developments should probably aim at building structures that are closer to real plant tissues in terms of geometry, permeability, elasticity, *etc.* For example, there is currently some effort to understand bubble penetration through the nanoporous pits separating xylem vessels and *e.g.* the role of surfactants; artificial systems with pore sizes similar to those of pits (typically 10–100 nm) would probably prove useful in these investigations. The synthetic structures based on porous silicon or hydrogels mentioned in this chapter currently have pores with diameters of only a few nanometers, which make bubble penetration extremely difficult and unlikely.

The second goal of artificial systems is to develop technologies inspired by plants to solve various problems in engineering, taking advantage of solutions already present in nature. This direction has already proven fruitful, with examples of biomimetic systems using negative pressure to pump water,[29,73] transport heat,[74] or desalinate water.[75] Improving these prototypes, but also explore other applications such as energy harvesting or mechanical actuation, should provide large inspiration for scientists in the near future.

Appendix A Effect of Air on the Saturation Vapor Pressure of Water

Equilibrium between liquid water and pure water vapor (in the absence of air) occurs when the two phases are at the saturation pressure, $P_{sat}(T)$, by definition. In this situation, the chemical potentials are equal in both phases ($\mu_\ell(P_{sat}) = \mu_v(P_{sat}) \equiv \mu_{sat}(T)$).

In the presence of air, the liquid and vapor phases are at a different pressure: due to mechanical equilibrium with air, liquid water is at atmospheric pressure, $P = P_{atm}$, while water vapor is characterized by its partial vapor pressure in air, p. By integration from the pure substance equilibrium (μ_{sat}), the corresponding chemical potentials are

$$\mu_\ell = \mu_{sat} + v_m(P - P_{sat}) \tag{4.19}$$

for liquid water (assuming an incompressible liquid), and

$$\mu_v = \mu_{sat} + RT \ln\left(\frac{p}{P_{sat}}\right) \tag{4.20}$$

for water vapor, assuming that it behaves as an ideal gas. Derivation of eqn (4.19)–(4.20) can be done using the thermodynamic relation equating the partial derivative of μ with respect to pressure to the molar volume, v, of the substance (constant $v = v_m$ for the liquid phase, $v = RT/p$ for the vapor phase).

We define p_{sat} as the equilibrium partial vapor pressure, p when the liquid is maintained at atmospheric pressure, $P = P_{atm}$. Using the equality of chemical potentials between eqn (4.19)–(4.20),

$$p_{sat} = P_{sat} \exp\left(\frac{v_m(P_{atm} - P_{sat})}{RT}\right) \tag{4.21}$$

which yields $p_{sat}/P_{sat} = 1.0007$ at 25 °C; this factor depends weakly on temperature.

At first sight, one could think that because $p_{sat} > P_{sat}$, water vapor at p_{sat} would be supersaturated, *i.e.* metastable with respect to condensation. However, this is not the case, because condensation is also affected by air pressure: the condensed liquid phase forms at P_{atm}, not at P_{sat}. This extra pressure makes condensation only favorable when $p > p_{sat}$. As a result, the equilibrium with liquid water at P_{atm} and water vapor at partial pressure p_{sat} in air is a true, stable equilibrium shifted from P_{sat} because of air pressure. Using this shifted equilibrium as the new reference state (μ_{sat}^{air}), the chemical potentials can be rewritten

$$\begin{cases} \mu_\ell &= \mu_{sat}^{air} + v_m(P - P_{atm}) \\ \mu_v &= \mu_{sat}^{air} + RT\ln\left(\dfrac{p}{p_{sat}}\right) \end{cases} \tag{4.22}$$

for the liquid and vapor phases, respectively.

We finally note that air has another potential effect on the liquid–vapor equilibrium of water. Indeed, by dissolving in liquid water at a concentration C_{air}, air results in an osmotic pressure $\Pi_{air} \simeq RTC_{air}$ with an associated shift in chemical potential $\Delta\mu_{air,sol} = -v_m\Pi_{air}$. From the solubility of air in water,[76] one can estimate $\Pi_{air} \simeq 2$ kPa when the water is saturated with air. Since $\Pi_{air} \ll P_{atm} \simeq 100$ kPa, $\Delta\mu_{air}^{sol} = -v_m\Pi_{air}$ is negligible compared to the shift in chemical potential introduced by air pressure, $\Delta\mu_{air}^{pressure} \simeq v_m P_{atm}$. In other words, the effect of dissolved air on the saturation vapor pressure is orders of magnitude smaller than the effect of the mechanical pressure of air, which is itself already quite small.

Appendix B Free Energy of a Confined Bubble

Here, we derive the free energy landscape of a vapor bubble (volume, V) in a liquid (V_ℓ), both enclosed in a container (V_c) that is elastic (potentially, infinitely stiff). Thus,

$$V_c = V + V_\ell \tag{4.23}$$

We use the state of the system at $V = 0$ (homogeneous liquid, no bubble) as the reference state, where the liquid pressure is P_0 (see Figure 4.7). Since the liquid occupies the whole container, $V_\ell = V_c$ in this reference state; we define the reference volume V_0 as being the value of V_ℓ and V_c in this situation. Note that the reference pressure P_0 is not atmospheric pressure here, but the initial pressure of the liquid, which is negative in the situations of interest in this chapter.

We consider an isothermal transformation, with a constant total number of water molecules (liquid + vapor) in the system. This assumption is justified by the fact that during the timescales of interest for bubble nucleation, the liquid does not have time to flow out of the cell (see Section 4.6.5). Pressure in the liquid varies as a function of bubble size: bubble expansion compresses the liquid (V_ℓ decreases) and makes its pressure P increase. This increase in pressure also makes the container expand (V_c increases if not infinitely stiff), and as we will see below, we can use knowledge about the variations of V_c with P to fully characterize the free energy variations of the *outside world* due to variations in bubble size.

For such a system, the natural thermodynamic potential is the Helmholtz free energy, F, and equilibrium is obtained when F is minimal. We use the notation $\Delta F = F - F_0$ to describe the Helmholtz free energy difference between the current state (bubble volume, V) and the reference state ($V = 0$).

Below we evaluate the different contributions to the total Helmholtz free energy variation ΔF of the system as a function of bubble size (V): liquid pressure (P) and compressibility (χ_ℓ), elastic deformation of the container (χ_c), surface tension of the liquid–vapor interface (γ), and evaporation of the liquid into the bubble (vapor pressure, P_{sat}). The derivation below follows a different and more general approach compared to earlier versions,[25,46] but yields the same final result.

a. Liquid pressure and compressibility.

We use the linearized equation of state of liquid water (eqn (4.1)), taking the homogeneous liquid at negative pressure as the reference state:

$$V_\ell - V_0 = -\chi_\ell V_0 (P - P_0) \tag{4.24}$$

We first consider that the liquid does not evaporate into the bubble and thus evolves at constant substance amount n when the bubble forms. In a second step, we will calculate the correction due to evaporation (see *Evaporation into bubble* paragraph below). Because $dF = -SdT - PdV + \mu dn$ for a pure substance (with parameters n, V, T), then $\Delta F = -\int P dV$ for constant T, n and we can directly obtain by integration of eqn (4.24) the free energy of the liquid as a function of its volume:

$$\Delta F_\ell = -P_0(V_\ell - V_0) + \frac{1}{2}\frac{1}{\chi_\ell V_0}(V_\ell - V_0)^2. \tag{4.25}$$

b. Evaporation into bubble. In reality, a small fraction of the liquid evaporates into the bubble (amount Δn). We make the simplifying assumption that this evaporation maintains the vapor in the bubble at the saturated vapor pressure, P_{sat}. There are subtleties in the consequences of this assumption but is impact on the results in classical nucleation theory is negligible.[43]

Since $F = \mu n - PV$, the total free energy of the vapor in the bubble is

$$F_{vap} = \mu_{sat}\Delta n - P_{sat}V \tag{4.26}$$

where μ_{sat} is the chemical potential of saturated vapor, which is also by definition the chemical potential of liquid water at pressure P_{sat}, because liquid and vapor are in equilibrium when $P = P_{sat}$.

In our situation, however, the liquid is not at P_{sat} but at P, so its chemical potential is $\mu_{sat} + v_m(P - P_{sat})$, neglecting second-order compressibility effects. Thus, using $F = \mu n - PV$ again, the change in free energy of the liquid due to evaporation is $\Delta F_{liq,evap} = -\Delta n(\mu_{sat} + v_m\,(P - P_{sat})) - P(-v_m\Delta n)$, or

$$\Delta F_{liq,evap} = \mu_{sat}(-\Delta n) + v_m \Delta n P_{sat} \tag{4.27}$$

As a result, the total effect of evaporation is the sum of the contributions of eqn (4.26) and (4.27)

$$\Delta F_{\text{evap}} = P_{\text{sat}}(v_{\text{m}}\Delta n - V) \simeq -P_{\text{sat}}V \tag{4.28}$$

where we neglect $v_{\text{m}}\Delta n$ compared to V because of the large difference between liquid and vapor densities; the term $v_{\text{m}}\Delta n$ should be kept when working close to the critical point, but we do not consider such situations here.

Note that it is also possible to add contributions from other gases present in the bubble,[46] but we neglect this effect here. In particular, we assume that dissolved air does not have time to significantly fill the bubble during the timescales of nucleation.

c. Elasticity of container. When pressure changes in the liquid, the container has an elastic response that we characterize using a linear approximation similar to the one we used for the compressibility of water (see Section 4.2.2):

$$V_{\text{c}} - V_0 = \chi_{\text{c}} V_0 (P - P_0) \tag{4.29}$$

For example, if the container is a spherical inclusion in an infinite, incompressible solid, $\chi_{\text{c}} = 3/(4G)$ where G is the shear modulus of the medium.[46]

When V_{c} changes by $\text{d}V_{\text{c}}$, the *outside world* (including the container itself) changes volume by $-\text{d}V_{\text{c}}$ and receives work from the fluid (pressure P) by an amount $\int P\text{d}V_{\text{c}}$. By definition, its free energy changes by the same amount, so that the contribution of the container deformation (and of the outside world) to the changes in the total free energy can be calculated from integration of eqn (4.29)

$$\Delta F_{\text{c}} = P_0 (V_{\text{c}} - V_0) + \frac{1}{2}\frac{1}{\chi_{\text{c}} V_0}(V_{\text{c}} - V_0)^2. \tag{4.30}$$

Note the similarity but also the difference in sign compared to eqn (4.25), due to the fact that while liquid contracts when pressure increases, the container expands. Note also that we do not need to know the details of the physical processes occurring in the outside world that result in how much the container deforms under pressure (*e.g.* entropic *vs.* mechanical, *etc.*): only knowing the relationship between container volume and inner pressure (eqn (4.29)) is sufficient to characterize the outside world contribution to the total free energy.

d. Total free energy. We can now write the total Helmholtz free energy ΔF as a function of the volumes V, V_ℓ, V_{c} by summing all the contributions above (eqn (4.25), (4.28), (4.30)) and the contribution of the surface tension γ of the liquid–vapor interface, which by definition is γA, where A is the surface area of the bubble:

$$\Delta F = \gamma A + (P_0 - P_{\text{sat}})V + \frac{1}{2V_0}\left[\frac{1}{\chi_\ell}(V_\ell - V_0)^2 + \frac{1}{\chi_{\text{c}}}(V_{\text{c}} - V_0)^2\right] \tag{4.31}$$

where we have used eqn (4.23) to simplify some terms.

We can further simplify eqn (4.31) by using the relationship between V, V_c and V_ℓ that is implied by the liquid and container equations of state. Indeed, from eqn (4.24) and (4.29), and using $V = V_c - V_\ell$, it is straightforward to show that

$$\begin{cases} V_\ell - V_0 = -\dfrac{\chi_\ell}{\chi_\ell + \chi_c} V \\[2ex] V_c - V_0 = \dfrac{\chi_c}{\chi_\ell + \chi_c} V \\[2ex] P - P_0 = \dfrac{1}{\chi_\ell + \chi_c} \dfrac{1}{V_0} V \end{cases} \tag{4.32}$$

These equations can also be found by direct minimization of ΔF with respect to V_ℓ and V_c for an imposed value of V. Injecting the volume relations from eqn (4.32) into eqn (4.31) yields

$$\Delta F = \gamma A + (P_0 - P_{sat})V + \frac{1}{2} \frac{1}{V_0(\chi_\ell + \chi_c)} V^2. \tag{4.33}$$

Note that we have not made assumptions about the shape of the bubble or container, and our calculations thus apply to arbitrary shapes.

We also note that eqn (4.33) is less general than eqn (4.31) because the former uses the equilibrium relations (4.32) and thus implicitly assumes that liquid and container volumes evolve in mechanical equilibrium given a constraint of bubble volume of size V. In other words, eqn (4.33) already uses a partial minimization of eqn (4.31) to relate V_ℓ and V_c to V and reduce the number of degrees of freedom to one (V). By contrast, eqn (4.31) has two degrees of freedom (three variables V_c, V_ℓ, V with the constraint $V + V_\ell = V_c$) and could in principle be used to describe situations where these quantities evolve independently (*e.g.* oscillation of the container and bubble that are not in phase). This may be confusing at first because we initially used equilibrium relations to derive the free energies of the liquid and of the elastic deformation of the container. However, this is only an illustration of the fact that reversible transformations can be used to probe the state variables of thermodynamic systems.

Acknowledgements

I am thankful to Teemu Hölttä, Benjamin Dollet, Yoël Forterre, Steven Jansen, Pono and others who have provided useful information during the preparation of this chapter through discussions and/or e-mail exchanges. I also thank Philippe Marmottant, Abraham Stroock, students and postdocs in their groups, and other colleagues with whom I have worked on the themes of negative pressure and cavitation in various contexts.

References

1. A. Ponomarenko, O. Vincent, A. Pietriga, H. Cochard, É. Badel and P. Marmottant, Ultrasonic emissions reveal individual cavitation bubbles in water-stressed wood, *J. R. Soc., Interface*, 2014, **11**, 20140480.
2. C. Llorens, M. Argentina, N. Rojas, J. Westbrook, J. Dumais and X. Noblin, The fern cavitation catapult: Mechanism and design principles, *J. R. Soc., Interface*, 2016, **13**, 20150930.
3. B. Choat, E. Badel, R. Burlett, S. Delzon, H. Cochard and S. Jansen, Noninvasive Measurement of Vulnerability to Drought-Induced Embolism by X-Ray Microtomography, *Plant Physiol.*, 2016, **170**, 273.
4. D. S. Meredith, Violent Spore Release in Some Fungi Imperfecti, *Ann. Bot.*, 1963, **27**, 39.
5. J. A. Milburn, Cavitation and Osmotic Potentials of Sordaria Ascospores, *New Phytol.*, 1970, **69**, 133.
6. M. T. Tyree and M. H. Zimmermann, *Xylem Structure and the Ascent of Sap*, Springer Science & Business Media, 2013.
7. H. Cochard, Cavitation in trees, *C. R. Phys.*, 2006, **7**, 1018.
8. A. D. Stroock, V. V. Pagay, M. A. Zwieniecki and N. M. Holbrook, The Physicochemical Hydrodynamics of Vascular Plants, *Annu. Rev. Fluid Mech.*, 2014, **46**, 615.
9. J. Dumais and Y. Forterre, "Vegetable Dynamicks": The Role of Water in Plant Movements, *Annu. Rev. Fluid Mech.*, 2012, **44**, 453.
10. X. Noblin, N. O. Rojas, J. Westbrook, C. Llorens, M. Argentina and J. Dumais, The Fern Sporangium: A Unique Catapult, *Science*, 2012, **335**, 1322.
11. N. P. Money and M. W. F. Fischer, Biomechanics of Spore Release in Phytopathogens, in *Plant Relationships V, The Mycota*, ed. H. B. Deising, Springer Science & Business Media, 2009, vol. 5, pp. 115–133.
12. M. Versluis, B. Schmitz, A. von der Heydt and D. Lohse, How Snapping Shrimp Snap: Through Cavitating Bubbles, *Science*, 2000, **289**, 2114.
13. S. N. Patek, W. L. Korff and R. L. Caldwell, Deadly strike mechanism of a mantis shrimp, *Nature*, 2004, **428**, 819.
14. W. M. Kier and A. M. Smith, The Structure and Adhesive Mechanism of Octopus Suckers, *Integr. Comp. Biol.*, 2002, **42**, 1146.
15. P. Nobel, *Physicochemical and Environmental Plant Physiology*, Elsevier, Cambridge, 2020.
16. K. H. Jensen, K. Berg-Sørensen, H. Bruus, N. M. Holbrook, J. Liesche, A. Schulz, M. A. Zwieniecki and T. Bohr, Sap flow and sugar transport in plants, *Rev. Mod. Phys.*, 2016, **88**, 035007.
17. P. G. Debenedetti, *Metastable Liquids: Concepts and Principles*, Princeton University Press, 1996.
18. C. E. Brennen, *Cavitation and Bubble Dynamics*, Cambridge University Press, 2014.
19. F. Caupin and E. Herbert, Cavitation in water: A review, *C. R. Phys.*, 2006, **7**, 1000.

20. F. Caupin and A. D. Stroock, The Stability Limit and other Open Questions on Water at Negative Pressure, in *Advances in Chemical Physics*, ed. H. E. Stanley, John Wiley & Sons, Inc., Hoboken, NJ, USA, 2013, pp. 51–80.
21. R. J. Speedy, Stability-limit conjecture. An interpretation of the properties of water, The, *J. Phys. Chem.*, 1982, **86**, 982.
22. O. Vincent, J. Zhang, E. Choi, S. Zhu and A. D. Stroock, How Solutes Modify the Thermodynamics and Dynamics of Filling and Emptying in Extreme Ink-Bottle Pores, *Langmuir*, 2019, **35**, 2934.
23. J.-F. Louf, G. Guéna, E. Badel and Y. Forterre, Universal poroelastic mechanism for hydraulic signals in biomimetic and natural branches, *Proc. Natl. Acad. Sci.*, 2017, **114**, 11034.
24. J. Irvine and J. Grace, Continuous measurements of water tensions in the xylem of trees based on the elastic properties of wood, *Planta*, 1997, **202**, 455.
25. O. Vincent, Dynamique de Bulles de Cavitation Dans de l'eau Micro-Confinée Sous Tension. Application à l'étude de l'embolie Dans Les Arbres, Ph.D. thesis, Université Joseph Fourier Grenoble 1, 2012.
26. O. Vincent, D. A. Sessoms, E. J. Huber, J. Guioth and A. D. Stroock, Drying by Cavitation and Poroelastic Relaxations in Porous Media with Macroscopic Pores Connected by Nanoscale Throats, *Phys. Rev. Lett.*, 2014, **113**, 134501.
27. C. Scognamiglio, F. Magaletti, Y. Izmaylov, M. Gallo, C. M. Casciola and X. Noblin, The detailed acoustic signature of a micro-confined cavitation bubble, *Soft Matter*, 2018, **14**, 7987.
28. W. M. Deen, Hindered transport of large molecules in liquid-filled pores, *AIChE J.*, 1987, **33**, 1409.
29. T. D. Wheeler and A. D. Stroock, The transpiration of water at negative pressures in a synthetic tree, *Nature*, 2008, **455**, 208.
30. T. D. Wheeler and A. D. Stroock, Stability Limit of Liquid Water in Metastable Equilibrium with Subsaturated Vapors, *Langmuir*, 2009, **25**, 7609.
31. O. Vincent, P. Marmottant, S. R. Gonzalez-Avila, K. Ando and C.-D. Ohl, The fast dynamics of cavitation bubbles within water confined in elastic solids, *Soft Matter*, 2014, **10**, 1455.
32. I.-T. Chen, D. A. Sessoms, Z. Sherman, E. Choi, O. Vincent and A. D. Stroock, Stability Limit of Water by Metastable Vapor–Liquid Equilibrium with Nanoporous Silicon Membranes, *J. Phys. Chem. B*, 2016, **120**, 5209.
33. J.-P. Franc and J.-M. Michel, Fundamentals of Cavitation, *Fluid Mechanics and Its Applications*, Springer, Netherlands, 2005.
34. K. A. Mørch, Reflections on cavitation nuclei in water, *Phys. Fluids*, 2007, **19**, 072104.
35. N. M. Holbrook, M. J. Burns and C. B. Field, Negative Xylem Pressures in Plants: A Test of the Balancing Pressure Technique, *Science*, 1995, **270**, 1193.
36. H. Cochard, A technique for measuring xylem hydraulic conductance under high negative pressures, *Plant, Cell Environ.*, 2002, **25**, 815.

37. H. Cochard, E. Badel, S. Herbette, S. Delzon, B. Choat and S. Jansen, Methods for measuring plant vulnerability to cavitation: A critical review, *J. Exp. Bot.*, 2013, **64**, 4779.
38. M. Berthelot, Sur quelques phénomènes de dilatation forcée des liquides, *Ann. Chim. Phys.*, 1850, **30**, 232.
39. Q. Zheng, D. J. Durben, G. H. Wolf and C. A. Angell, Liquids at Large Negative Pressures: Water at the Homogeneous Nucleation Limit, *Science*, 1991, **254**, 829.
40. M. E. M. Azouzi, C. Ramboz, J.-F. Lenain and F. Caupin, A coherent picture of water at extreme negative pressure, *Nat. Phys.*, 2013, **9**, 38.
41. E. Herbert, S. Balibar and F. Caupin, Cavitation pressure in water, *Phys. Rev. E*, 2006, **74**, 041603.
42. K. Ando, A.-Q. Liu and C.-D. Ohl, Homogeneous Nucleation in Water in Microfluidic Channels, *Phys. Rev. Lett.*, 2012, **109**, 044501.
43. M. Blander and J. L. Katz, Bubble nucleation in liquids, *AIChE J.*, 1975, **21**, 833.
44. G. Menzl, M. A. Gonzalez, P. Geiger, F. Caupin, J. L. F. Abascal, C. Valeriani and C. Dellago, Molecular mechanism for cavitation in water under tension, *Proc. Natl. Acad. Sci.*, 2016, **113**, 13582.
45. P. M. Wilt, Nucleation rates and bubble stability in water-carbon dioxide solutions, *J. Colloid Interface Sci.*, 1986, **112**, 530.
46. O. Vincent and P. Marmottant, On the statics and dynamics of fully confined bubbles, *J. Fluid Mech.*, 2017, **827**, 194.
47. P. S. Epstein and M. S. Plesset, On the Stability of Gas Bubbles in Liquid-Gas Solutions, *J. Chem. Phys.*, 1950, **18**, 1505.
48. A. A. Atchley, The Blake threshold of a cavitation nucleus having a radius-dependent surface tension, *J. Acoust. Soc. Am.*, 1989, **85**, 152.
49. E. N. Harvey, D. K. Barnes, W. D. McElroy, A. H. Whiteley, D. C. Pease and K. W. Cooper, Bubble formation in animals. I. Physical factors, *J. Cell. Comp. Physiol.*, 1944, **24**, 1.
50. R. E. Apfel, The Role of Impurities in Cavitation-Threshold Determination, *J. Acoust. Soc. Am.*, 1970, **48**, 1179.
51. D. H. Trevena, *Cavitation and Tension in Liquids*, A. Hilger, Bristol, England, Philadelphia, 1987.
52. L. Kaack, M. Weber, E. Isasa, Z. Karimi, S. Li, L. Pereira, C. L. Trabi, Y. Zhang, H. J. Schenk, B. Schuldt, V. Schmidt and S. Jansen, Pore constrictions in intervessel pit membranes provide a mechanistic explanation for xylem embolism resistance in angiosperms, *New Phytol.*, 2021, **230**, 1829.
53. M. Kanduč, E. Schneck, P. Loche, S. Jansen, H. J. Schenk and R. R. Netz, Cavitation in lipid bilayers poses strict negative pressure stability limit in biological liquids, *Proc. Natl. Acad. Sci.*, 2020, **117**, 10733.
54. H. J. Schenk, K. Steppe and S. Jansen, Nanobubbles: A new paradigm for air-seeding in xylem, *Trends Plant Sci.*, 2015, **20**, 199.

55. H. J. Schenk, S. Espino, D. M. Romo, N. Nima, A. Y. Do, J. M. Michaud, B. Papahadjopoulos-Sternberg, J. Yang, Y. Y. Zuo, K. Steppe and S. Jansen, Xylem Surfactants Introduce a New Element to the Cohesion-Tension Theory, *Plant Physiol.*, 2017, **173**, 1177.
56. H. F. Wang, *Theory of Linear Poroelasticity with Applications to Geomechanics and Hydrogeology*, Princeton University Press, 2000.
57. C. Drysdale, Radiation dynamics of a cavitation bubble in a liquid-filled cavity surrounded by an elastic solid, *Phys. Rev. E*, 2017, **95**, 053104.
58. O. Vincent, P. Marmottant, P. A. Quinto-Su and C.-D. Ohl, Birth and Growth of Cavitation Bubbles within Water under Tension Confined in a Simple Synthetic Tree, *Phys. Rev. Lett.*, 2012, **108**, 184502.
59. M. Gallo, F. Magaletti, D. Cocco and C. M. Casciola, Nucleation and growth dynamics of vapour bubbles, *J. Fluid Mech.*, 2020, **883**, A14.
60. M. A. Bruning, M. Costalonga, J. H. Snoeijer and A. Marin, Turning Drops into Bubbles: Cavitation by Vapor Diffusion through Elastic Networks, *Phys. Rev. Lett.*, 2019, **123**, 214501.
61. L. De Roo, L. L. Vergeynst, N. J. F. De Baerdemaeker and K. Steppe, Acoustic Emissions to Measure Drought-Induced Cavitation in Plants, *Appl. Sci.*, 2016, **6**, 71.
62. T. Leighton, *G.,* The Acoustic Bubble, secondoftwo publisher Elsevier, 1994.
63. W. Lauterborn and T. Kurz, Physics of bubble oscillations, *Rep. Prog. Phys.*, 2010, **73**, 106501.
64. T. Hölttä, T. Vesala and E. Nikinmaa, A model of bubble growth leading to xylem conduit embolism, *J. Theor. Biol.*, 2007, **249**, 111.
65. E. Zwaan, S. Le Gac, K. Tsuji and C.-D. Ohl, Controlled cavitation in microfluidic systems, *Phys. Rev. Lett.*, 2007, **98**, 254501.
66. M. Pellegrin, Nucléation et Dynamique de Bulles de Cavitation Dans Des Liquides Confinés Sous Tension : Expériences Dans Des Systèmes Microfabriqués et Simulations de La Dynamique Moléculaire, Ph.D. thesis, Université Nice Sophia Antipolis, 2015.
67. A. A. Doinikov, Cavitation in a liquid-filled cavity surrounded by an elastic medium: Intercoupling of cavitation events in neighboring cavities, *Phys. Rev. E*, 2018, **98**, 013108.
68. A. El Amri, Propagation d'une Onde de Cavitation En Milieux Confinés, Ph.D. thesis, Université Côte d'Azur, 2017.
69. A. R. G. Lang, B. Klepper and M. J. Cumming, Leaf Water Balance During Oscillation of Stomatal Aperture, *Plant Physiol.*, 1969, **44**, 826.
70. K. Steppe, S. Dzikiti, R. Lemeur and J. R. Milford, Stomatal Oscillations in Orange Trees under Natural Climatic Conditions, *Ann. Bot.*, 2006, **97**, 831.
71. T. J. Brodribb, D. Bienaimé and P. Marmottant, Revealing catastrophic failure of leaf networks under stress, *Proc. Natl. Acad. Sci.*, 2016, **113**, 4865.
72. T. J. Brodribb, M. Carriqui, S. Delzon and C. Lucani, Optical Measurement of Stem Xylem Vulnerability, *Plant Physiol.*, 2017, **174**, 2054.

73. W. Shi, R. M. Dalrymple, C. J. McKenny, D. S. Morrow, Z. T. Rashed, D. A. Surinach and J. B. Boreyko, Passive water ascent in a tall, scalable synthetic tree, *Sci. Rep.*, 2020, **10**, 1.
74. I.-T. Chen, A. Pharkya and A. D. Stroock, Analysis of superheated loop heat pipes exploiting nanoporous wick membranes, *AIChE J.*, 2014, **60**, 762.
75. Y. Wang, J. Lee, J. R. Werber and M. Elimelech, Capillary-driven desalination in a synthetic mangrove, *Sci. Adv.*, 2020, **6**, eaax5253.
76. R. Sander, Compilation of Henry's law constants (version 4.0) for water as solvent, *Atmos. Chem. Phys.*, 2015, **15**, 4399.

CHAPTER 5

Root–Soil Interaction

E. KOLB,*[a] M. QUIROS,[a] G. J. MEIJER,[b] M. B. BOGEAT-TRIBOULOT,[c] A. CARMINATI,[d] E. ANDÒ,[e] L. SIBILLE[e] AND F. ANSELMUCCI[e]

[a] PMMH, CNRS UMR 7636, ESPCI Paris, Université PSL, Sorbonne Université, Université Paris Cité, F-75005, Paris, France; [b] Department of Architecture and Civil Engineering, University of Bath, Claverton Down, Bath BA2 7AY, United Kingdom; [c] Université de Lorraine, AgroParisTech, INRAE, UMR Silva, 54000 Nancy, France; [d] Physics of Soils and Terrestrial Ecosystems, Institute of Terrestrial Ecosystems (ITES), Department of Environmental Systems Science (D-USYS), ETH Zürich, Universitätstrasse 16, 8006 Zürich, Switzerland; [e] Univ. Grenoble Alpes, CNRS, Grenoble INP, 3SR, F-38000, Grenoble, France

*Emails: evelyne.kolb@sorbonne-universite.fr; Gjm36@bath.ac.uk

5.1 Introduction

The physiology, mechanics and hydraulics of plant roots are a complex and stimulating challenge for soft matter scientists. Understanding their mechanics and growth properties at different spatial scales as well as their complex interplay with the environment requires multidisciplinary approaches and interests many research fields ranging from soil science[1] and civil engineering to biomechanics[2] and biophysics[3,4] and even extends to applied domains like food processing[5] or bio-composite engineering.[6]

Plant roots absorb the water and nutrients required to satisfy the shoot's demand and ensure the mechanical anchorage and stability of the whole plant against wind blowing, water flows or soil movements.[7,8] They may also provide functions other than those directly beneficial to the plant itself, as

Soft Matter Series No. 15
Soft Matter in Plants: From Biophysics to Biomimetics
Edited by Kaare H. Jensen and Yoël Forterre

Published by the Royal Society of Chemistry, www.rsc.org

roots interact with and alter the properties of the surrounding soil in many ways. Vegetation may be used as an ecologically friendly and sustainable alternative to more traditional hard engineering solutions for soil reinforcement in civil engineering but also for deep C sequestration through increased root development at large depths.[9]

Soil-property-altering effects of roots can be grouped into two categories. Firstly, roots add additional strength to the soil. This can help to reduce erosion[10–12] or help mitigate against streambank instabilities[13,14] or (shallow) landslides.[15–17] Secondly, plant transpiration changes the soil moisture balance in the soil, which for example can be used to control moisture levels in landfill covers.[18]

Clearly the intricate properties of root and soil raise fundamental questions related to soft matter principles traditionally addressed in complex fluids, granular and porous media, composite and fibrous materials, cellular foams or mechanics of thin and inflatable structures. In this chapter we present some aspects of the root–soil interaction by focusing on the material properties of roots and their physical effects on soils. In the first part, we detail the specificity of the growth and mechanics of a single root and how the growth of a root is affected by the mechanical strength of the soil. In the second part, we introduce its counterpart, the soil, and the mechanical and hydric modifications provoked by the growing root at the local scale. In the last part, we extend to the whole root system and present the complex interplay between soil and roots, especially in the context of soil management and civil engineering.

5.2 The Single Root: An Interesting Material for Soft Matter Studies

5.2.1 How Does a Root Grow?

A growing root is not simply an advancing thin rod.[19] Root elongation occurs in the root apex, by the two fundamental processes of growth in plants: cell proliferation and cell expansion. In the root apex, these two processes are spatially separated, in the meristem and in the elongation zone, respectively (see Figure 5.1). The spatial extent of these two zones is controlled by different molecular gradients (hormones, reactive oxygen species) maintained by cell activity and molecular flows.[20] New cells are produced in the meristem by cell division and some cell expansion, bringing cell size over a threshold before the next division. The cells reaching the transition zone between the meristem and the elongation zone leave their meristematic state and start to expand rapidly, increasing their volume by up to 200 times by vacuolisation (increase in size of the vacuole, an organelle consisting of a membrane containing an aqueous solution). In addition, cell expansion is strongly anisotropic, giving the root a cylindrical shape. Finally, while the root apex structure is rather constant, it is a dynamic structure in which cells flow from the tip to the maturation zone and go through significant morphological changes.[21,22]

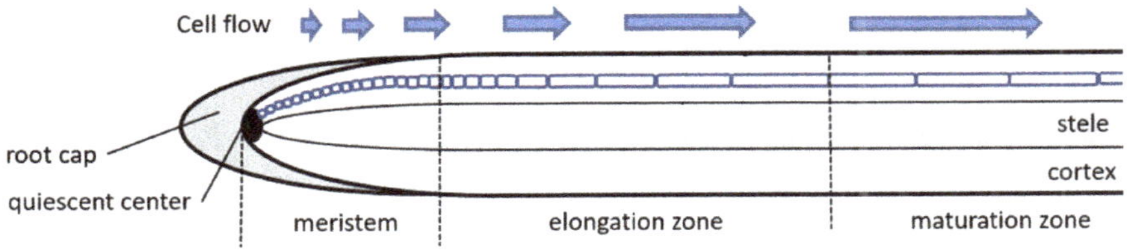

Figure 5.1 Cell flow in the constant root apex structure. The root apex is covered by a root cap that protects the meristem. The quiescent centre, which is an organisational centre, is surrounded by stem cells that divide and produce cells that proliferate in the meristem. Cells elongate and divide at the same rate in the meristem. In the elongation zone, cells only elongate.

Quantifying organ growth requires a detailed characterisation of the motion velocity field, which can be monitored using video or time-lapse photography. Since root growth occurs mainly in the longitudinal direction, most studies considered only this longitudinal axis, reducing growth analysis to one dimension. However, roots grow also in diameter, and longitudinal and diametric growths do not respond similarly to constraints.[23,24] While the root apex is pushed into the soil by expanding cells, cells within the root apex are moved away from the root tip (Figure 5.1) and it is convenient to analyse growth as a function of the distance to the root tip. Pioneering growth kinematics studies include those by Goodwin and Stepka,[25] who analysed growth of the transparent root of *Phleum pratense* by following cell displacement under a microscope, and Erickson and Sax,[21] who used streak photography of corn roots marked with lamp black in water solution. By tracking the displacement velocity of cells or of artificial marks at the root surface, they characterised the velocity field.

Figure 5.2A presents a typical velocity field along a poplar root apex, the referential being the quiescent centre (defined in Figure 5.1). The plateau highlights the maturation zone, and the maximal velocity V_{max} corresponds to the overall root growth rate (RGR). Taking the first spatial derivative of this velocity field provides a quantitative characterisation of the relative elemental growth rate (REGR) or strain rate along the root apex (Figure 5.2B and C). The longitudinal profile of REGR shows that material expansion is very low in the meristem, increases strongly when the cells leave their meristematic state, reaches up to 30% h^{-1} in the middle of the elongation zone and then decreases progressively to zero at the distal end of the elongation zone, when cells reach their final size and enter in maturation.

Since these pioneer works, technological advances in computer, imaging and image processing have promoted high-throughput analyses. For instance, tracking the natural marks generated at the surface of the roots by infrared light (Figure 5.2D) instead of carbon particles deposited on the root surface has allowed high spatial resolution and long-term monitoring.[26,27] Particle image velocimetry (PIV), which permits the quantification of the instantaneous speed of a flow from the analysis of image sequences, is at the basis of several applications dedicated to kinematic analyses.[28–30] The development of these

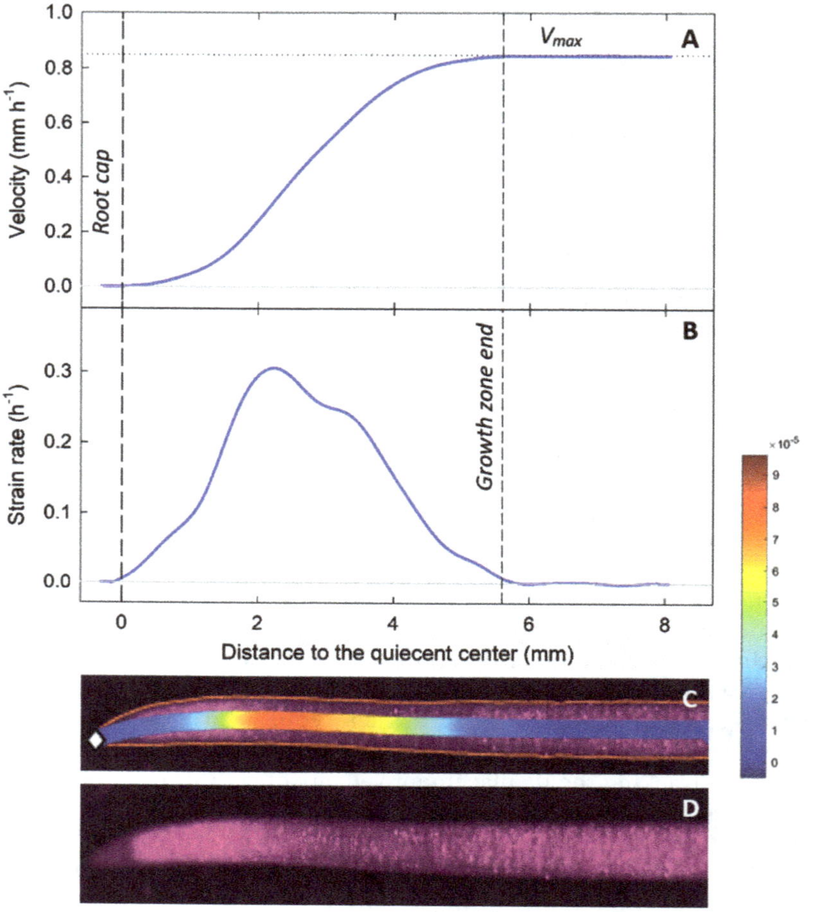

Figure 5.2 Growth parameters along a poplar root apex, the referential being the quiescent centre. (A): velocity field. (B): strain rate in h^{-1}. (C): visualisation of the strain rate by Kymorod.[30] The amplitude of the strain rate is given by the colour code on the right, with a scale in s^{-1}. (D): picture of the root apex with the natural marks generated by the infrared light (wavelength 850 nm).

tools contributed to many growth kinematic studies aiming to understand the regulation of root growth in response to its environment.[24,31–33]

While cell proliferation in the meristem is indispensable for root growth, most of the increase of tissue volume is due to rapid cell expansion in the elongation zone.[34] Plant cells are surrounded by a rigid pecto-cellulosic cell wall, allowing cells to live in hypo-osmotic media without bursting, by opposing the internal hydrostatic pressure, called the turgor pressure (P). Cell expansion requires water and solute inflows, and the cell wall to expand. The visco-elasto-plastic nature of the primary cell wall and the production of new cell wall elements make the volume expansion of the growing cells possible.[35] If water and solute flows are assumed not to be limiting, root growth depends on cell turgor pressure and on the cell wall rheological and biologically regulated properties of elongating cells. In fact, phenomenological analyses of root growth showed that, under steady state, the root growth velocity (or RGR, $\mathrm{mm\,h^{-1}}$) is observed to be close to an affine function of cell turgor pressure P (Figure 5.3A):

$$\mathrm{RGR} = m\cdot(P - Y) \tag{5.1}$$

where m is an extensibility coefficient characterising the ease of axial expansion and Y is a minimal turgor threshold.[36]

Attempts are made to relate this macroscopic phenomenological growth law (eqn (5.1)) to the irreversible expansion rate at the cellular level established for the expansive growth of cells with walls, like algal, fungal or plant cells (see Chapter 3). These Lockhart laws couple the net rate of water uptake that maintains the turgor pressure P to the rate of cell wall deformation[37] (see Chapter 1). Under turgor pressure, the rigid walls of the cell are loaded in tension, like the walls of a vessel under pressure (Figure 5.3B). Balance of forces implies that the hoop (orthoradial) stress $\sigma_{\theta\theta}$ and axial stress σ_{zz} in the cell wall are directly related to the turgor pressure P by a geometrical amplification factor $\frac{R}{h}$ that depends on the thickness h of the cell wall and on the radius R of the cell (with $R \gg h$), such that:

$$\sigma_{\theta\theta} = 2\sigma_{zz} = \frac{R}{h}\cdot P \tag{5.2}$$

Although currently debated, the cell wall has often been considered as a Bingham fluid, which deforms irreversibly above a yield stress σ_Y with a plastic viscosity η_P: while turgor pressure P exceeds a given threshold P_Y, cell walls flow and expand. Whereas P is isotropic, the primary cell walls are mechanically anisotropic. They behave like composite reinforced materials made of (rigid) cellulose microfibrils embedded in a (soft) gel of pectin and hemicelluloses. Due to a preferential orientation of the microfibril deposition, the walls of a growing cell are softer in the axial (longitudinal) direction, and cells elongate mainly in the axial direction. Thus the relative variation of the length ℓ of a cell with time t, or strain rate $\dot{\varepsilon}$, can be written as:

$$\dot{\varepsilon} = \frac{1}{\ell}\frac{\mathrm{d}\ell}{\mathrm{d}t} = \frac{\sigma_{zz} - \sigma_Y}{\eta_P} = \frac{R}{2h}\cdot\frac{(P - P_Y)}{\eta_P} = \Phi(P - P_Y) \tag{5.3}$$

with $P_Y = \frac{2h}{R}\sigma_Y$ and $\Phi = \frac{R}{2h}\cdot\frac{1}{\eta_P}$ is the extensibility defined at the cellular level (see Chapter 1). A crude (and formally not exact) extrapolation at the tissue scale of eqn (5.3) consists in replacing the length ℓ in the denominator with L_{GZ}, the typical length over which the root grows. This approximation gives information on the scaling of the root growth velocity or $\mathrm{RGR} \sim \Phi(P - P_Y)\cdot L_{GZ}$.

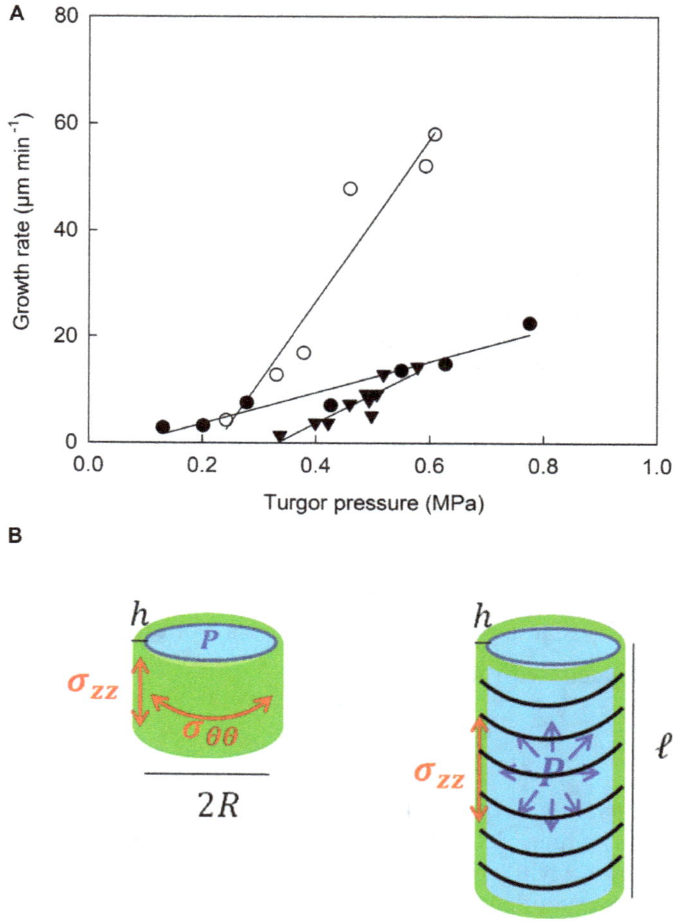

Figure 5.3 (A) Velocity of root growth as a function of turgor pressure P (at the root scale). Maize (open circles) and wheat (solid circles) roots were bathed in a series of mannitol solutions for adjusting the turgor pressure. Pea roots (solid triangles) were grown in soils of different density. Reproduced from ref. 36 with permission from John Wiley & Sons, Copyright 1994 New Phytologist Trust. (B) Mechanisms of expansive growth for a cell of radius R with rigid cell wall under turgor pressure P. The surrounding cell wall (in green) withstands the axial and orthoradial (hoop) tensile stresses associated with the internal (isotropic) turgor pressure P (in blue). The mechanical anisotropy of the cell wall due to a particular deposition of cellulose microfibrils (schematically symbolised by the hoop fibres in black) leads to a longitudinal increase of the cell length ℓ when the turgor pressure increases.

The similarity between eqn (5.1) and (5.3) helps to understand the physical meaning of m and Y entering eqn (5.1). The minimal turgor pressure for growth $Y = P_Y$, is in fact associated with a tensorial quantity, the axial yield stress of the cell wall. On its side, the extensibility parameter m scales like $m \approx \frac{R}{2h} \cdot \frac{1}{\eta_P} \cdot L_{GZ} \approx \Phi \cdot L_{GZ}$, where L_{GZ} is taken as the root growing length and Φ is its average extensibility, usually expressed in $MPa^{-1}\,h^{-1}$ units.

While cell turgor pressure P is observed to be uniform longitudinally over the growth zone,[23,38] the strain rate is not uniform and shows a bell-shaped profile along the root apex (Figure 5.2B). Thus, as cells are moved along the growth zone, strain rate variations must be controlled by variations in cell wall properties through parameters such as η_P or σ_Y. At the tissue scale, the root growth velocity dependency with turgor pressure (eqn (5.1)) does not provide a direct measurement of $\frac{1}{\eta_P}$ or σ_Y, but of m and Y. Determination of the two latter parameters along the root growth zone revealed that m is relatively constant and that Y varies according to the variations of the strain rate.[39]

5.2.2 How Does a Root Behave Mechanically?

One of the first steps towards a better understanding of the root penetration mechanisms or for implementing mechanical models for the stability of root-reinforced soils starts with the knowledge of the mechanical properties of individual roots. Even at this scale, the derivation of mechanical properties of a single root is challenging because it deals with a multicellular, inhomogeneous and anisotropic tissue under turgor pressure. Therefore it is sometimes difficult to find consistent values of the mechanical parameters for plant roots, like Young's moduli or ultimate strengths.

Most experiments carried out tensile tests on roots, either *in situ* by pulling out roots from the soil or *ex situ* by performing tensile tests with universal testing machines on excised segments of roots. In most works, the tensile modulus E and the tensile strength σ_U of roots are observed to decay with increasing root diameter d. Roots with smaller diameters appeared to be stronger and stiffer. Usually some negative power-law relationships of the form $E \propto d^{-\alpha}$ or $\sigma_U \propto d^{-\beta}$ are proposed, but the exponents α and β greatly vary among works. In the meta-analysis of Mao *et al.*,[40] the β values for root strength are ranging from 0.51 to 1.77, depending on the species. This difference in β values covers a distinction according to the functional group of the species: large woody roots are mainly found in trees or shrubs, whereas fines roots are common to all species. To be incorporated into models of soil stability, phenomenological laws have been derived by averaging over many species from the same functional group. For instance, an averaged relationship $\sigma_U = 28.97\mathrm{d}^{-0.52}$ is obtained for tree roots, the diameter d being in mm and the strength σ_U in MPa. Another law $\sigma_U = 21.05\mathrm{d}^{-1.15}$ is obtained for grass or herbaceous roots. These results have practical importance for interpreting the stability of root-reinforced soils: the thinner roots that have high tensile strength would slip while the thicker roots would break during soil shear[41] (see Section 5.4.3).

The power-law relation between tensile strength and diameter has been mainly explained by changes in root chemical composition, either by an increase of cellulose content[42] or alternatively by a decrease in the lignin/cellulose ratio as the diameter increases.[43] Clearly the description in terms of diameter is not sufficient to describe the variety of behaviours. By performing tensile tests on excised segments of barley (cereal), Loades *et al.*[44] have shown that the correlation of the power-decay regression law is improved when the roots are classified according to their type (either seminal, nodal or lateral). For root diameters ranging from 0.4 to 0.6 mm, seminal roots can be five times stiffer than lateral roots (with a Young's modulus of around 7 MPa). This difference according to the root type shows the importance of root internal structure and anatomy with possible changes in cell wall structure, composition or tissue density. These differences are also observed for different ages or locations from the apex.[41,42,45] The growth environment also affects the root's mechanical properties. For instance, in waterlogged conditions, roots are weaker and more compliant. This difference is probably due to the formation of aerenchyma, a spongy tissue with increasing air spaces within roots to help the plant survive in anoxic conditions.

In fact, the former mechanical description in terms of diameter implicitly assumes that roots within the same diameter class have a similar function. However, root morphology varies greatly among and within species. According to McKormack *et al.*,[46] the broadly defined fine root group should be split into two distinct classes: absorptive fine roots and transport fine roots. Absorptive fine roots approximately represent the first-, second-, and often third-order roots in the branching hierarchy, while transport fine roots are those of higher developmental orders. The transport fine roots are characterised by secondary development, with the appearance of a cork periderm (secondary dermal tissue that replaces the epidermis, the outermost layer of cells, along older roots), the thickening of cell walls and the deposition of additional suberin, a biopolymer which reduces movement of water and ions radially across the root and changes the tissue chemistry and therefore the mechanics of the roots. In contrast, absorptive fine roots grow only in length (primary growth), have a shorter life and show no secondary growth.

Whatever its fate (fine short-lived or coarser long-lived), any root undergoes compression while growing in length or elongating in soil. This prompts compression tests to be carried out on these different root types, paying particular attention to their water status.

Interestingly, in the context of food engineering, the water status of the plant tissues is known to have a drastic influence on their mechanical and rheological properties. In this field, improved knowledge is needed to maintain product quality and tissue firmness during harvesting, handling, transport and storage of processed foods like bread or fruits and vegetables. The long-term storage of tubers like potatoes[47–49] and the drying of vegetables and fruits like apples[50] or pumpkins[51] have been characterised in food technology by classical mechanical tests like compression and tensile cycles or oscillatory shear tests, at different ages or with different osmotic adjustments to control the moisture content of the plant tissue. Changes in water content lead to differences in turgor pressure at the cell scale and have a tremendous effect on macroscopic mechanical properties like elastic modulus, fracture toughness and yield strength. Models have been proposed to take into account the hierarchical cellular structure of plants for explaining the macroscopic mechanical tissue behaviour from the microscopic scale with pressurised cells.[52,53]

As for plant parenchyma, the hydric status of the root tissue, which is linked to the turgor pressure, is a key parameter to take into account when performing mechanical tests on roots. Yet its effect has often been disregarded. By testing excised segments of chick-pea radicules with a universal testing machine, we have observed that the Young's modulus $E(t)$ in compression dramatically evolves with time t after excision. For example, E increases by a factor of 35 from $E_1 = 8.0 \pm 0.3$ MPa (value obtained at $t = 13$ min after excision for the first test of compression) to $E_{10} = 285 \pm 20$ MPa after a duration of around $t = 2.5$ hour for a given root segment (of diameter $d = 1.490 \pm 0.025$ mm) compressed at a strain rate of $\dot{\varepsilon} = 3\%\,\text{min}^{-1}$ (upper curve (yellow) cross centre square in Figure 5.4A). The amplification factor on Young's modulus was only 16 for another root tested at the same strain rate but with a larger initial diameter $d = 1.860 \pm 0.025$ mm (half-full diamond (yellow) in Figure 5.4A). Compared with segment of roots exposed to air, the Young's modulus of roots placed in mannitol baths at the isotonic concentration of 0.15 M (such that there is no net flux of water between the root and the bath) did almost not evolve with time, with average values around 2 MPa. Therefore, the evolution of the Young's modulus with time is clearly linked to a drying process of the excised root in air. But the drying process for roots with the same length is more rapid for thinner roots compared with thicker roots: in proportion they have a smaller reservoir of water in the bulk volume compared with their external surface exposed to drying. Thus plotting the evolution of the Young's modulus as a function of the root's diameter $d(t)$ or equivalently as a function of the area of the root section $S(t)$ at mid-length (Figure 5.4B) is a way to rescale the data coming from segment of roots with different initial diameter, thus evolving differently with time t of drying. The collapse of curves is rather good and shows power-law decays of the form $\frac{E(t)}{E_1} \propto \left(\frac{S(t)}{S_1}\right)^{-\alpha}$, with E_1 and S_1 being the corresponding values obtained for the first compression test.

This behaviour is reminiscent of that observed in solid foams or engineering cellular solids.[54] These models have also been adapted for some plant tissues, like parenchyma, where a power-law relation exists between the macroscopic Young's modulus E of the plant tissue, the microscopic Young's modulus E_S of the solid part (cellular walls) and the solid fraction (calculated without the protoplasm filling the cells), *i.e.*

$$\frac{E}{E_S} \approx \phi^n \tag{5.4}$$

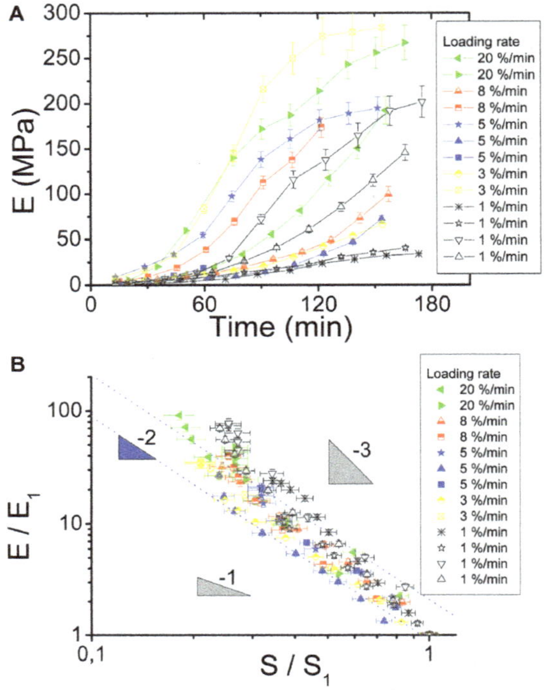

Figure 5.4 Evolution of the Young's modulus E in compression in air for segments of roots subjected to a series of k loading–unloading cycles (from $k=1$ to $k=10$) separated by rest periods during which the root is drying. The colours correspond to different loading rates. One root experiencing a loading rate of 5% min^{-1} broke between cycles $k=4$ and $k=5$. (A) E as a function of time t. (B) rescaling of the Young's modulus E as a function of the root section S for the different experiments in log-log scales. The Young's modulus $E(t)$, as well as the section $S(t)$ which depend on time have been normalised by their corresponding values E_1 and S_1 at time $t=t_1$ (values at the first cycle). The two dotted (blue) lines are guides for the eye and correspond to power-laws with exponent -2. For comparison, the slopes of -1 and -3 are also represented by the slopes of the grey triangles. Image courtesy of E. Kolb.

Depending on the turgor pressure, different exponents $n>0$[55] are observed.

In the case of root drying, the solid fraction of the root tissue increases with time due to water losses of the protoplasm (plant cell without the cell walls), which leads to the solid part being radially shrunk into a smaller section. Thus the solid fraction $\phi(t)$ of the cell walls in the tissue increases with time t. According to eqn (5.4), the cellular foam model might explain the increase of the Young's modulus $E(t)$ with time of drying, at least phenomenologically. Thus $E(t)$ might progressively increase from the Young's modulus of the tissue under full turgor $E_0=E(t\rightarrow 0)$ until the one $E_S=E(t\rightarrow \infty)$ of the solid part (cell wall) only.

Finally, these results might provide a possible alternative or complementary explanation for the observed decaying power relationship relating

Young's modulus (and also ultimate strength) to root diameter in soil science tests. Roots of smaller diameter dry more rapidly, and consequently their Young's moduli increase faster with time. When testing roots of different diameters, this drying could have a drastic effect on the mechanical results: the cutting and fixation of roots in a mechanical setup require some incompressible time delay during which the roots dry if directly exposed to air. If the roots initially have an identical turgor pressure, thin roots tested after a given time delay will show a higher Young's modulus than thicker ones. Therefore, the dependency of the Young's modulus with diameter might be in certain cases an artefact that comes from different drying rates. This approach might rationalise some mechanical data obtained in soil science and explain a part of the huge variability of Young's modulus and strengths of roots or shoots reported in the literature. In some cases, this variability could be due to an evolution of the root moisture content, which is very sensitive to air drying, even for small time exposures (a few minutes).

Note that the same results have been obtained by Boldrin *et al.*[56] in tensile tests on woody roots during progressive and controlled dehydration moisture equilibration in soil with contrasting water contents. The authors also concluded that root water status is a factor that can cause (inappropriately) high strength values and the large variability reported in literature for thin roots. They recommended testing all root diameter classes with consistent moisture for fair comparison.

An important final point to mention is that mechanical stimulation is known to affect plant growth and material properties by changing the development of shoots and roots.[57] This response is termed "thigmomorphogenesis" and is observed for trees[58,59] as well as for herbaceous plants. For instance, in the work of Goodman and Ennos,[60] the roots of flexed shoots of sunflowers and maize plants were not only thicker and more numerous than those of plants that received no mechanical stimulation, but their own mechanical properties were also affected by external mechanical stimulation: the flexed primary nodal roots of maize were stiffer and stronger. These types of studies show that roots can respond to changes in soil strength[61] and therefore one can consider reducing plant lodging by manipulating soil mechanical properties.[60]

5.2.3 How do the Mechanical Stresses of the Soil Affect Root Growth?

While growing in the soil, a root tip experiences a variety of biological and physico-chemical cues from its close environment. Root growth and morphology are affected by these spatial and temporal fluctuations of soil properties. This root response is called "phenotypic plasticity".

Among the soil physical properties, compaction is known to have a drastic effect on root growth. As a direct consequence of compaction, root penetration is mechanically restricted. The compacted state of a soil is measured by mini penetrometers, with a probe diameter ranging from 0.1 to 10 mm depending on experiment (laboratory or field). Usually the penetrometer consists of a metal probe with a conical tip attached to a cylindrical shaft of smaller diameter to avoid lateral friction. The force required to push this metal cone into the soil at a constant velocity (around 1 mm/min) divided by its cross-sectional area gives the mechanical strength of the soil (in MPa), or "mechanical impedance", as it is known in the soil science community. Compared with non-impeding soils, root elongation rate is typically halved in repacked soils, with penetrometer resistances from 0.8 MPa (for cotton) to 2 MPa (for maize and peanut) in the absence of water deficit.[62] So increased soil strength induces a decrease of the axial growth rate, resulting in a decrease of the individual root length and of the whole root system size.[61]

To model the growth biomechanics of a root in a resisting soil characterised by a mechanical strength σ, Greacen and Oh (1972)[63] suggested that the elongation rate of an impeded root can be described by a modified form of the Lockhart equation: in the axial direction, the soil resistance acts against the effective pressure $P-Y$ needed for axial growth and pushing. If water is not limiting cell expansion, the root growth velocity (and following, the RGR) is assumed to be proportional to $(P-Y-\sigma)$. Thus a modified version of the Lockhart equation was proposed in the following form (which may be subject to discussion depending on the definition of σ):[64]

$$\frac{1}{L}\cdot\frac{\mathrm{d}L}{\mathrm{d}t}=\Phi\cdot(P-Y-\sigma) \tag{5.5}$$

with the same notations as in eqn (5.1) or eqn (5.3): P is the cell turgor pressure, Y is the minimum pressure threshold depending on cell wall biochemistry, t is the time. In eqn (5.5), L is simply defined as the length of the "elongating tissue under consideration" and Φ is the corresponding extensibility. The term $Y+\sigma$ is considered as a new effective threshold for growth. Note that the frictional stresses at the root flanks are not explicitly taken into account in eqn (5.5) where only an axial stress at the tip is considered.[65]

In a strong soil, the root is completely impeded when the growth stops. According to the modified Lockhart eqn (5.5), the condition $\frac{\mathrm{d}L}{\mathrm{d}t}=0$ gives the maximum soil strength which roots can overcome:

$$\sigma_{\mathrm{Max}}=P-Y \tag{5.6}$$

Due to the quasi-static conditions of growth this value can be viewed as the maximum axial stress a root is able to exert on a soil. Thus the cell turgor pressure P (of the order of 0.5 MPa) gives the right order of magnitude of the "maximum growth pressure" σ_{Max} a root is able to exert in a soil.[19]

Laboratory model experiments provide a direct way to measure the axial growth force generated by an impeded root while growing against a force sensor. With force techniques such as shear beam apparatus, force transducer, dead-load system or digital balance,[66] the axial force was observed to increase gradually with time until a plateau was reached after a typical duration of 10 to 24 hours. Then the maximum axial force F_{Max} the root exerts is divided by its cross-sectional area to obtain the maximum growth pressure σ_{Max}. For instance, for primary roots of pea[66] with a root diameter d of the order of 1.5 mm, the maximum force was around 1 N and the corresponding $\sigma_{Max} = \dfrac{F_{Max}}{(\pi/4) \cdot d^2}$ was of the order of $\sigma_{Max} \approx 0.5$ MPa. Note that this huge value of 5 atmospheres is completely consistent with the equation relating σ_{Max} to the turgor pressure P (eqn (5.6)).

The precise value of σ_{Max} depends on the species and on the apparatus, but also varies with the determination of the root diameter d, as $\sigma_{Max} \propto \dfrac{1}{d^2}$. It should be stated how the root diameter was measured, either *in situ* or after having extracted the root from the impeding system, because the latter method might lead to tissue relaxation and/or root drying. In addition, when a root is subjected to high axial stress or long-term stress, there is often a bulging of the root axis[67] with an increase in diameter that appears to occur in the elongation zone.[68] Therefore, the exact location along the root axis and the time at which root's diameter was measured must be accurately considered when comparing σ_{Max} data.

Despite these possible discrepancies, due mainly to uncertainties in diameter measurements, the σ_{Max} values reported by successive reviews are always in the range of 0.1–1 MPa, regardless of species.[65,66,69] For example, in the work of Clark *et al.*[70] (1999), the mean value of σ_{Max} for dicots was 0.41 MPa (pea, lupin, sunflower), while the mean for monocots (wheat, maize, barley, rice) was 0.44 MPa. Similar values were also obtained for the axial growth of lateral roots of agricultural plants. Even for tree seedlings, $\sigma_{Max} = 0.22$ MPa for the primary root axes of young tree seedlings of acacia and $\sigma_{Max} = 0.15$ to 0.25 MPa for lateral roots of 3–4-month-old eucalyptus.[71] Interestingly, the radial stress obtained for primary root growth experiencing radial constriction is also found in the same range. By using a technique of photoelasticity, we measured an average lateral stress of 0.3 MPa for chickpea roots of millimeter diameter growing in the gap between two photoelastic discs ensuring lateral constriction.[72] But radial pressures in excess of 2 MPa were also derived from indirect measurements where roots were capable of breaking stiff rigid chalk in Misra *et al.*[73] (1986).

In brief, the "maximum growth pressures" of various species are consistent with each other and are close to the cell turgor pressure. However, the σ_{Max} values measured on roots in laboratory experiments are observed to be smaller than the maximum soil strength σ_{Soil} fully impeding root growth and measured by penetrometry in soils: roots are able to grow in soils of strength higher than predicted by σ_{Max}. This shift may be explained by method artefacts, or by the shape and frictional characteristics of the root tip. In particular, a more acute root tip seems to better penetrate strong soils.[74] Furthermore, the release of border cells and mucilage by the root cap is often viewed as a way to lubricate the root–soil contact[75,76] and to favour the penetration of the root inside the soil.

Another explanation comes from a time-dependent biological response of the plant root or acclimatation to the mechanical stress. Indeed the cell wall properties (Y and Φ) were shown to change in response to variations in cell water relations[77] and in response to mechanical impedance.[67] An increase in cell turgor pressure could also be an efficient way to counteract the soil mechanical resistance but there is no conclusive evidence, as it was observed that the turgor pressure P in the impeded roots was either increased[78,79] or unchanged.[67]

Besides a global increase in the bulk density, at a finer scale, compaction leads to the alteration of the soil structure and of the connectivity of the pore network with a decrease of the number of coarser pores. These changes in the pore network limit the nutrient and water transport to roots as well as the gas diffusion within the soil.[80] So in addition to the mechanical effect that physically reduces the effective pressure for cell expansion, soil compaction could affect root growth through a limited rate of oxygen supply to the root or a thigmomorphogenetic response linked to the accumulation of ethylene within the root. Root growth is known to be sensitive to hypoxia, and the similarity of root growth responses to soil compaction and hypoxia was nicely demonstrated recently.[81] Another recent research highlights that soil compaction favours ethylene production and accumulation within the root and that this hormonal signal is necessary and sufficient to restrict root growth.[80] Disentangling the mechanical and the biochemical parts of the root growth reduction will need to consider the temporal aspects of the responses more thoroughly. Following, a refined version of the Lockhart equation is better suited to account for the root thigmomorphogenetic response or accommodation to the mechanical stress σ:

$$\frac{1}{L}\cdot\frac{\mathrm{d}L}{\mathrm{d}t} = \Phi(\sigma)\cdot(P(\sigma) - Y(\sigma) - \sigma) \quad (5.7)$$

P, Φ, Y are not constants, but are physiological properties that are actively regulated and depend on the actual stress σ, on the root environment and also on the mechanical stress history of the root.[82]

In addition to axial growth modification due to a direct effect (change of the effective threshold in Lockhart equation through eqn (5.5)) or an indirect effect (change in the biomechanical parameters involved in the axial elongation, eqn (5.7)), mechanical impedance commonly causes roots to become thicker at a longer timescale.[83] Most of the diameter increase is due to cell radial enlargement in the outer cortex layer but can also be produced by an increase of the number of cell layers in the stele and in the cortex[84] (see Figure 5.1). The increase in diameter is viewed as a mechanical advantage to penetrate hard soil. According to the review of Chen *et al.* (2021)[85] on penetration processes, this radial expansion can create fractures and produce deformation-induced softening ahead of the tip in cohesive soils (*i.e.* clays) or relaxation of the soil's effective stresses ahead of the root tip in noncohesive soils (*i.e.* sands).[86,87] Moreover, roots with larger diameter might have a larger bending stiffness, which helps in stabilising the root against buckling events.[19] Buckling is a mechanical instability of elastic rods that occurs for an axial compressive force at the end that increases linearly with the rod bending stiffness ($E{\cdot}I$). E is the rod Young's modulus and I its second moment of area. Assuming that roots behave like rods and have a circular cross-section $E{\cdot}I \propto E{\cdot}d^4$, d being the root's diameter. For instance, a 60% increase in root diameter, which is observed for a heavily compacted sandy loam soil[65] will result in a 240% increase in the root's bending stiffness. Then thicker roots (either intrinsically or due to a morphological response) have a higher probability of straight penetration into the soil. Conversely, thinner roots are more prone to buckle and to reorient their growth trajectory, resulting in a more tortuous root system architecture. More recent works showed that root anatomical phenes like cortex thickness are better predictors of bending strength than root diameter *per se*.[88]

Root curvature and reorientation of the growth axis may be caused by a passive phenomenon such as buckling but could also result from an active process involving differential growth across the root section as a result of a touch-avoidance response at the root tip. In some situations, reorientation is considered a positive adaptive trait that allows the root to bypass obstacles without being blocked. More subtle effects such as circumnutation[89] (helical movement of the root apex of varying amplitude and frequency) also "help" the root to bypass obstacles.[90]

All that has been said so far concerns elongating roots (young, non-suberised): that is, primary growth. In comparison, the forces exerted by large tree roots experiencing growth in diameter (secondary growth) have not been so much investigated. Yet the costs of tree root damage on infrastructure such as sidewalks and pipes are considerable.[91,92] Most of the works in this field characterised the pavement displacements and cracks and focused on the root morphology in size, branch occurrence, or cross-sectional shape.[93] Indeed, mechanical stimulation (due, for instance, to the loading produced by wind sway or due to the restricted soil displacement imposed by the above pavement) can modify the secondary growth of stems[94] but also of roots[95,96] and causes adaptive growth[97] such as the development of an eccentric growth ring or the evolution from a circular to an I- or T-shaped root section. In rare cases only, forces produced by secondary growth of roots were measured: either indirectly by measuring *a posteriori* the induced deformation of solid foams sandwiching the roots[98] or *in situ* by using load cells on cherry tree roots[99] (Note that in this case, the authors only provided a measurement in N/m without clearly specifying the reference length. We have deduced a pressure by taking a root diameter of 10 cm). In both cases, the pressure was close to 0.4 MPa, again of the order of the turgor pressure.

5.3 The Impact of the Growing Root on the Physical Properties of Soil

So far, we have focused on the growth and mechanics of single roots growing in impeding soils. We will now turn our attention to the effects of roots on the soil. After recalling some basics of soil mechanics, we will describe the mechanical and hydric effects produced by the growing root on the surrounding soil.

5.3.1 The Mechanical Strength of Soil

Soil—the geologically deposited material below the organic humus—is a three-phase material composed of a solid phase (grains or particles), a liquid phase (most often water) and a gaseous phase (air plus water vapour). Liquid and/or gas fill the spaces between the solid particles, called "pores". Below the water table, the soil is saturated: water fills the pores and the gaseous phase is almost missing. Above the water table, in the vadose zone, all three phases are present.

Soil can be considered at various spatial scales; for each scale different physical characteristics of the soil are pertinent:

- at the bulk scale: soil density and water content
- at the scale of an assembly of particles: the closeness of the packing (local packing fraction) and the particle contact orientations resulting from the particle shape and form anisotropy
- finally, at the scale of a single particle: particle size, shape and roughness (insofar as it affects the friction coefficient between particles).

Starting from the smallest scale, soil particles can be split by size into mainly coarse cohesionless grains and fine cohesive particles. Fine cohesive particles are typically smaller than 80 μm and include silts and clays. Putting aside the physico-chemical adhesive forces between such fine particles, the mechanical interaction between soil particles can be considered, for the sake of simplicity, as essentially frictional in nature. At the local scale, friction between particles and their shape (which can cause interlocking when interparticle friction is overcome, and acts as a further barrier to overcome) are two key parameters to assess the mechanical properties of soils such as the macroscopic friction coefficient μ.

Moving onto particle assemblies, for a given soil (*i.e.* for a given grain size distribution), particle packing fraction (dry density) is a key parameter affecting the soil's mechanical properties. Thereafter, the distribution of particle orientations and interparticle contacts (soil anisotropy) can play an important role.

When sheared (*i.e.* made to undergo a non-isotropic deformation) a soil can change in density—if the soil is initially dense then it will be loosened by shearing, and *vice versa*: if loose, it can densify. There exists an intermediate value of density ("critical density") at, or after which, there is no change in volume when shearing. A kinematically constrained dense sand (for instance under isochoric conditions such as an undrained saturated sand) is very hard to deform as it cannot dilate.

Since the soil can be thought of as particles interacting frictionally, the friction force between particles needs to be overcome when the soil is deformed. In the simplest model, the interparticle friction force depends linearly on the interparticle normal force. The latter depends positively on the external stresses applied on the soil volume. This means that the more highly stressed or deeper a soil is, the "stronger" (*i.e.* harder to deform) it becomes.

Thinking now about the effect of water, in the vadose zone, the risen water above the water table is explained by capillary rise—the same effect that sucks coffee into the suspended sugar cube in contact with the surface. This is due to the geometry of the pores (although chemical effects can also cause this) and is a stronger effect for smaller pores. This "suspended" water above the water table has a pressure lower than atmospheric pressure. The corresponding negative difference of pressure will be referred to as water pressure later on. These negative water pressures suck soil particles together; thus suction increases the normal forces between particles and consequently the frictional resistance between particles. It is for this reason that sand castles can be built with vertical walls with partially wet sand. Study of this "water retention" phenomenon, for a given material, reveals that suction increases with soil drying. This suction can be particularly strong for fine soils and is still present even at rather low water content. A positive (hydrostatic) water pressure in saturated soils takes away some strength in the same way that a negative water pressure (the suction) increases strength. The result is the fundamental concept of "effective stress" σ' carried by the connected soil particles, which drives deformations of the solid skeleton.

Finally moving back to the bulk scale, and summarising, the resistance to shearing of a given soil (defined previously as "strength") can be roughly depicted as: soil strength depends on particle friction and shape, soil density and effective stress. The classical Mohr–Coulomb failure criterion includes the effective stress dependent strength $\mu \cdot \sigma'$ described above, plus a non-stress dependent strength-cohesion c, resulting directly from physico-chemical forces between fine soil particles:

$$\tau = \mu \cdot \sigma' + c \tag{5.8}$$

where τ is the shear stress at failure.

A complete introduction to soil mechanics can be found in Muir Wood.[100]

5.3.2 How Does Root Growth Affect Soil at the Particle Scale?

As plants grow, and consequently roots develop in the soil searching for water, nutrients and anchorage, root volume must increase, which in turn must affect the soil. This complex mechanical interaction is difficult to study for a number of reasons, but principally because soil is not transparent (although some experimentalists use synthetic transparent soils[101]) and the root system is a fine, networked structure, growing within it.

A key and early paper on the subject of root soil interaction is that of Dexter[102] (1987), which presents a mathematical model for the compression of soil around roots, assuming the root volume increase is balanced by a loss of pore volume in the surrounding soil. The paper clearly states the fact that the necessary soil parameters are difficult to measure. The model presented is that of an exponential fall-off of densification with radial distance from the root.

Although there are many standard ways of testing soils, and of analysing root systems, almost all are either not fully representative (laboratory test on an elementary volume of rooted soil) or too destructive (pulling up a root system, or epoxying and slicing a rooted specimen) to be of use to study the root's effect on the particle-scale organisation of a real soil. Studying a root growing in soil against a glass wall is a popular technique.[103] Although this technique yields some interesting observations, how well these represent the interactions deep in the soil is hard to know—the glass–soil interface is a singularity for the root and the soil is imposed to deform in plane strain condition (which is far from the 3D field case).

Less intrusive imaging techniques are beginning to show promise for the observation of roots and soil interaction at the particle scale.[104] Most notably are 3D imaging techniques that use penetrating radiation to see through samples and reconstruct a 3D field of attenuation. This means that a given state of the soil–root system can be measured at a given point in time, and subsequent measurement can be used to monitor evolutions. With the advent of microtomography, X-ray Computed Tomography (CT) is now routinely capable of imaging a granular system composed of grains of 100 μm in size at a resolution sufficient to identify individual grains, which provides an ideal way to directly measure root–soil interaction at the particle scale (Figure 5.5). It is worth noting that repeated CT scans may affect root growth, especially when plants are exposed to X-rays for a longer time, and this needs to be considered when interpreting the measurements.

X-ray tomography measures a 3D field of a "CT-value" roughly proportional to the material density and (as usual in imaging) there is a trade-off between field of view and spatial resolution, meaning that either a small volume can be studied with a lot of detail or a larger volume with less detail. Current capabilities are resolutions down to 1 μm and samples as large as 50 cm in diameter (when using a medical scanner). A specific challenge in moist soil is that the X-ray attenuation of water and of the young root can be quite similar, although there are methods to circumvent this identification problem.

In the study of geomaterial deformations, this tool has yielded extremely interesting results of local, 3D strain fields accompanying strain localisation upon mechanical loading.

The particle-scale interactions of a growing root and a soil can be classified into three regimes of interaction depending on the relative size of root diameter close to the tip and the size of the pores between the soil particles,

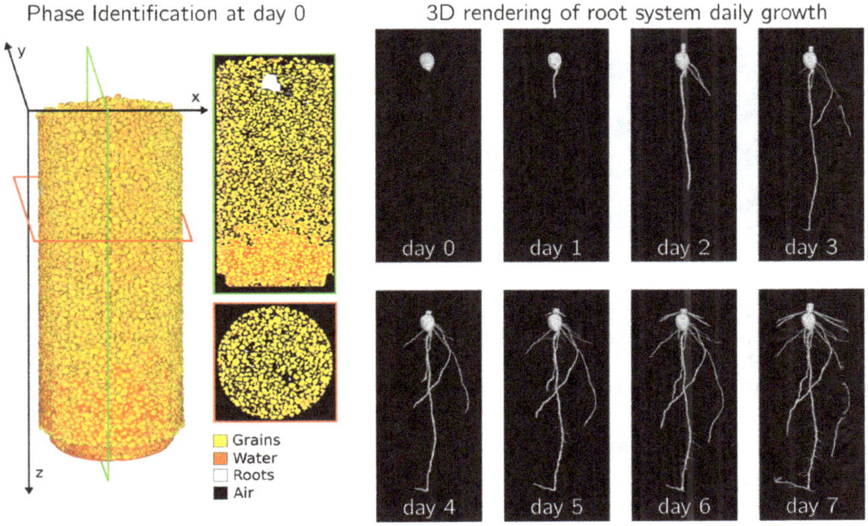

Figure 5.5 3D rendering of maize root growth in Hostun sand with colours corresponding to the identified phases in the system (solid, water, air and roots) at a given time on the left; extraction of the root phase and time evolution of the maize root system 7 days after the seeding in the sand sample (resolution 40 μm per pixel), adapted from ref. 125.

along the lines of Bodner *et al.*[105] (2014) and Lucas *et al.*[106] (2019). Figure 5.6 presents the three different regimes—on the left the "soil displacement" regime occurs when the root size is much larger than the pore size. In this case (in a homogeneous clay, for example) the root must displace soil particles to be able to advance. The weight and friction of the soil must necessarily be overcome, but as is common in these soils, some cohesion due to suction is to be expected too. At the other extreme (Figure 5.6, right) the "pore invasion" regime is possible if the diameter of the growing root is much smaller than the pore sizes between the grains in the growing medium (coarse sand or gravel, for example); then the root can grow into the pores without having to move any grains. Between these two regimes, in the "intermediate" regime (Figure 5.6, middle) where roots, grains and pores are all of the same characteristic size, displacement of the soil is still needed for the root to advance.

Using this as a guide, the relevant literature can be seen in this light. Referring back to Dexter[102] (1987), the model of soil densification implies implicitly a sufficiently loose soil that will densify upon shearing, but the soil is not necessarily in such a state. Besides, there is very little imaging work where the soil is homogeneous (at the scale observed), which might correspond to the soil displacement regime, which appears to be a gap needing to be filled to understand root penetration into homogeneous clays. Although particles cannot be imaged if the field of view should meaningfully

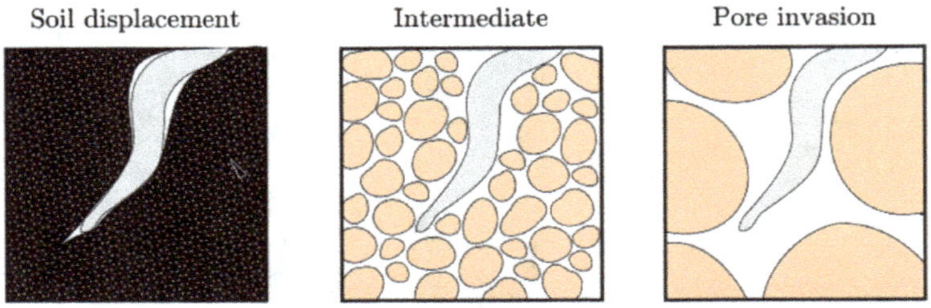

Figure 5.6 Illustration of the three regimes of interaction of a growing root with soil. From left to right, the relative size of root diameter compared with pores between particles decreases: when soil particles are much smaller than the root (left) they will be displaced; when the root is much smaller than the particle and pores (right), the root will initially invade pores.

contain the root, changes in soil density (when corrected for water migration) should be measurable in the CT value.

When considering the "pore invasion" regime, few published works with imaging situate themselves in this case regarding root soil interaction—since this is quite an unlikely topsoil—however, there is some important work worth mentioning here concerning fibre-reinforced soils. Work from the University of Bristol[107,108] and further work with X-ray imaging[109] reveal that the presence of small and short pore-filling fibres, randomly oriented and only a few percent by volume, can increase the shear strength of sand. Significantly, the presence of these fibres makes sand behave as if it is denser, the fibres effectively "stealing" some void space from the sand. The jump from the system studied to a root system is not to be underestimated, but this presents an important bridge for the study of root–soil interaction.

Most work can be classified in the intermediate regime, which is doubtlessly the most attractive for tomographic imaging. Pores can be individually measured with imaging at this scale, which is of importance since many studies are principally interested in changes of porosity and the resulting changes in permeability. Since the effect of a root pushing soil particles cannot be isotropic, some shear deformation is expected. This shear produces either increase of soil volume (dilation) or decrease of soil volume (compaction) depending on the initial density of the soil. The supplementary materials coming with Keyes *et al.*[110] (2017) give an insight about this effect. More recently, Anselmucci *et al.*[111] (2021) measured both shear- and volumetric-induced deformations and obtained values ranging from a few percent or less, at few millimetres away from the root surface, up to almost 10% at the direct vicinity of the root. By considering the very simple model for soil behaviour outlined in the previous section, much of the seemingly contradictory soil behaviour that is discussed in literature may be clarified.

For instance, it is stated in the recent work of Lucas *et al.*[106] (2019), that "Contradictory evidence exists regarding whether and to which extend [sic] roots change soil structure in their vicinity". Nevertheless, the authors conclude their study by saying that this can be reconciled by integrating the mechanics of root growth and plant interaction with the existing soil structure.

This is a key point, but in our opinion it could be taken further. This can be done by completing the picture with a thorough description of the behaviour of the soil itself (in the form of a suitable model), which is an additional key information to explain and predict the mechanical effects of the disturbance by the root.

Helliwell *et al.*[112] (2017) use X-ray tomography to study changes of soil porosity due to root growth on sandy and clay soils systematically, revealing a small zone of increased porosity close to the root—an expected effect of geometry in the form of "steric exclusions", reflecting the fact that grains can be packed less densely against a locally flat object such as a root. Further afield from the root, porosity may still either increase or decrease, depending on the soil nature, according to the authors, but it may well be explained by the initial soil density. A very complete study from these authors[113] varying soil type and density, as well as plant type, systematically shows a local increase in porosity next to the root, and sometimes a densification away from it. Koebernick *et al.*[114] (2019) use synchrotron X-ray imaging on soils finer than usual and present a model for porosity around to a growing root including the effect of steric exclusions.

Going forwards, it is worth keeping in mind the possibility of more complex interactions between soil and a growing root than only the porosity modification induced directly *via* mechanical interaction. Beyond the porosity, the pore size distribution may be affected.[105] Besides, the constitutive hydro-mechanical coupling in partially saturated soils (as discussed previously in terms of suction) may also contribute to additional soil deformation of crack opening in the root vicinity due to the drying of soil by the root's water uptake.[113] The role played by root hairs is also poorly investigated because it concerns a low scale hardly reachable by direct imaging. Root hairs seems to contribute to the soil deformation induced *via* the hydro-mechanical coupling;[115] but could the root hairs rearrange soil particles directly, at least for fine soils (as per Champion & Barley[116])?

X-ray tomography has proven to be an important and valuable tool for the direct observation of root soil interaction with minimum disturbance. Although the focus on porosity with the aim of revealing changes in permeability is laudable in the current studies, this paints only a limited picture of the key parameters relevant to the mechanics of the soil, ignoring soil microstructure, as well as subtle effects such as small-scale particle segregation.

Further input from soil mechanics in the interpretation of soil deformation behaviour could be around the description of deformation itself as a tensor, with invariants that merit study in themselves (roughly volumetric strain—well studied and "shear" strain—unfortunately often missed—but not in Keyes *et al.* (2017)[110] or Anselmucci *et al.* (2021),[111] for example). Another interesting area probably not yet studied for practical reasons is the post-root-penetration stage, where thickening and stiffening of the root happens, with possibly different effects on the soil.

There is a bright future for the meeting of soil mechanics and plant physics around these imaging tools, and among others the effect that the soil state (grain size, density, microstructure) has on the root system that eventually develops.

5.3.3 How Does Root Mucilage Affect the Hydric Properties of the Soil?

Root growth is facilitated by the secretion of mucilage from the root cap.[117] Mucilage is a polymeric substance secreted at the root tip that is believed to act as a lubricant facilitating root growth in dry, hard soils. However, mucilage has other functions, which have just started to be understood and which go beyond root growth. These functions are crucial to understand the physical properties of the soil near the root, called the rhizosphere. In this section we review the physical interactions between mucilage and the soil and their implications for root and rhizosphere processes.

Mucilage is a gel mainly composed of polysaccharides, but also containing a small fraction of lipids.[118] It has a large water adsorption capacity (it can hold an amount of water a hundred times greater than its dry weight[119]), and a significant fraction of this water can be retained as the soil dries.[120] As a consequence, mucilage increases the capacity of the rhizosphere to retain water. Besides simply increasing the volume of water in the rhizosphere, mucilage also changes the spatial configuration of the liquid phase. Mucilage decreases the surface tension and increases the viscosity of the soil solution.[120,121] These properties avoid the break-up of liquid bridges, connecting soil particles due to capillary forces.

An example of mucilage distribution in a sandy soil during drying is shown in Figure 5.7. The image was obtained using a light transmission microscope after drying of mucilage from chia seed in a sandy soil. The picture shows long filaments connecting distant grains as well as large bridges between grains in contact with each other. Note that the bridges are hollow, indicating that the polymers were deposited while the capillary bridges between soil particles were still large. Further drying caused water cavitation inside the hollow cylinders.

A theory of the physics of the deposition of polymer solution in porous media was proposed in Carminati *et al.* (2017)[121] and Benard *et al.* (2019).[120] Compared to water in porous media, where capillary forces (together with adsorption in clay soils) are the dominant forces for describing the liquid configuration,[122] in polymer solutions viscosity has to be taken into consideration. As the soil dries, the concentration of the polymer solution increases, the surface tension decreases and the viscosity increases. The thinning rate of a viscous liquid bridge between two soil particles is proportional to the ratio between surface tension and viscosity.[123] As during drying this number becomes increasingly smaller, long-living filaments can be found after drying of mucilage in soils. Additionally, drying of mucilage in porous media results in the formation of two-dimensional surfaces, such as

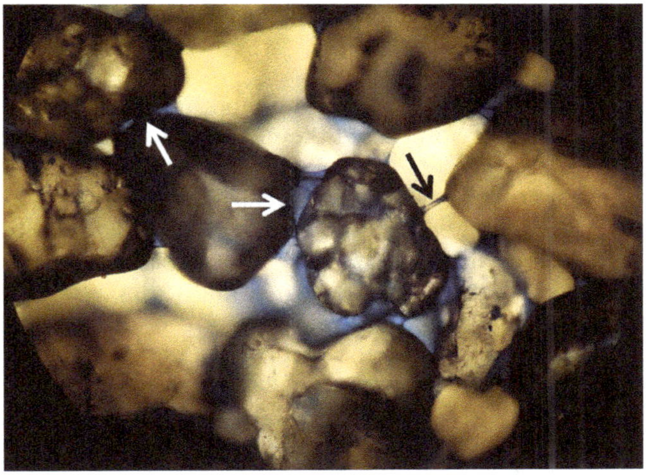

Figure 5.7 Chia seed mucilage deposited in a sandy soil upon drying. Thin bridges (black arrow) connecting distant particles, and thicker bridges (white arrows) between particles in touch with each other are visible. The sand grains have a diameter of 100–200 μm. The picture was obtained with a light transmission image. The method is described in ref. 120. Image courtesy of P. Benard.

hollow cylinders up to more complex interconnected surfaces.[120] These surfaces form when the viscosity of the polymer solution impedes the retreat of the polymers together with water. This critical point depends on viscous forces (which impede the retreat of the polymers) and water adsorption, which pulls water toward the regions with higher polymer concentrations. When water adsorption is no longer sufficient to overcome the viscous forces, the polymers are deposited and their concentration rapidly increases. The deposited surfaces act as a new solid matrix.

The consequences of these processes are manifold. The connection of the liquid phase is preserved during soil drying. This means that the root surface remains in physical contact with the soil matrix during soil drying. This is fundamental for the uptake of water and nutrients by roots during drought spells,[120,124] as well as for the activity of microorganisms in the rhizosphere.

Additionally, the formation of viscous bridges between soil particles also has the effect of stabilising the porous medium. This is visible when excavating the roots out of the soil. Typically, one observes a sheath of soil particles adhering to the root surface. This layer of particles, including root hairs, root exudates and microorganisms, is referred to as rhizosheath. The porosity of the rhizosheath is expected to be impacted by the presence of mucilage, which (1) attenuates the compacting effect of capillary forces and (2) limits the capacity of soil particles to be displaced during drying. In summary, mucilage is a key physical element used by plant roots to engineer the soil microenvironment.

5.4 The Complex Interplay of the Root System and Soil

In the previous section we focused on the modification at the root scale of the physical properties of soil due to the presence of a single root. We now turn to a global scale to assess how soils are altered by roots.

In an engineering context, soil is treated as a continuum material despite consisting of different materials (soil particles, water, air, organics, *etc.*) in various phases (solids, liquids, gases). This approach is continued when extended to rooted soil; any local effects near individual roots are "smeared out" over the continuum, and the engineering properties of the "root-reinforced soil" as a single, homogenous material are investigated.

These properties are changed by roots in several interconnected ways. Firstly, roots alter the structure of the soil by filling pores or dislocating particles during the root growth process (see Figure 5.6). The mechanical behaviour of soil (*e.g.* strength, stiffness, volume change during deformation) is affected by changes in soil structure and porosity, as described in Section 5.3.2 and observed by use of X-ray CT.[125] In addition, changing the pore structure alters the ability to maintain suction pressures (see Section 5.3.1), resulting in changes to the soil–water retention curve[126] and increases the saturated hydrological permeability.[127] The effect of the latter on the initiation of landslides is contested: on the one hand, increasing the permeability leads to water reaching potential failure planes earlier.[128] The increased pore water pressure reduces the soil strength and increases the likelihood of soil failure. On the other hand, the increased lateral drainage may channel more water downslope before the water is able to percolate to potential failure planes,[129] therefore increasing slope stability.

A second pathway through which the presence of roots alters the mechanical behaviour of soil is through changes in soil water content. Evapotranspiration reduces the soil moisture content and increases suction pressure,[130] resulting in additional soil strength. This is known as hydrological root-reinforcement. Increased levels of suction may increase soil shrinkage and soil cracking,[15,131] which may be undesirable, for example near infrastructure such as railway lines.[132,133] Different plant species will have different suction-increasing characteristics,[130,134] so careful species selection is important.

A third pathway through which roots alter the mechanical behaviour of soil is through adding mechanical reinforcement. Roots have their own mechanical stress–strain (constitutive) behaviour, which is very different from that of soil. While soils are generally loaded in compression or shear and are weak in tension due to their particulate nature, roots have significant tensile strengths. Together, roots and soil form a composite material, the behaviour of which must be a combination of the material behaviours of both and therefore be different from that of unrooted soil.

Considerable debate has been given to whether hydrological or mechanical root-reinforcement is most important for slope stability applications.

Hydrological reinforcement can be much larger but is more variable throughout the year. For slope stability applications in temperate regions, where landslides often coincide with cold and wet seasons with minimal evapotranspiration and near-saturated ground conditions, mechanical root-reinforcement is generally considered more important.[135]

5.4.1 Shear Strength of Rooted Soil

The ability of the soil to resist shear deformations is often key for engineering applications. Soils are generally loaded in compression due to self-weight but have high compressive strength, making failure in shear the most likely. Landslides are a typical example; here a mass of soils shears off the underlying stationary soil or rock, and so is an example of erosion. Because of these applications, historically research has focused on finding the increase in the peak shear strength of rooted soil. Starting in the late 1970s, this has often been quantified as an additional cohesion term in the Mohr–Coulomb failure criterion (defined in eqn (5.8)) often used for soils.[136,137]

Small-diameter roots reinforce the soil through mobilising tensile resistance.[135] When roots are sufficiently anchored, either through soil–root interface friction or lateral branching,[138] their full tensile capacity may be reached and the roots may break. When anchorage is insufficient, the roots will slip ("pull-out failure"). Thicker roots can mobilise significant resistance through bending or shear effects and may reinforce a slope in a way similar to soil nails, a common slope stabilisation technique used in civil engineering.[15] While these mechanisms are widely recognised, no clear guidelines are available to help differentiate between "thin" and "thick" roots based on a simple criterion of root diameter. Rather a distinction between absorptive and transport fine roots might be useful for fair comparison of mechanical properties (see Section 5.2.2).

While soil typically requires small strains (shear strains in the order of percentages) to mobilise peak resistance, roots are much more flexible with tensile failure strains in the order of 15–20%. Experiments have shown that significantly larger shear displacements are required to mobilise the full capacity, much larger than those in unrooted soil.[139–141] While this does not pose much of a problem when analysing the stability of natural slope or erosion problems (where deformation does not matter; only peak reinforcement does), it may be more important when vegetation is used for stabilising road or railway embankments or cuttings that may come with strict deformation tolerances.

5.4.2 Laboratory Measurement of Root-reinforcement

Measuring the mechanical properties of rooted soil is a major challenge. Using standard soil mechanics laboratory tests to measure the stress–strain behaviour of rooted soil, such as the direct simple shear test or the triaxial test, comes with two major limitations. Test samples are relatively small, making them more prone to sample disturbance, which is a major concern in rooted soil because of the need to sever roots during sampling. Small samples, furthermore, may lack sufficient root length for realistic root anchorage within the soil. Secondly and more importantly, standard testing equipment can apply only limited strains. While these are sufficient to get most soils beyond peak resistance and towards the critical state, at this point strains in roots are still relatively small and peak rooted soil strength may not be reached.

Because of these limitations, the most commonly used test for measuring the strength of rooted soil is the direct shear test (or "shear box") (Figure 5.8), in which one half of a sample is pulled laterally, forcing a failure plane to localise near the interface between the moving and stationary halves. This test is easily adaptable for tall samples (*e.g.* 500 mm[142,143]), allowing tests with more realistic root lengths. Displacements may be large, allowing the full tensile capacity of roots to be reached. This test gives results only about displacement and forces applied to the outside of the sample, while (internal) stresses and strains (which tend to localise near the failure plane) remain unknown. Recent X-ray testing on rooted direct shear samples shows that these deformation patterns may be complicated and different from those in unrooted soil.[143] This makes this test less suitable for characterising the soil constitutive behaviour, but it remains useful for testing root-reinforcement in discrete sliding problems such as landslides.

Root architecture (*e.g.* root tortuosity, orientations, branching) has been recognised as affecting mechanical reinforcement. It is, however, hugely variable and difficult to control in laboratory conditions, making studying its effect in isolation challenging. Root analogues can be used instead (see Figure 5.9 for an example), allowing the use of reproducible architectures[144,145] at the expense

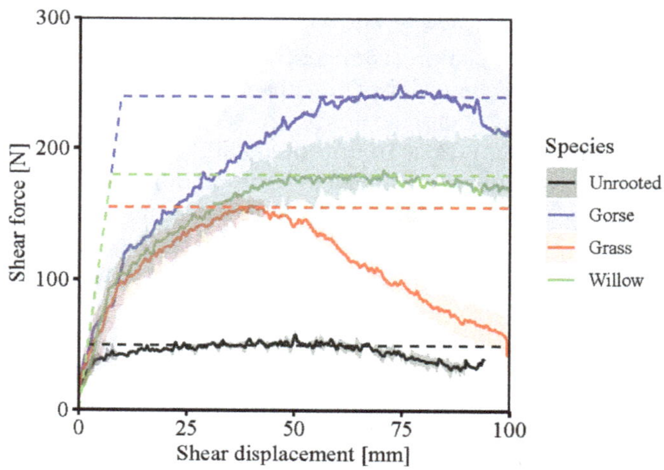

Figure 5.8 Shear box tests of unroooted soils and rooted soils with different species. Image courtesy of G. J. Meijer.

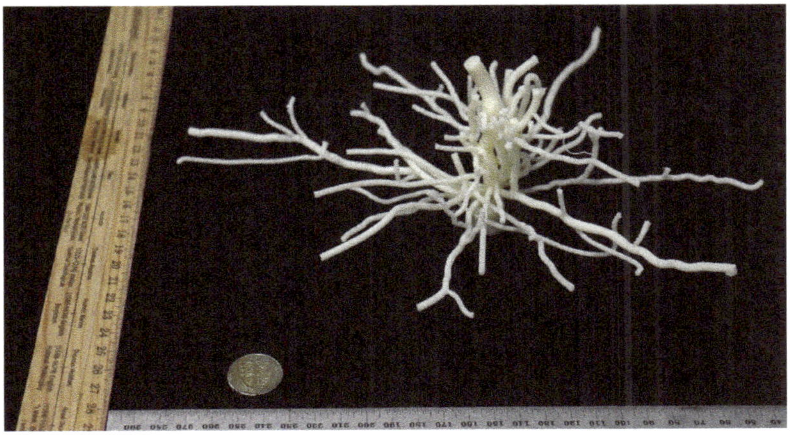

Figure 5.9 3D-printed scale model of a tree root system. The root system is shown upside down and printed from Acrylonitrile Butadiene Styrene plastic. Image courtesy of X. Y. Zhang and J. A. Knappett, University of Dundee.

of accuracy in root mechanical properties. The advance of 3D printing has enabled the printing of more complicated architectures, allowing for the study of entire root systems, for example of scale models of trees,[146] and has enabled the production of root analogues that simultaneously reinforce through mechanical and hydrological mechanisms.[147]

Methods to study changing soil behaviour due to an increase in matric suction are well established. Typically, this is measured using a triaxial test, in which the air and water pressure within the pores of the soil sample can be independently controlled.

The strength and stiffness of soil is strongly dependent on the magnitude of mechanical stress in the soil. A geotechnical centrifuge can be used to test scale models. By increasing the gravity, realistic soil stresses are achieved in the scale model. This methodology has recently been used to investigate the stability of vegetated slopes.[148–150] Careful selection of the centrifuge acceleration is required to find a good balance between accurate scaling of root diameters, strengths and growth depth.[140]

5.4.3 Field Measurement of Root-reinforcement

Measuring root-reinforcement in field conditions is equally challenging. The most common method to test the strength of rooted soil is again the direct shear test, which has been successfully used for decades.[137,139,151] The field version is generally larger than its laboratory equivalent, with square shear areas typically around 300-500 mm wide. A block of soil is excavated around the sides and encased in a four-sided rectangular frame, which is translated laterally while measuring the resistance. Additional overburden may be placed on top to maintain realistic confining pressures. To isolate the effect of roots, measurements in similar but unrooted soil are required.

A second commonly used approach is to collect measurements of both the root quantity and root strength to predict the additional reinforcement using predictive models. The number of roots and their diameters can be measured either in the field by counting root intersections with a vertical plane ("trench wall" method[17,152]) or by removing them from soil samples in the lab by means of washing with water.[153] As mentioned in Section 5.2.2, the root strength is commonly measured in the lab using uniaxial tensile tests on individual roots.[154] Alternatively, field pull-out testing can be used, in which individual roots are pulled out of the soil while measuring the force required.[155] This will lead to more conservative strength values since roots may slip out rather than break.

Recently, new experimental techniques have been trialled to develop more portable and faster tests that can be used to characterise rooted soils on sites with limited accessibility or when many tests are required due to spatial or temporal variability. The "corkscrew test" (Figure 5.10A) limits soil disturbance by rotational insertion. Subsequent upwards translational

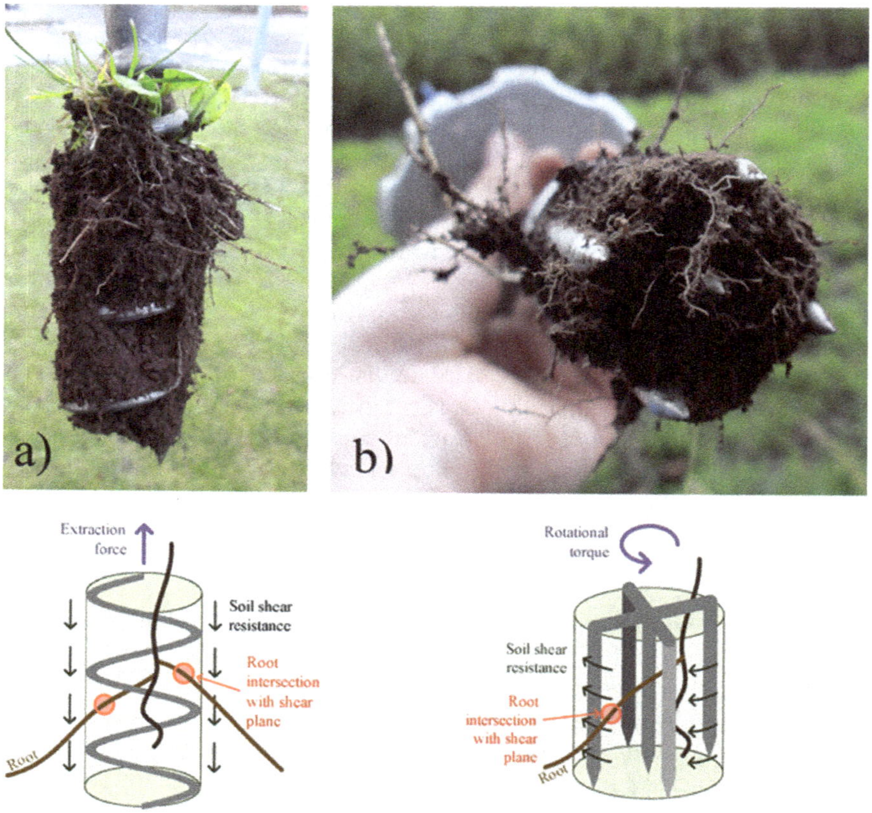

Figure 5.10 (a) "Corkscrew" and (b) "pin vane" device to measure the shear strength of root-reinforced soil. Image courtesy of G. J. Meijer.

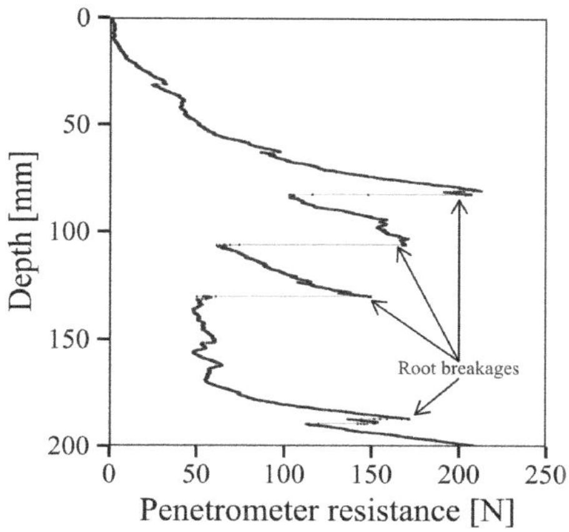

Figure 5.11 Example depth-resistance trace measured using the "blade penetrometer" in the surface soil layer of a mature Pedunculate oak plantation. More details about this site, test methodology and results can be found in ref. 157.

extraction mobilizes rooted soil strength along the outside of the soil plug caught within the helix, measured with a load cell.[141] The "pin vane test" (Figure 5.10B) measures the resistance against rotation of a soil cylinder.[156] Alternatively, the "blade penetrometer test" infers the properties of individual roots from the measured change in penetrometer resistance with depth caused by individual roots (gradual increase in resistance followed by a sudden decrease associated with root failure underneath the penetrometer tip, see Figure 5.11).[157]

The change in soil strength due to root water uptake is often quantified by measuring the penetration resistance of the soil (*e.g.*[134]). Techniques for measuring soil water contents (*e.g.* theta probes) and matric suctions (*e.g.* field tensiometer) in the field are well established and will therefore not be further discussed.

5.4.4 Root-reinforcement Modelling Techniques

Dedicated modelling of mechanical root-reinforcement models started in the 1970s, with what is now commonly referred to as the Wu/Waldron model.[136,137] The peak root-reinforcement, expressed as an additional soil cohesion, is calculated using the sum of the tensile strength of all roots crossing a soil plane and a simple correction factor accounting for the effects of root orientation, often set to 1.2. Its assumption that all roots break simultaneously has often been criticised, resulting in the advance of Fibre Bundle Models,[158] which account for sequential root breakage and therefore predict lower but more realistic reinforcements. This approach has since been refined into the Root Bundle Model, making sequential mobilisation a function of root stiffness[159] and, in some versions, a statistical distribution of root strength.[160] A significant amount of work has tried to use such an approach to find the most realistic Wu/Waldron correction factor, now often thought to be much nearer 0.4.

Root-reinforcement by individual, thicker roots requires accounting for additional bending and shear resistances. Such effects have been investigated using Discrete Elements Methods[161] or Finite Element Methods by incorporating roots as discrete elements in the mesh.[162,163] Alternatively, such roots can be modelled using spring-supported beam bending theory, commonly used in the field of structural mechanics.[164–166]

Recently, some research has been directed towards modelling the full stress–strain behaviour of homogenous, rooted soil, by developing appropriate constitutive models. One can either develop a single constitutive relation[167,168] or use constitutive modelling techniques developed for soil–fibre composites. In the latter approach, the individual constitutive behaviours of both roots and soil are combined based on their volumetric fractions.[169]

5.5 Concluding Remarks

The many interactions between soil and roots and their effects on the mechanical behaviour of rooted soil are complex. The large number and diversity of recent publications in this field show how this field of research lies at the crossroads of different disciplines.

In the context of climate change and food security, breeding programs on the most worldwide food crops (such as wheat, maize or soybean) federate research teams working in physiology, soil science or plant genetics.[170] Bad rooting due to hard layers or compacted soil limits the access of roots to water and nutrient resources. As plants have to invest more carbon in root growth, the amount of carbohydrates available to the above-ground parts is reduced and the crop yield is lower, leading to significant economic impacts. Adapting root systems to specific soil conditions and identifying the root traits enabling for efficient growth at minimal carbon costs could improve the crop rooting in the extending compacted arable lands.[171] But breeding approaches would gain from cross-talks with other fundamental research fields. Although the specific and genotypic diversity could be a source of improvement,[81] understanding the root growth and morphological responses to soil physical properties at the root scale is also needed. Among others, this approach involves fields such as biomechanics, biophysics and soft matter to identify how roots respond to spatial and temporal variations in soil properties such as moisture, gas diffusion, temperature and compaction.

The observed ongoing climate change is also associated with stronger and more frequent windstorms, and the sustainability and development of forest ecosystems depend on their capacity to survive and adapt to multiple environmental cues. The stability of trees depends on the development of their root system and thus on the interactions between root growth and soil properties to resist uprooting caused by wind. But intense rainfall events, by soaking the soil, affect the root–soil complex and the anchorage of large trees.

On the other hand, the use of plant root systems in soil science and civil engineering for stabilising soils against landslides, erosion or earthquakes does not necessarily meet the same requirements as for agriculture and forest management.[172,173] Despite the potential of vegetation to be used as an engineering material, widespread application remains limited to date, as engineers require reliable material parameters to ensure soil failure does not occur. But the biologically, spatially and temporally variable nature of vegetation provides some major challenges. To facilitate wider future uptake, the behaviour of rooted soil for engineering applications needs to be better understood and uncertainty margins reduced, requiring interdisciplinary approaches.

Plant roots are also a fabulous source of inspiration for biomimetic approaches. The way the root penetrates a soil is a model of efficient digging that soft robotics aims to reproduce. In particular, during primary root growth, the portion of the root moving relative to the soil is limited to the elongation zone, which minimises friction with the soil, unlike a penetrating rod. One of the engineering challenges is to minimise the energy consumption of the burying process by optimising the shape and mechanical properties of the probe and the strategies of actuation for the penetration.[85]

Materials of the roots and more generally of the plants themselves are the subject of intense biomimetic research in material science. For example, a growing cell wall is a thin, strong and pliant extracellular layer.[35] It forms a powerful composite material, capable of extension without decohesion of the rigid cellulose microfibrils from the soft polymeric hydrated matrix. Reproducing these peculiar properties from scratch is a challenge in chemistry. The strong, non-electrostatic chemical interaction between cellulose and xyloglucan (one type of hemicellulose found in the cell wall soft matrix) is commonly reproduced for applications in the textile, food, pulp and paper industries, and even at the nanoscopic scale for "green" nanocomposite films made entirely from renewable resources of such biopolymers.[174] At another scale, the mucilage released by the root tip with its particular wetting and rheological properties forms a peculiar gel network in contact with water[175] and might be also a source of inspiration from a physico-chemical point of view. But its special properties also come from the fact that the gel is trapped in the pores between the soil particles and is a part of the rhizosphere, the small volume of soil modified by roots.

Clearly, studying the mechanical interactions between growing roots and rearrangements of soil particles is part of a wider research issue involving the spatio-temporal complexity of the rhizosphere's functioning, with physical, chemical and biological feedbacks between roots, particles, rhizodeposition and microbial activity.[176]

Acknowledgements

EK is particularly grateful to all the members of the ROSOM project through the program of Agropolis Foundation (reference ID 1202-073) (Labex Agro:

ANR-10-LABX-001-01), to the Board of Directors and the Bureau of ALERT Geomaterials and to the GDR PhyP ≪ Biophysique et biomécanique des plantes ≫ (n°2007) of the CNRS, that allowed to organise interdisciplinary meetings, courses and workshops on the root-soil system. EK particularly thanks C. Hartmann, P. Genet, P. Kurowski and R. Peralta y Fabi for starting the adventures on the roots together, and L. Dupuy and V. Legué for continuous exchanges. Thanks also to F. Danjon for providing us with some references on trees.

References

1. J. Read and A. Stokes, *Am. J. Bot.*, 2006, **93**, 1546–1565.
2. C. Coutand, J. L. Julien, B. Moulia, J. C. Mauget and D. Guitard, *J. Exp. Bot.*, 2000, **51**, 1813–1824.
3. O. Hamant and J. Traas, *New Phytol.*, 2010, **185**, 369–385.
4. V. Mirabet, P. Das, A. Boudaoud and O. Hamant, in *Annual Review of Plant Biology*, 2011, ed. S. S. Merchant, W. R. Briggs and D. Ort, vol. 62, pp. 365–385.
5. R. L. Stroshine, *Physical Properties of Agricultural Materials and Food Products*, R. Stroshine, West Lafayette, Ind., 2004.
6. D. Shah, *Mater. Des.*, 2014, **62**, 21–31.
7. B. Gardiner, P. Berry and B. Moulia, *Plant Sci.*, 2016, **245**, 94–118.
8. C. J. Stubbs, D. D. Cook and K. J. Niklas, *J. Exp. Bot.*, 2019, **70**, 3439–3451.
9. M.-F. Dignac, D. Derrien, P. Barré, S. Barot, L. Cécillon, C. Chenu, T. Chevallier, G. T. Freschet, P. Garnier, B. Guenet, M. Hedde, K. Klumpp, G. Lashermes, P.-A. Maron, N. Nunan, C. Roumet and I. Basile-Doelsch, *Agron Sustainable Dev.*, 2017, **37**, 14.
10. G. Gyssels, J. Poesen, E. Bochet and Y. Li, *Prog. Phys. Geogr.*, 2005, **29**, 189–217.
11. N. Pollen-Bankhead and A. Simon, *Geomorphology*, 2010, **116**, 353–362.
12. E. Gasser, P. Perona, L. Dorren, C. Phillips, J. Hübl and M. Schwarz, *Water*, 2020, **12**, 893.
13. B. Abernethy and I. D. Rutherfurd, *Geomorphology*, 1998, **23**, 55–75.
14. B. B. Docker and T. C. T. Hubble, *Geomorphology*, 2008, **100**, 401–418.
15. N. Coppin and I. Richards, *Use of Vegetation in Civil Engineering, {CIRIA} Book 10*, Butterworths, Kent, 1990.
16. D. H. Gray and R. B. Sotir, *Biotechnical and Soil Bioengineering Slope Stabilization, a Practical Guide for Erosion Control*, John Wiley & Sons Inc, New York, 1996.
17. C. Moos, P. Bebi, F. Graf, J. Mattli, C. Rickli and M. Schwarz, *Earth Surf. Processes Landforms*, 2016, **41**, 951–960.
18. C. W. W. Ng, A. K. Leung and J. Ni, *Plant–Soil Slope Interaction*, CRC Press, Boca Raton, 2019: Taylor & Francis, a CRC title, part of the Taylor & Francis imprint, a member of the Taylor & Francis Group, the academic division of T&F Informa, plc.

19. E. Kolb, V. Legué and M.-B. Bogeat-Triboulot, *Phys. Biol.*, 2017, **14**, 065004.
20. D. De Vos, K. Vissenberg, J. Broeckhove and G. T. S. Beemster, *PLoS Comput. Biol.*, 2014, **10**, e1003910.
21. R. O. Erickson and K. B. Sax, *Proc. Am. Philos. Soc.*, 1956, **100**, 487–498.
22. W. K. Silk, *Int. J. Plant Sci.*, 1992, **153**, S49–S58.
23. R. E. Sharp, W. K. Silk and T. C. Hsaio, *Plant Physiol.*, 1988, **87**, 50–57.
24. X. L. Yang, G. Dong, K. Palaniappan, G. H. Mi and T. I. Baskin, *Plant, Cell Environ.*, 2017, **40**, 264–276.
25. R. H. Goodwin and W. Stepka, *Am. J. Bot.*, 1945, **32**, 36–46.
26. A. Walter, H. Spies, S. Terjung, R. Kusters, N. Kirchgessner and U. Schurr, *J. Exp. Bot.*, 2002, **53**, 689–698.
27. F. Bizet, I. Hummel and M. B. Bogeat-Triboulot, *J. Exp. Bot.*, 2015, **66**, 1387–1395.
28. P. Basu, A. Pal, J. P. Lynch and K. M. Brown, *Plant Physiol.*, 2007, **145**, 305–316.
29. N. Wuyts, A. G. Bengough, T. J. Roberts, C. J. Du, M. F. Bransby, S. J. McKenna and T. A. Valentine, *Planta*, 2011, **234**, 769–784.
30. R. Bastien, D. Legland, M. Martin, L. Fregosi, A. Peaucelle, S. Douady, B. Moulia and H. Hofte, *Plant J.*, 2016, **88**, 468–475.
31. R. E. Sharp, V. Poroyko, L. G. Hejlek, W. G. Spollen, G. K. Springer, H. J. Bohnert and H. T. Nguyen, *J. Exp. Bot.*, 2004, **55**, 2343–2351.
32. M. Royer, D. Cohen, N. Aubry, V. Vendramin, S. Scalabrin, F. Cattonaro, M. B. Bogeat-Triboulot and I. Hummel, *J. Exp. Bot.*, 2016, **67**, 5961–5973.
33. C. Youssef, F. Bizet, R. Bastien, D. Legland, M. B. Bogeat-Triboulot and I. Hummel, *J. Exp. Bot.*, 2018, **69**, 5157–5168.
34. T. I. Baskin, *Wiley Interdiscip. Rev.: Dev. Biol.*, 2013, **2**, 65–73.
35. D. J. Cosgrove, *Nat. Rev. Mol. Cell Biol.*, 2005, **6**, 850–861.
36. J. Pritchard, *New Phytol.*, 1994, **127**, 3–26.
37. A. Geitmann and J. K. E. Ortega, *Trends Plant Sci.*, 2009, **14**, 467–478.
38. M. B. Triboulot, J. Pritchard and D. Tomos, *New Phytol.*, 1995, **130**, 169–175.
39. J. Frensch and T. C. Hsiao, *Plant Physiol.*, 1995, **108**, 303–312.
40. Z. Mao, L. Saint-Andre, M. Genet, F.-X. Mine, C. Jourdan, H. Rey, B. Courbaud and A. Stokes, *Ecol. Eng.*, 2012, **45**, 55–69.
41. S. Mickovski and L. P. H. Beek, *Plant Soil*, 2009, **324**, 43–56.
42. M. Genet, A. Stokes, F. Salin, S. Mickovski, T. Fourcaud, J. F. Dumail and R. van Beek, *Plant Soil*, 2005, **278**, 1–9.
43. C. B. Zhang, L. H. Chen and J. Jiang, *Geomorphology*, 2014, **206**, 196–202.
44. K. W. Loades, A. G. Bengough, M. F. Bransby and P. D. Hallett, *Plant Soil*, 2013, **370**, 407–418.
45. K. W. Loades, A. G. Bengough, M. F. Bransby and P. D. Hallett, *Plant Soil*, 2015, **395**, 253–261.
46. M. L. McCormack, I. A. Dickie, D. M. Eissenstat, T. J. Fahey, C. W. Fernandez, D. Guo, H.-S. Helmisaari, E. A. Hobbie, C. M. Iversen, R. B. Jackson, J. Leppälammi-Kujansuu, R. J. Norby, R. P. Phillips,

K. S. Pregitzer, S. G. Pritchard, B. Rewald and M. Zadworny, *New Phytol.*, 2015, **207**, 505–518.
47. T.-T. Lin and R. E. Pitt, *J. Texture Stud.*, 1986, **17**, 291–313.
48. M. G. Scanlon, C. H. Pang and C. G. Biliaderis, *Food Res. Int.*, 1996, **29**, 481–488.
49. M. Bentini, C. Caprara and R. Martelli, *Biosyst. Eng.*, 2009, **104**, 25–32.
50. M. L. Oey, E. Vanstreels, J. De Baerdemaeker, E. Tijskens, H. Ramon, M. L. A. T. M. Hertog and B. Nicolai, *Postharvest Biol. Technol.*, 2007, **44**, 240–247.
51. L. Mayor, R. L. Cunha and A. M. Sereno, *Food Res. Int.*, 2007, **40**, 448–460.
52. K. J. Niklas, *Am. J. Bot.*, 1989, **76**, 929–937.
53. L. J. Gibson, *J. R. Soc., Interface*, 2012, **9**, 2749–2766.
54. L. J. Gibson and M. F. Ashby, *Adv. Polym. Technol.*, 1989, **9**, 165–166.
55. M. Warner, B. L. Thiel and A. M. Donald, *Proc. Natl. Acad. Sci.*, 2000, **97**, 1370–1375.
56. D. Boldrin, A. K. Leung and A. G. Bengough, *Plant Soil*, 2018, **431**, 347–369.
57. A. M. Goodman and A. R. Ennos, *Int. J. Plant Sci.*, 2001, **162**, 691–696.
58. A. Stokes, B. Nicoll, M. Coutts and A. Fitter, *Can. J. For. Res.-Rev. Can. Rech. For. - CAN J For. RES*, 1997, **27**, 1049–1057.
59. M. P. Coutts, *Plant Soil*, 1983, **71**, 171–188.
60. A. M. Goodman and A. R. Ennos, *J. Exp. Bot.*, 1996, **47**, 1499–1507.
61. J. Correa, J. A. Postma, M. Watt and T. Wojciechowski, *J. Exp. Bot.*, 2019, **70**, 6019–6034.
62. A. G. Bengough, B. M. McKenzie, P. D. Hallett and T. A. Valentine, *J. Exp. Bot.*, 2011, **62**, 59–68.
63. E. L. Greacen and J. S. Oh, *Nat. New Biol.*, 1972, **235**, 24–25.
64. B. M. McKenzie, C. E. Mullins, J. M. Tisdall and A. G. Bengough, *Plant, Cell Environ.*, 2013, **36**, 1085–1092.
65. S. R. Tracy, C. R. Black, J. A. Roberts and S. J. Mooney, *J. Sci. Food Agric.*, 2011, **91**, 1528–1537.
66. L. J. Clark, A. G. Bengough, W. R. Whalley, A. R. Dexter and P. B. Barraclough, *Plant Soil*, 1999, **209**, 101–109.
67. C. Croser, A. G. Bengough and J. Pritchard, *Physiol. Plant.*, 2000, **109**, 150–159.
68. P. S. Kuzeja, P. M. Lintilhac and C. F. Wei, *J. Plant Physiol.*, 2001, **158**, 673–676.
69. I. Potocka and J. Szymanowska-Pulka, *Ann. Bot.*, 2018, **122**, 711–723.
70. L. J. Clark and P. B. Barraclough, *J. Exp. Bot.*, 1999, **50**, 1263–1266.
71. G. Azam, C. D. Grant, R. K. Misra, R. S. Murray and I. K. Nuberg, *Plant Soil*, 2013, **368**, 569–580.
72. E. Kolb, C. Hartmann and P. Genet, *Plant Soil*, 2012, **360**, 19–35.
73. R. Misra, A. Dexter and A. Alston, *Plant Soil*, 1986, **95**, 315–326.
74. T. Colombi, N. Kirchgessner, A. Walter and T. Keller, *Plant Physiol.*, 2017, **174**, 2289–2301.

75. A. G. Bengough and B. M. McKenzie, *J. Exp. Bot.*, 1997, **48**, 885–893.
76. M. Iijima, S. Morita and P. W. Barlow, *Plant Prod. Sci.*, 2008, **11**, 17–27.
77. J. Frensch and T. C. Hsiao, *Plant Physiol.*, 1994, **104**, 247–254.
78. L. J. Clark, W. R. Whalley, A. R. Dexter, P. B. Barraclough and R. A. Leigh, *Plant, Cell Environ.*, 1996, **19**, 1099–1102.
79. L. J. Clark, W. R. Whalley and P. B. Barraclough, *J. Exp. Bot.*, 2001, **52**, 167–171.
80. B. K. Pandey, G. Huang, R. Bhosale, S. Hartman, C. J. Sturrock, L. Jose, O. C. Martin, M. Karady, L. A. C. J. Voesenek, K. Ljung, J. P. Lynch, K. M. Brown, W. R. Whalley, S. J. Mooney, D. Zhang and M. J. Bennett, *Science*, 2021, **371**, 276–280.
81. H. Sjulgard, D. Iseskog, N. Kirchgessner, A. G. Bengough, T. Keller and T. Colombi, *Environ. Exp. Bot.*, 2021, **188**, 104494.
82. A. G. Bengough, C. Croser and J. Pritchard, *Plant Soil*, 1997, **189**, 155–164.
83. S. A. Materechera, A. M. Alston, J. M. Kirby and A. R. Dexter, *Plant Soil*, 1992, **144**, 297–303.
84. K. P. Barley, *Aust. J. Biol. Sci.*, 1965, **18**, 499.
85. Y. Chen, A. Khosravi, A. Martinez and J. DeJong, *Bioinspiration Biomimetics*, 2021, **16**, 046012.
86. D. Hettiaratchi, *Philos. Trans. R. Soc., B*, 1990, **329**, 343–355.
87. J. M. Kirby and A. G. Bengough, *Eur. J. Soil Sci.*, 2002, **53**, 119–127.
88. J. G. Chimungu, K. W. Loades and J. P. Lynch, *J. Exp. Bot.*, 2015, **66**, 3151–3162.
89. M. Riviere, J. Derr and S. Douady, *Phys. Biol.*, 2017, **14**, 051001.
90. I. Taylor, K. Lehner, E. McCaskey, N. Nirmal, Y. Ozkan-Aydin, M. Murray-Cooper, R. Jain, E. W. Hawkes, P. C. Ronald, D. Goldman and P. N. Benfey, *Proc. Natl. Acad. Sci. U. S. A.*, 2021, **118**, e2018940118.
91. J. Mullaney, T. Lucke and S. J. Trueman, *Landsc. Urban Plan.*, 2015, **134**, 157–166.
92. G. W. Watson, A. M. Hewitt, M. Custic and M. Lo, *Arboric. Urban For*, 2014, **40**, 249–271.
93. J. Grabosky and N. Gucunski, *Urban For. Urban Green.*, 2019, **41**, 238–247.
94. B. Niez, J. Dlouha, B. Moulia and E. Badel, *Trees-Struct. Funct*, 2019, **33**, 279–291.
95. A. Stokes, A. H. Fitter and M. P. Coutts, *J. Exp. Bot.*, 1995, **46**, 1139–1146.
96. A. Stokes, B. Nicoll, M. Coutts and A. Fitter, *Can. J. For. Res.*, 1997, **27**, 1049–1057.
97. F. Danjon, T. Fourcaud and D. Bert, *New Phytol.*, 2005, **168**, 387–400.
98. J. C. Grabosky, E. Thomas Smiley and G. A. Dahle, *Arboric. Urban For.*, 2011, **37**, 35–40.
99. S. Ishihara, K. Tanaka and Y. Shinohara, *RILEM Bookseries*, 2012, **4**, 317–326.
100. D. M. Wood, *Soil Behaviour and Critical State Soil Mechanics*, Cambridge University Press, 1990.
101. H. Downie, N. Holden, W. Otten, A. J. Spiers, T. A. Valentine and L. X. Dupuy, *PLoS One*, 2012, **7**, e44276.

102. A. R. Dexter, *Plant Soil*, 1987, **97**, 401–406.
103. A. V. Vollsnes, C. M. Futsaether and A. G. Bengough, *Eur. J. Soil Sci.*, 2010, **61**, 926–939.
104. S. J. Mooney, T. P. Pridmore, J. Helliwell and M. J. Bennett, *Plant Soil*, 2012, **352**, 1–22.
105. G. Bodner, D. Leitner and H.-P. Kaul, *Plant Soil*, 2014, **380**, 133–151.
106. M. Lucas, S. Schlüter, H.-J. Vogel and D. Vetterlein, *Sci. Rep.*, 2019, **9**, 16236.
107. A. Diambra, E. Ibraim, D. M. Wood and A. R. Russell, *Geotext. Geomembr.*, 2010, **28**, 238–250.
108. A. Diambra and E. Ibraim, *Géotechnique*, 2015, **65**, 296–308.
109. I. Soriano, E. Ibraim, E. Andò, A. Diambra, T. Laurencin, P. Moro and G. Viggiani, *Granul. Matter*, 2017, **19**, 75.
110. S. D. Keyes, L. Cooper, S. Duncan, N. Koebernick, D. M. McKay Fletcher, C. P. Scotson, A. van Veelen, I. Sinclair and T. Roose, *J. R. Soc., Interface*, 2017, **14**, 20170560.
111. F. Anselmucci, E. Andò, G. Viggiani, N. Lenoir, R. Peyroux, C. Arson and L. Sibille, *Geotech. Lett.*, 2021, **11**, 96–102.
112. J. Helliwell, C. Sturrock, S. Mairhofer, J. Craigon, R. Ashton, A. Miller, W. Whalley and S. Mooney, *Sci. Rep.*, 2017, **7**, 14875.
113. J. R. Helliwell, C. J. Sturrock, A. J. Miller, W. R. Whalley and S. J. Mooney, *Plant, Cell Environ.*, 2019, **42**, 1974–1986.
114. N. Koebernick, K. R. Daly, S. D. Keyes, A. G. Bengough, L. K. Brown, L. J. Cooper, T. S. George, P. D. Hallett, M. Naveed, A. Raffan and T. Roose, *New Phytol.*, 2019, **221**, 1878–1889.
115. N. Koebernick, K. R. Daly, S. D. Keyes, T. S. George, L. K. Brown, A. Raffan, L. J. Cooper, M. Naveed, A. G. Bengough, I. Sinclair, P. D. Hallett and T. Roose, *New Phytol.*, 2017, **216**, 124–135.
116. R. A. Champion and K. P. Barley, *Soil Sci.*, 1969, **108**, 402.
117. A. G. Bengough and B. M. McKenzie, *J. Exp. Bot.*, 1997, **48**, 885–893.
118. D. B. Read and P. J. Gregory, *New Phytol.*, 1997, **137**, 623–628.
119. M. E. McCully and J. S. Boyer, *Physiol. Plant.*, 1997, **99**, 169–177.
120. P. Benard, M. Zarebanadkouki, M. Brax, R. Kaltenbach, I. Jerjen, F. Marone, E. Couradeau, V. J. M. N. L. Felde, A. Kaestner and A. Carminati, *Vadose Zone J.*, 2019, **18**, 180211.
121. A. Carminati, P. Benard, M. Ahmed and M. Zarebanadkouki, *Plant Soil*, 2017, **417**, 1–15.
122. M. Tuller, D. Or and L. M. Dudley, *Water Resour. Res.*, 1999, **35**, 1949–1964.
123. R. Sattler, S. Gier, J. Eggers and C. Wagner, *Phys. Fluids*, 2012, **24**, 023101.
124. M. Zarebanadkouki, T. Fink, P. Benard and C. Banfield, *Vadose Zone J.*, 2019, **18**, 190021.
125. F. Anselmucci, E. Andó, L. Sibille, N. Lenoir, R. Peyroux, C. Arson, G. Viggiani and A. G. Bengough, *E3S Web Conf.*, 2019, **92**, 12011.
126. C. W. W. Ng, J. J. Ni, A. K. Leung and Z. J. Wang, *Géotechnique Lett.*, 2016, **6**, 106–111.

127. A. K. Leung, D. Boldrin, T. Liang, Z. Y. Wu, V. Kamchoom and A. G. Bengough, *Géotechnique*, 2017, 1–7.
128. A. J. C. Collison, M. G. Anderson and D. M. Lloyd, *Proc. Inst. Civ. Eng. – Water Marit. Energy*, 1995, **112**, 168–175.
129. F. Preti, *Ecol. Eng.*, 2013, **61**(Part C), 633–645.
130. D. Boldrin, A. K. Leung and A. G. Bengough, *Géotechnique*, 2020, 1–15.
131. S. Bordoloi, J. Ni and C. W. W. Ng, *Sci. Total Environ*, 2020, **729**, 138760.
132. J. Scott, F. Loveridge and A. S. O'Brien, ed. V. Cuellar, E. Dapena, E. Alonso, J. M. Echave, A. Gens, J. L. de Justo, C. Oteo, J. M. Rodriguez-Ortiz, C. Sagaseta, P. Sola and A. Soriano, IOS Press, 2007, pp. 659–664.
133. J. A. Smethurst, K. M. Briggs, W. Powrie, A. Ridley and D. J. E. Butcher, *Géotechnique*, 2015, **65**, 869–882.
134. D. Boldrin, A. K. Leung and A. G. Bengough, *Plant Soil*, 2017, **416**, 437–451.
135. A. Stokes, C. Atger, A. G. Bengough, T. Fourcaud and R. C. Sidle, *Plant Soil*, 2009, **324**, 1–30.
136. L. J. Waldron, *Soil Sci. Soc. Am. J.*, 1977, **41**, 843–849.
137. T. H. Wu, W. P. McKinnell III and D. N. Swanston, *Can. Geotech. J.*, 1979, **16**, 19–33.
138. L. Dupuy, T. Fourcaud and A. Stokes, *Eur. J. Soil Sci.*, 2005, **56**, 319–327.
139. J. C. Ekanayake, M. Marden, A. J. Watson and D. Rowan, *N. Z. J. For. Sci*, 1997, **27**, 216–233.
140. T. Liang, A. G. Bengough, J. A. Knappett, D. MuirWood, K. W. Loades, P. D. Hallett, D. Boldrin, A. K. Leung and G. J. Meijer, *Ecol. Eng.*, 2017, **109**, 207–227.
141. G. J. Meijer, A. G. Bengough, J. A. Knappett, K. W. Loades and B. C. Nicoll, *Can. Geotech. J.*, 2018, **55**, 1372–1390.
142. K. W. Loades, A. G. Bengough, M. F. Bransby and P. D. Hallett, *Ecol. Eng.*, 2010, **36**, 276–284.
143. D. J. Bull, J. A. Smethurst, I. Sinclair, F. Pierron, T. Roose, W. Powrie and A. G. Bengough, *Proc. R. Soc. Math. Phys. Eng. Sci.*, 2020, **476**, 20190838.
144. S. B. Mickovski, A. G. Bengough, M. F. Bransby, M. C. R. Davies, P. D. Hallett and R. Sonnenberg, *Eur. J. Soil Sci.*, 2007, **58**, 1471–1481.
145. O. Hamza, A. G. Bengough, M. F. Bransby, M. C. R. Davies and P. D. Hallett, in *Eco- and Ground Bio-engineering: The Use of Vegetation to Improve Slope Stability*, ed. A. Stokes, I. Spanos, J. E. Norris and E. Cammeraat, Springer, Dordrecht, The Netherlands, 2007, pp. 13–20.
146. T. Liang, J. A. Knappett and A. G. Bengough, ed. *C. Gaudin and D.* White, 2014, CRC, vol. 1, pp. 361–366.
147. V. Kamchoom, A. K. Leung and C. W. W. Ng, *Jpn. Geotech. Soc. Spec. Publ.*, 2016, **2**, 236–240.
148. C. W. W. Ng, V. Kamchoom and A. K. Leung, *Landslides*, 2016, **13**, 925–938.
149. R. Sonnenberg, M. F. Bransby, P. D. Hallett, A. G. Bengough, S. B. Mickovski and M. C. R. Davies, *Can. Geotech. J.*, 2010, **47**, 1415–1430.

150. A. Askarinejad and S. Springman, *Computer Methods and Recent Advances in Geomechanics*, CRC Press, 2014, pp. 1339–1344.
151. E. Comino, P. Marengo and V. Rolli, *Soil Tillage Res.*, 2010, **110**, 60–68.
152. P. J. Gregory, *Plant Roots: Growth, Activity and Interaction with Soils*, Blackwell Publishing, Oxford, 2006.
153. M. Genet, N. Kokutse, A. Stokes, T. Fourcaud, X. Cai, J. Ji and S. Mickovski, *For. Ecol. Manag.*, 2008, **256**, 1517–1526.
154. D. Boldrin, A. K. Leung and A. G. Bengough, *Ecol. Eng.*, 2017, **109**, 196–206.
155. J. E. Norris, *Plant Soil*, 2005, **278**, 43–53.
156. G. J. Meijer, A. G. Bengough, J. A. Knappett, K. W. Loades and B. C. Nicoll, *Géotechnique*, 2016, **66**, 27–40.
157. G. J. Meijer, G. Bengough, J. Knappett, K. Loades and B. Nicoll, *Geotechnique*, 2018, **68**, 320–331.
158. N. Pollen and A. Simon, *Water Resour. Res.*, 2005, **41**, W07025.
159. M. Schwarz, P. Lehmann and D. Or, *Earth Surf. Processes Landforms*, 2010, **35**, 354–367.
160. M. Schwarz, F. Giadrossich and D. Cohen, *Hydrol. Earth Syst. Sci.*, 2013, **17**, 4367–4377.
161. F. Bourrier, F. Kneib, B. Chareyre and T. Fourcaud, *Ecol. Eng.*, 2013, **61**, 646–657.
162. L. X. Dupuy, T. Fourcaud, P. Lac and A. Stokes, *Am. J. Bot.*, 2007, **94**, 1506–1514.
163. S. B. Mickovski, A. Stokes, R. van Beek, M. Ghestem and T. Fourcaud, *Ecol. Eng.*, 2011, **37**, 1523–1532.
164. T. H. Wu, R. M. McOmber, R. T. Erb and P. E. Beal, *J. Geotech. Eng. ASCE*, 1988, 114, 1351–1375.
165. T. Liang, J. A. Knappett and N. Duckett, *Géotechnique*, 2015, **65**, 995–1009.
166. G. J. Meijer, D. Muir Wood, J. A. Knappett, G. A. Bengough and T. Liang, *Int. J. Numer. Anal. Methods Geomech.*, 2019, **43**, 684–707.
167. B. M. Świtała and W. Wu, *Géotechnique*, 2018, **68**, 481–491.
168. B. M. Świtała, A. Askarinejad, W. Wu and S. M. Springman, *Géotechnique*, 2018, **68**, 375–385.
169. D. Muir Wood, A. Diambra and E. Ibraim, *Soils Found.*, 2016, **56**, 765–778.
170. T. Colombi and T. Keller, *Soil Tillage Res.*, 2019, **191**, 156–161.
171. A. Bishopp and J. P. Lynch, *Nat. Plants*, 2015, **1**, 15117.
172. M. Yang, P. Defossez, F. Danjon, S. Dupont and T. Fourcaud, *Plant Soil*, 2017, **411**, 275–291.
173. C. Mattheck, M. Teschner and J. Schafer, *J. Theor. Biol.*, 1997, **184**, 261–269.
174. B. Jean, L. Heux, F. Dubreuil, G. Chambat and F. Cousin, *Langmuir*, 2009, **25**, 3920–3923.
175. P. Benard, M. Zarebanadkouki, M. Brax, R. Kaltenbach, I. Jerjen, F. Marone, E. Couradeau, V. J. M. N. L. Felde, A. Kaestner and A. Carminati, *Vadose Zone J.*, 2019, **18**, 180211.
176. D. Vetterlein, A. Carminati, I. Kögel-Knabner, G. P. Bienert, K. Smalla, E. Oburger, A. Schnepf, T. Banitz, M. T. Tarkka and S. Schlüter, *Front. Agron.*, 2020, **2**, DOI: 10.3389/fagro.2020.00008.

CHAPTER 6

Invasive Processes in the Life Cycle of Plants and Fungi

KARUNA KAPOOR AND ANJA GEITMANN*

Department of Plant Science, McGill University, Macdonald Campus, 21111 Lakeshore, Ste-Anne-de-Bellevue, Quebec, Canada
*Email: geitmann.aes@mcgill.ca

6.1 Introduction

The cells of multicellular organisms generally occupy specific locations within tissues and organs to serve specialized metabolic or structural functions. In animal bodies, there are numerous exceptions to this spatial constancy: blood and lymph cells are transported through the entire body by the respective circulatory systems; neurons elongate their axons through other tissues to innervate them; and cancer cells migrate through tissues and change location *via* the circulatory system. In plants, no circulation of cells occurs, although the position of individual nuclei may change across substantial distances, in particular in large, multinucleate syncytia or coenocytes. The nuclear migration is similar to that occurring in many filamentous fungi and slime molds, where nuclei can move over large distances within multinucleate cells.[1] Because of the absence of cell migration proper, however, spatial constancy of cells within the plant body is more pronounced than in animals. This is even more so since plant cells are encased in an extracellular matrix and glued to each other by the middle lamellae, dramatically limiting movement of cells relative to each other.[2] Collectively, the middle lamella and the encasing cell walls in the plant body are called the apoplast, a scaffold that confers structural stability to plant organs and cements cells in space. However, even within the

Soft Matter Series No. 15
Soft Matter in Plants: From Biophysics to Biomimetics
Edited by Kaare H. Jensen and Yoël Forterre

Published by the Royal Society of Chemistry, www.rsc.org

plant body, certain cells produce extensions either to reach distant regions in the same organism or to explore the external environment. Just like in animal organisms, these motile activities in plants often require the cells to invade or squeeze between their neighbors and through narrow spaces.

Invasive cell types in plants include root hairs, pollen tubes, sclerenchyma fibers, and laticifers. The invasive lifestyle and associated elongated cell shape can serve a variety of purposes. Elongated cells can provide structural stability to the organ, analogous to the steel rods reinforcing a concrete structure (sclerenchyma fibers), facilitate procurement of nutrients and water from distant sources (root hairs),[3] or transport cargo (pollen tubes). A second type of invasion to which plant bodies are subjected is that by symbionts and pathogens of fungal and bacterial origin. Some of the invasive activities performed by other organisms exploit structural openings in the plant body, such as stomata or intercellular air spaces, whereas others see the invader drill into the apoplast or even into the lumen of individual cells. As long as cellular extension or invasion exploits openings in the tissue traversing gas or liquid spaces, the invading cell does not encounter any mechanical obstacle. However, when invasion or extension occurs against or through a solid or viscous matrix, typically the apoplast, the invading cell has to overcome mechanical obstacles.

The forces required for the invasive and migratory behavior of animal cells such as neurons, cancer cells, and fibroblasts are generated by the cytoskeletal system and have been covered in numerous reviews.[4,5] In the following we focus on the invasive lifestyle of walled cells in plants, fungi, and oomycetes. We elaborate on the key structural parameters involved in maintaining an intrusive activity in walled cells—the turgor pressure and the regulation of cell wall mechanical properties.[6] The cytoskeleton in these cells is important in its role as regulator of intracellular trafficking and cell wall assembly.[7] We discuss how invasive cells may facilitate their activities by secreting agents that soften the invaded substrate and we review experimental techniques and numerical methods developed to measure the invasive force.

6.2 Invasive Growth Serves a Diverse Range of Functions

6.2.1 Elongated Cells Can Confer Structural Stability

Among the longest cells in the plant body are fibers with lengths up to 120 mm in *Boehmeria nivea*, 33 mm in *Linum usitatissimum*.[8] These sclerenchymatous cells (cells with thick lignified secondary cell wall and typically dead at maturity) serve to stabilize the plant body against mechanical stress.[6,9] This mechanical role depends on structural and geometrical parameters that involve the accumulation of cellulose and lignin in significant amounts at mechanically critical positions within organs.[6] Fibers are formed in various plant organs, including roots, shoots, leaves[10,11] and are particularly abundant in the phloem or secondary xylem of eudicotyledon plants and surrounding the vascular bundles in the leaves of monocotyledons.[6,12] The structural function of fibers relies on their resistance to compressive, bending, and tensile forces all of which may occur in the same organ. Fibers have also been considered to act akin to animal muscles, as in some situations they are thought to have contractile properties.[11,13,14] This action is displayed, for instance, during gravitropic responses, climbing and underground positioning of geophytes.[11] Because of their mechanical, structural and biochemical properties, sclerenchyma fibers are key for the mechanical properties of timber-based construction materials and constitute an energy-rich component for the fuel wood industry. The flexibility combined with tensile resistance of bast fibers in flax, ramie, hemp, jute, and kenaf are also exploited in the textile industry.[10,11,15,16]

The extreme length of sclerenchyma fibers is generated starting from relatively short precursor cells formed in meristems—the stem cell niches in plants. Fiber differentiation and morphogenesis, therefore, require the cells to expand in a highly anisotropic manner and through surrounding, slower-growing tissues (Figure 6.1a). This occurs through intrusive growth that penetrates the apoplast connecting neighboring cells.[6,10,11,17] Plant fibers are thus, in principle, an excellent model to study plant invasive growth, cell wall formation, and cell wall mechanics. However, despite their structural properties and economic importance, the mechanics underlying their intrusive behavior is still poorly understood. This is mainly due to the fact that fibers are formed within the depth of the organ and surrounded by other tissues, which renders their isolation or *in situ* live cell observation difficult.[6,11]

6.2.2 Invasion for Cargo Delivery Across Tissues

Elongated cells provide an excellent catheter-like system that enables the transport of cargo from one region of the organ to another, even across tissue boundaries. One such *trans*-tissue transfer is required for successful fertilization in flowering plants. The sperm cells must be shuttled from the pollen grains upon their arrival on the landing platform of the pistil—the stigma—through the stylar transmitting tissue to the female gametophyte located in the ovary. This transfer is accomplished by a cellular protuberance formed by the germinating pollen grain, the pollen tube. This protuberance emerges from an opening in the hard outer wall of the pollen grain and undergoes localized elongation at the very apex. This tip growth process is reflected in the extremely polar organization of the cytoplasm[18,19] and the spatially confined expansion of the cell surface.[20,21] Depending on plant species, the diameter of a pollen tube varies between 5 and 20 μm and the length can extend up to tens of centimeters depending on the length of the pistil. While the entire pollen tube consists of a single vegetative cell, the active portion of the pollen tube cytoplasm is restricted to the apical portion of the cell only, and the distal region is plugged off and eventually degenerates.[22] When invading the stigma and transmitting tissue of the style, pollen tubes have to overcome multiple mechanical constraints (Figure 6.1b–d). The elongating tip of the tube, therefore, has to exert sufficient penetration force in

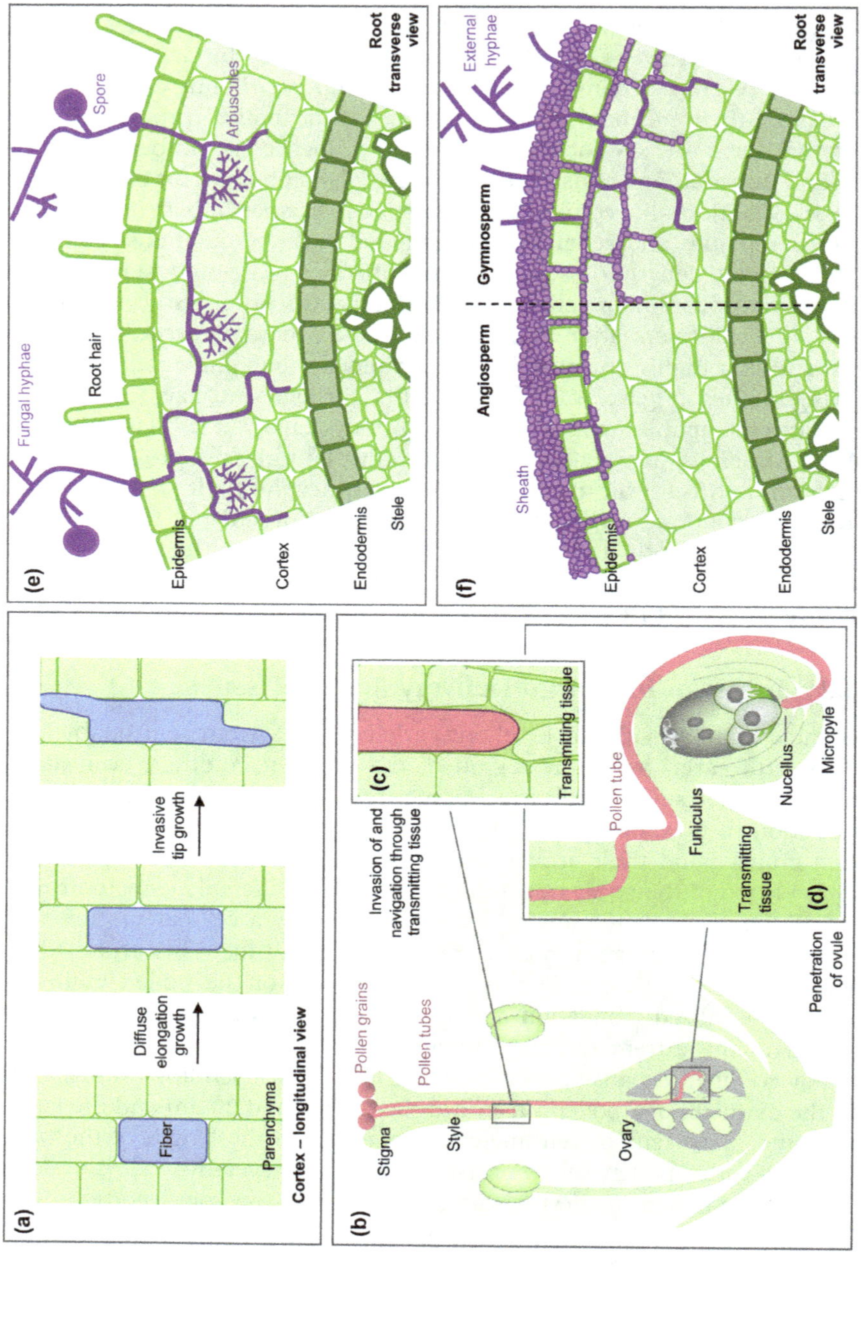
(a)
Fiber
Parenchyma
Cortex – longitudinal view
Diffuse elongation growth
Invasive tip growth
(b)
Stigma
Pollen grains
Pollen tubes
Style
Ovary
Invasion of and navigation through transmitting tissue
(c)
Transmitting tissue
Penetration of ovule
(d)
Transmitting tissue
Funiculus
Pollen tube
Nucellus
Micropyle
(e)
Fungal hyphae
Root hair
Spore
Arbuscules
Epidermis
Cortex
Endodermis
Stele
Root transverse view
(f)
Sheath
Angiosperm
Gymnosperm
External hyphae
Epidermis
Cortex
Endodermis
Stele
Root transverse view

order to withstand the external compressive stress generated by transmitting tissue while maintaining the direction towards its target, the ovule.[23,24]

6.2.3 Spreading out for Procurement of Nutrients and Water

Cells elongating beyond the perimeter of the organism, into the external substrate, confer the ability to explore a larger space on the search for nutrients or water. In order to fulfill their nutritional needs from diverse biological and synthetic substrates, filamentous fungi extend colonies by forming branched hyphae that have the ability to penetrate solid substrates of considerable mechanical stiffness, including rocks. Hyphae are commonly formed by yeasts, mushroom-forming fungi, and also by oomycete water molds.[6,25,26] Hyphae are tube-shaped cells with a diameter ranging from 2 to 20 μm, elongating at the tip region, similar to pollen tubes. The hyphae of a fungal organism often grow in a direction centrifugal with regards to the center of the mycelium to ensure the most efficient exploration of the substrate.[6] Although hyphae usually elongate individually, they sometimes aggregate in parallel to form a structure called a rhizomorph. These rhizomorphs can have a diameter of several millimeters and can elongate several meters in length.[6] They have the invading capacity to penetrate through soil or wood in search of nutrients, which are then transferred to developing fruiting bodies.[27,28] When penetrating and elongating, fungal hyphae secrete enzymes that digest polymers to sugars and other molecules that can be taken up through the plasma membrane.[25] The enzymatic digestion is also thought to soften the physical impedance of the surrounding substrate.[25]

Certain fungal hyphae establish a symbiotic relationship with plant root tissues, where the plant provides organic molecules such as sugar to the fungus and the fungal partner provides water and minerals absorbed from the soil to the plant.[29] This association is highly intimate, since it involves the invasion of the fungal hyphae into the apoplast of the root epidermal layer

Figure 6.1 Various types of invasive cells: (a) Development of sclerenchyma fiber cell from meristem precursor. Initial elongation occurs by diffuse growth in lockstep with surrounding parenchyma cells, but once these cease elongation the continuation of fiber growth occurs by tip growth. (b) Mechanical obstacles in the pathway of the pollen tube towards the ovary include the stigmatic cuticle, the apoplast of the transmitting tissue, the micropyle, and the nucellus. (c) Pollen tube making its way through the apoplast of the stylar transmitting tract. (d) Pollen tube emerging from the transmitting tissue elongating on the surface of the funiculus, turning into the micropyle of the ovule. (e) Fungal hyphae of arbuscular mycorrhiza, a type of endomycorrhiza, invade the epidermis and cortex of the plant root by penetrating the apoplast and building branched structures inside the lumen of root cortex cells. (f) Fungal hyphae of ectomycorrhiza cover the root surface forming a sheath and penetrate the root epidermis forming a Hartig net. Occasionally, in particular in contact with gymnosperms, the hyphae of ectomycorrhiza reach external cortex layers.

(ectomycorrhiza) or the root cortex (endomycorrhiza)[6,29] (Figure 6.1e and f). Endomycorrhiza not only invade the root apoplast but actually grow into the root cell lumen, where they form arbuscular structures that augment the interaction surface between the fungus and the protoplast of the plant cell.

The plant root is not only subject to invasion, it is an active invader and exerts this action at different scales. At the cell scale, roots produce their own extensions that explore the substrate beyond the surface of the organ. This is done by root hairs—tube-shaped, tip-growing extensions emerging from the epidermal cells of the root that range from 5 to 17 μm in diameter and about 0.1 to 1.5 mm in length.[30,31] Similar to fungal symbionts, root hairs serve to increase the interaction surface between the absorbant region of the root and the soil, thus facilitating nutrient and water uptake.[3,32,33]

Root invasion also occurs at the scale of the organ. In their interaction with the growth substrate, roots grow against and interact mechanically with soil particles. The soil is not the first or only obstacle for root growth, as newly forming roots (radicles) encounter mechanical impedance in the form of the seed coat[34] or, in the case of lateral roots, the outer layers of the primary root.[35] Since lateral roots are formed from the pericycle—a cell layer in the central stele of the primary root—reaching the outside requires the exertion of invasive forces for the young lateral root primordium to break through the outer tissue layers of the primary root (Figure 6.2). The mechanical interaction of the root with an external matrix influences distinct morphological and developmental changes in the root system.[36] Studies done on the roots of cereal crop species have revealed irregular cortical cell growth, increased root diameter, and bending and buckling of the root tip as a result of mechanical resistance from the soil particles.[36,37] This mechanical impedance created as a result of compacted soil layers or soil drying can be a major limiting factor to root elongation and hence nutrient uptake for the plant. Increased soil strength requires roots to exert higher forces to ensure successful soil penetration.[38] To minimize the effect of soil mechanical impedance, roots of maize (*Zea mays*) and soybean (*Glycine max*) use natural or artificial macropores in the soil or growing matrix. Invading these openings in the otherwise compact material allows the roots to follow the path of least resistance.[38–40] Another common response to increased soil strength is the thickening of roots, which decreases penetrative stresses and stabilizes the root.[41,42] Root hairs have been suggested to play a supportive role during root penetration by anchoring the root to the surrounding soil.[33,43]

Given the fundamental importance of root growth through various soil types for yield and drought resistance, numerous biomechanical frameworks have been established for soil penetration mechanics. Root growth forces have been quantified with the help of technologies such as photoelastic discs,[44] or cantilever sensors[45] and 3D living imaging of the roots in transparent soil,[46] as well as X-ray based imaging,[47] have greatly advanced the field. Since roots are multicellular tissues of macroscopic size, we refer to excellent reviews on the topic of invasive root growth[36,48] as well as Chapter 5, and focus on the single-celled growth of root hairs in the present chapter.

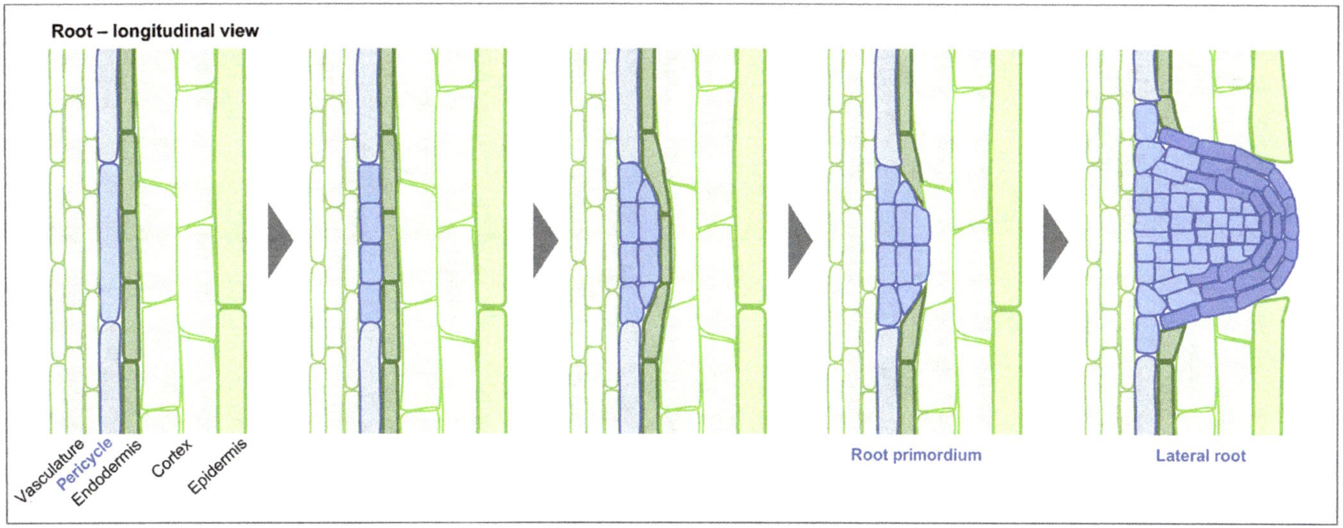

Figure 6.2 Lateral root formation initiates with cell divisions in a spatially confined region of the pericycle located in the central steele of the primary root. The resulting lateral root primordium develops a new root apical meristem whose continuous cell divisions produce the elongating new root, which in turn breaks through the cortex and epidermis of the primary root to continue growing through the soil.

6.3 Cell Mechanics of Intrusive Growth

6.3.1 Highly Polarized Cell Extension Directs Force Generation

The growth pattern of invasive cells is unique, as the cells are highly polarized and typically extend exclusively at the very apex of the cell.[18,20,21,49] The morphogenetic process generating these cylindrical cells has been subject to multiple efforts to characterize the cell-mechanical underpinnings through modeling.[50,51] The apical cytoplasm of tip-growing cells is often densely populated by vesicles, both exocytotic and endocytotic (Figure 6.3). Exocytotic vesicles deliver the material required for cell expansion, notably cell wall precursors and membrane material. In pollen tubes, the apical vesicle population forms an inverted cone-shaped region that is fed by Golgi-derived vesicles delivered to the apex by an array of actin filaments[52] (Figure 6.3b), whereas fungal hyphae typically feature a structured vesicle aggregate called Spitzenkörper organized by microtubules[53] (Figure 6.3a). This interaction of Spitzenkörper and microtubules is also known to function as a 'compass' that serves in preserving the 'directional memory' of the fungal hyphae when they navigate around the obstacles.[142] Bigger organelles such as mitochondria and endoplasmic reticulum sometimes share the apical space but more typically remain in more distal regions of the tubular cell. Vesicle movement in hyphae occurs towards the Spitzenkörper and from there radiates to the apical plasma membrane, whereas in pollen tubes, vesicles are delivered through a circular movement which in angiosperm pollen tubes occurs in an inverse fountain-shaped pattern controlled by the actin cytoskeleton.[54–56] In all tip-growing cell types, the spatial organization of the cytoplasm ensures that both delivery of new cell wall material and the expansion of the existing cell wall are confined to the apical region of the elongating cell. This extreme polarization of tip-growing cells is very different from the growth behavior of most other plant cells, which display more global deformations across the cell surface, also termed diffuse growth.[57] The continuous addition of new cell wall material is controlled by cellular feedback mechanisms[51] and involves breaking and forming crosslinks between newly added and existing cell wall polymers.[58,59] Once the cell wall material at the tip is excreted it starts to stiffen, which locks in the cylindrical diameter. In pollen tubes, this stiffening is caused by the gelation of pectin polymers whose secreted methyl esterified configuration becomes de-esterified *in muro* (in the wall) through the action of pectin methylesterase.[60,61] In fungal hyphae, the main wall components are chitin and β-glucans and the distal stiffening involves hydrogen and covalent bonds.[62,63] The tip-focused maintenance of the cylindrical geometry is a self-similar morphogenetic process that is regulated through a fine-tuning between internal turgor pressure and biochemical cell wall properties at the apical growing region.[6,21]

While pollen tubes, fungal hyphae, and root hairs display a clearly discernable pattern of tip growth, the growth pattern of sclerenchyma fibers may be more complex. In secondary xylem fibers, true tip growth seems to prevail, as these fibers develop in the portions of the stem in which the

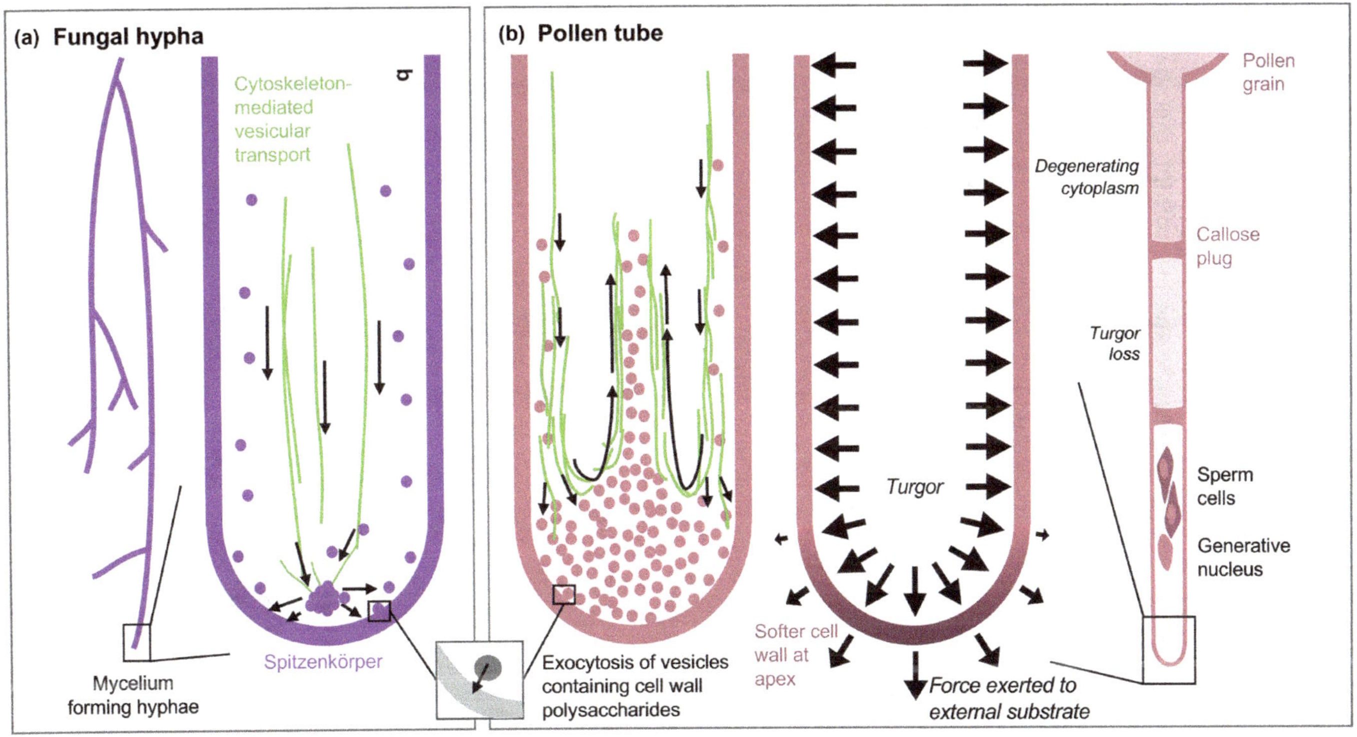

Figure 6.3 Cellular tip growth is fuelled by a continuous supply of secretory vesicles delivered by cytoskeleton-mediated transport. (a) Many fungal hyphae feature a distinct vesicle aggregate, the Spitzenkörper, which regulates vesicle transport to the apical plasma membrane. (b) In angiosperm pollen tubes, vesicles and other organelles move in a reverse fountain shaped pattern. The apical cell wall is softer than the wall in the distal region. Yielding of the cell wall to the turgor pressure in the apical region enables the cell to exert forces onto an external substrate. The cytoplasm of the pollen tube is segmented by callose plugs into the apical, viable region containing the sperm cells and generative nucleus, and distal regions that lose turgor and degenerate.

tissues surrounding the fibers ceased elongation.[11] The initial elongation of primary flax phloem fibers on the other hand, seems to occur through diffuse growth as the neighboring cells continue elongating in lockstep at least during the initial developmental phase of the organ. During this early developmental phase, the entire cell surface enlarges,[64] followed by an intrusive elongation of the fiber tips once growth in the neighboring cells has ceased (Figure 6.1a). Distinguishing the different growth patterns is aided by monitoring strain patterns of the cell surface and by profiling the fiber cell wall and fiber tips *via* biochemical and mechanical parameters.[14,65,66]

6.3.2 Turgor Pressure Generates the Invasive Force in Walled Cells

Cell growth in plants involves the expansion of the existing cell wall driven by the turgor pressure,[67] but whether the exertion of invasive forces is equally dependent on turgor or whether other cellular features are involved has been a matter of discussion. Extension of sclerenchyma fibers has been proposed to rely on elevated turgor based on the assumption that the soft and thin-walled growing fibers would be squished if they were less turgescent than the adjacent cells, but experimental quantification of turgor is elusive.[9] The increase in turgor pressure at the initiation of intrusive growth and the maintenance of turgor during the fiber elongation require the movement of water into the developing fiber. Gene expression patterns suggest that regulation of this water movement during fiber growth involves aquaporins—protein channels facilitating water movement across membranes.[9,68,69]

The turgor pressure of growing pollen tubes has been measured in lily and was found to range between 0.1 and 0.4 MPa.[70] Consistent with this, the maximum force that a pollen tube can produce when overcoming a mechanical barrier has been measured to be approximately 10 μN for lily[71] and 1.5 μN for *Camellia japonica.*[72] Since force is the product of the pressure and the interaction surface between the pressurized vessel and the substrate,[6] the latter value was calculated to correspond to a pressure of 0.19 MPa, consistent with the magnitude of turgor. While the measured invasive forces in pollen tubes are consistent with the notion that turgor is the driving force of their invasive activity, this does not necessarily mean that the turgor pressure is also the parameter that is tuned to regulate the magnitude of growth speed or invasive force. In fact, different growth rates do not seem to be correlated with different turgor values, and even non-growing pollen tubes can have a turgor similar to that of growing pollen tubes.[70] Variations in growth rate seem to rely instead on a modulation of the biomechanical properties of the cell wall, which in turn can oscillate through the effect of exocytosis of new cell wall material and cell-wall-modifying enzymes.[73] This modulation takes place at the apical cell wall, which is substantially more compliant than the cylindrical portion,[21] thus enabling the exertion of forces against an outside substrate at this site of the cell surface (Figure 6.3b).[6] Even if the apical wall is relatively compliant, a threshold turgor is necessary for pollen tube elongation and invasion as the plasmolyzed pollen tubes are unable to grow.[58] On the other hand, excessive turgor pressure can result in tube bursting,[74] and turgor clearly must be carefully calibrated to be within a particular range.[75]

The maintenance and rapid regulation of turgor in pollen tubes is likely to be facilitated by the segmentation of the protoplast that separates the continuously elongating growing region from degenerating distal regions where turgor is gradually lost. This segmentation is achieved by the deposition of plugs made through localized centripetal invagination of the cell wall built from callosic wall material (Figure 6.3b). These plugs are produced repeatedly once the male germ unit has moved forward through a particular tubular segment and ensures that the volume of the living portion of the cytoplasm remains within a relatively constant range.[74,76]

The values of maximum force generated by pollen tubes in *in vitro* setups does not necessarily allow one to deduce the actual penetration force exerted inside the pistil, as this force is dependent on both growth rate and the stiffness and texture of the surrounding matrix.[6,24] The use of differently stiffened growth matrices allows the determination of whether the penetrative ability in a given cell type is influenced by the substrate and what the optimal stiffness is.[23,24,77] Pollen tube species such as *Arabidopsis* actually grow better in a stiffened medium compared to a liquid medium, in line with the tissue architecture of the pistil in this species.[24] A systematic comparison of plant species with solid style (transmitting tissue consisting of densily packed cells) or hollow style (transmitting tissue lining a cell-free canal filled with a viscous extracellular matrix) revealed a consistently different behavior of the respective pollen tubes when confronted with a stiff artificial matrix *in vitro*.[24] Tubes adapted to a solid style not only display greater ability to penetrate a stiffer matrix when compared with the pollen tubes from species with hollow style, but also prefer a stiffer medium when presented with a choice, a phenomenon termed durotropism.[24] The growth matrix influences pollen tube growth not only through its stiffness properties, but structural anisotropy can also cause pollen tubes to display significantly altered growth directions. In *Arabidopsis* mutants with altered cellulose orientation in the stigmatic papilla, pollen tubes formed coils around the papillae rather than growing straight along these longitudinal cells.[143]

Fungal hyphae are exposed to highly variable substrates and, therefore, have to be able to efficiently adjust to changing external osmolyte concentration. Certain substrates that hyphae are able to penetrate are phenomenally stiff, such as rock. Some fungal species can create specialized structures that are able to produce the pressures that are required to invade particularly resistant surfaces. One such example is the plant pathogen *Magnaporthe grisea*, which produces infection pegs from flattened, enlarged hyphal tips called appressoria. Appressoria can establish a turgor pressure of up to 8 MPa that enable the invasive structure to penetrate the plant epidermis.[78] A tight control of the turgor pressure therefore seems to be crucial for fungal organisms. Upon hyperosmotic shock, fungal hyphae of *Neurospora crassa* display reduced turgor pressure, hyphal growth, and decrease in hyphal volume, but all parameters were found to rapidly return to their original values through regulatory mechanisms.[79] The second crucial element in addition to high turgor is the capacity of the appressorium to firmly adhere to the surface of the structure to be invaded. This adherence is key to prevent pushback caused by the invading infection peg. While the invasion angle of the *Magnaporthe grisea* infection peg is normal to the plant leaf surface, a different strategy is employed by *Phytophthora*, a plant pathogenic oomycete. *Phytophthora* hyphae do not form appressoria, but they do adhere to the surface of the plant orgen to then assume an oblique angle to breach the surface. The angled approach has been likened to the slicing principle of single-beveled Japanese kitchen knifes and was accordingly named the 'naifu-mechanism'.[80] Whatever the angle of attack, it is generally assumed that the infection peg or invading hypha requires substantial turgor pressure, although whether this truly applies to all types of hyphae and under all conditions remains unclear, since certain oomycete hyphae seemingly grow even in the absence of any measurable turgor pressure.[81,82]

6.3.3 Cytoskeletal Elements Regulate Tip Growth and Invasion Through Cell Wall Assembly

Force generation in animal cells involves cytoskeleton-based actions such as polymerization of cytoskeletal arrays and contractile mechanisms. Similar principles have been proposed for some walled cells[83,84] since they were observed to elongate under low or absent turgor pressure.[84] In pollen tubes, pharmacological interference with actin polymerization reduces the cell's ability to invade and penetrate a stiffened medium;[22,23] however, whether this effect can be ascribed directly to any force generation by the cytoskeletal arrays is unclear. Interestingly, moderate interference with the actin cytoskeleton also abolishes growth oscillations in these tubes.[85] This indicates that the role of actin in regulating the invasive force of pollen tube growth may be mediated by the delivery of cell wall material to the expanding apex and thus the dynamics of the supply of the building material that is required for the cell to elongate.[22,60,86,87]

Studies in oomycetes *Achlya bisexualis* and *Phytophthora cinnamomi* provide similar evidence for such an indirect role of actin. Comparing hyphae growing either invasively (through agar) or non-invasively (on the agar surface) revealed an 'actin-free zone' at the tip region of the former but not the latter.[88,144] The authors suggest that the depletion of actin at the tip region of invasive cells would ensure a greater yielding capacity of the cell wall, thus allowing for higher invasive force to be exerted.[88,89] Similar evidence for actin functioning as a 'restraint' for tip expansion has been found in zygomycete *Gilbertella persicaria*[90] and ascomycete *Geotrichum candidum.*[91] In other words, actin is suggested to be involved in tip yielding and thus the invasive growth in fungal hyphae, through its effect on the cell wall. Microtubules, on the other hand, do not have a significant effect on the pollen tubes' ability to invade a stiffened medium, but they seem to be involved in enabling pollen tubes to change growth orientation as demonstrated by pharmacological treatment.[23] Mutation in the microtubule-severing enzyme KATANIN1 confirms that invasive pollen tube growth is not affected by interfering with microtubule dynamics.[143] The helically bundled cortical microtubules localized at the sub apical region of a tip growing cell are, however, suggested to be involved in developing a 'structural memory' for the cell be able to restore growth direction after an encounter with an obstacle.[145,146]

6.4 Chemical and Enzymatic Tools Facilitating Invasion

The pollen tube encounters multiple types of obstacles on its way through the stigmatic and stylar tissues, starting with the stigmatic cuticle layer (if present), followed by the transmitting tissue, the micropyle of the ovule, the nucellus enveloping the female gametophyte, and the synergid cells adjacent to the egg cell. In species with solid transmitting tissue, the intercellular spaces are typically narrower than the tube diameter,[92] several additional mechanisms such as enzymatic lysis and chemical digestion are thought to be employed by the tube to facilitate invasion. Enzymes can help break down the molecules involved in cell adhesion or can degrade the invading tissue completely. Some pollen tubes, for example those of *Brassica napus* L., produce cutinase, an enzyme that digests the cuticle of the stigmatic papillae.[93] Agents with the potential to affect the apoplast are expansins,[94,95] polygalacturonase, glucanase, endoxylanase,[96] and pectin esterase.[97] While gene expression profiles show that these proteins are expressed by pollen tubes, it has been difficult to tease apart whether they serve to modulate the pollen tube's own wall or are targeted at the transmitting tissue. Another elegant way to soften the solid matrix is autodigestion *via* programmed cell death.[98] This has been shown to occur upon pollination in plant species such as petunia,[99] where the transmitting tissue undergoes cell death, leading to turgor loss and tissue softening. This both softens the pollen tube's path and may also provide additional nutrients for the elongating cell.

Enzymatic lysis is a crucial tool for fungal species to enable infection of other organisms. Plants use their external cuticle layer and the polysaccharide rich cell wall as the 'first line of defense' against invaders.[100] Fungi, therefore, produce a wide variety of enzymes that have the ability to depolymerize the plant cell wall polysaccharides.[100] Among these are pectinase,[101,102] cellulase,[102] arabinase,[103] xylanase,[104] and galactanase.[105] Enzyme-deficient fungal mutants display reduced ability to cause infection.[106,107] Cutinase is formed in particular by fungal species that do not form appressoria and hence are unable to generate similarly high physical forces.[77,108] The secondary cell walls of wood tissues are more resistant to degradation both physically and chemically, but some fungal species such as white rot and brown rot produce enzymes that degrade hemicellulose, lignin, and cellulose, core polysaccharides forming the compound middle lamella in wood.[109] Enzymatic digestion of cell walls is also often employed by invasive organisms that trigger the development of plant tumors.[110] For instance, in the case of corn smut disease, the infection by the biotrophic fungus *Ustilago maydis* alters the level of hemicelluloses in the infected plant cell wall, modifying the abundance of xylose and arabinose, a process that presumably facilitates piercing the organ surface.[111] Subsequent tumor formation results from a local plant cell enlargement and cell division triggered by the fungus. Unlike in animals, tumors are rarely fatal for plants, since metastasis is impossible due to the lack of a circulatory system.[110]

6.5 Biomechanical Approaches to Quantify Invasive Forces

Invasive growth clearly relies on cell mechanical features and the cells' ability to generate forces. To better understand invasive growth, it is therefore essential to quantitatively characterize these features. Significant effort has been put into designing experimental devices that are able to determine the mechanics of the cell wall, to quantify turgor, and to measure the invasive force of individual cells. Advancement in micro-measurement technology in recent years has been instrumental, notably microfluidics (microdevices with controlled fluid flow) and microelectromechanical system (MEMS)-based platforms.[112–116] These Lab-on-Chip (LoC) devices allow the creation of micro-environments that mimic aspects of the natural growth environment, such as varying degrees of mechanical resistance, chemical gradients, or patterned physical obstacles (Figure 6.4a–c). Importantly, LoC allow observation and manipulation of individual cells, confining their growth to a single focus plane and thus enabling high-resolution microscopy and extended time-lapse imaging.[117] The fabrication of microfluidic and MEMS devices allows for micrometer precision but requires engineering expertise (*e.g.* direct-write lithography and cleanroom facilities),[118,119] but technically simpler and more affordable alternatives can be used if spatial resolution of the design is less critical.[120] Exploiting LoC technology, a platform called the TipChip was developed to study the growth pattern of pollen tubes, for example, to characterize their chemotropic behavior and response to electrical fields[113,121] (Figure 6.4a). These studies are conducted with the aim of understanding how the pollen tube orients its growth in the maze of the female tissues.[122] The electric LoC is fabricated from two separate bondable modules: a PDMS (polydimethylsiloxane)-based microfluidic module for accommodating the suspension of cells in liquid medium and a micro electrode module with a metallic layer that serves to apply the electric field.[113] As in all variations of the TipChip, the height of the microfluidic channel network is determined by the size of the pollen grains (80 μm for *Camellia japonica*),[113] whereas other microfluidic platforms place the grains outside of the microchannel network proper, allowing for a smaller vertical dimension to fit the growing pollen tubes more snugly.[116] Depending on the experimental design and the needs for continuous fluid flow, the channel design must allow fluid-flow-mediated placement of pollen grains to locations or traps where they are immobilized and from where the tubes grow into or towards the testing setup.[123,124] The microchannel

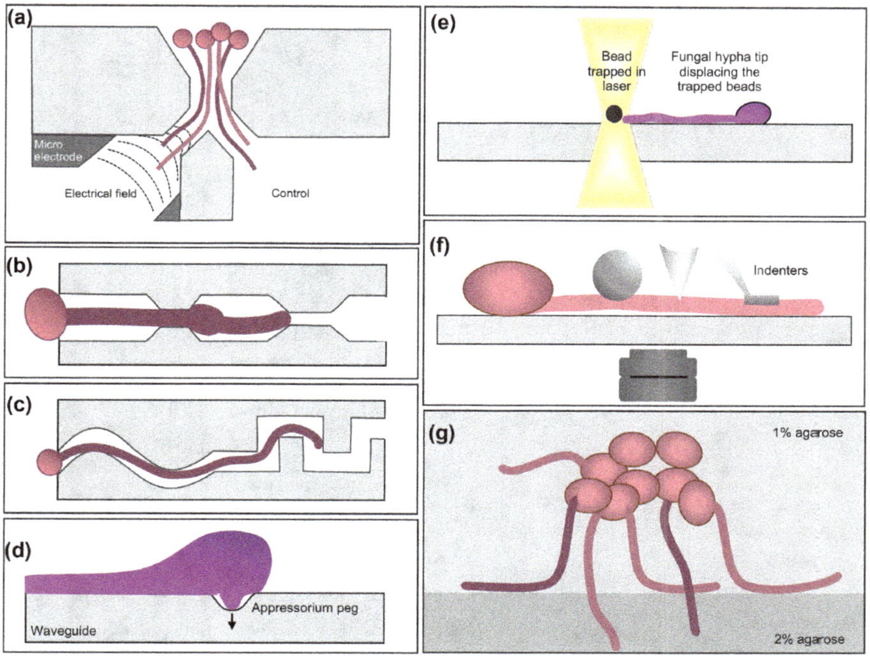

Figure 6.4 Biomechanical approaches to quantifying invasive and oriented tip-growth behavior. (a) Lab-on-chip device offering pollen tubes a choice to grow toward or away from an electrical field. (b) Microfluidic design to challenge elongating pollen tubes with narrow gaps. The pollen tube diameter transiently widens following gap passage. (c) Microchannels with complex geometrical patterns designed to investigate the pollen tube's ability to cope with mechanical obstacles and reorient growth. (d) Measuring penetrative forces of fungal appressorium using waveguide deformation. (e) Optical tweezers to measure the force exerted by hyphal tips by way of displacement of trapped beads. (f) Microindentation using indenters of different shape and size to determine cellular stiffness. The indenters are attached to cantilevers, the deflection of which is monitored by optical sensors. (g) Assay exposing growing pollen tubes to media stiffened to different degrees by varying the concentration of agarose.

architecture must also be designed to avoid clogging to allow for effective liquid exchange.

To quantify the pollen tube's invasive forces, LoC have been employed to expose them to a variety of narrow spaces and complex mazes[114] with sophisticated microscopic design features, such as elastic cantilevers serving as strain gauges (Figures 6.4b–c and 6.5c and d).[72] Microchannels featuring consecutive narrow gaps were designed to mimic the microarchitecture of the pistillar tissue (Figures 6.4b and 6.5b).[114,116,125] An elongating pollen tube deformed the PDMS sidewalls of the gap, allowing for the calculation of the force exerted to maintain its diameter against compressive stress.[125] Intriguingly, the pollen tube diameter changed transiently after it made its way

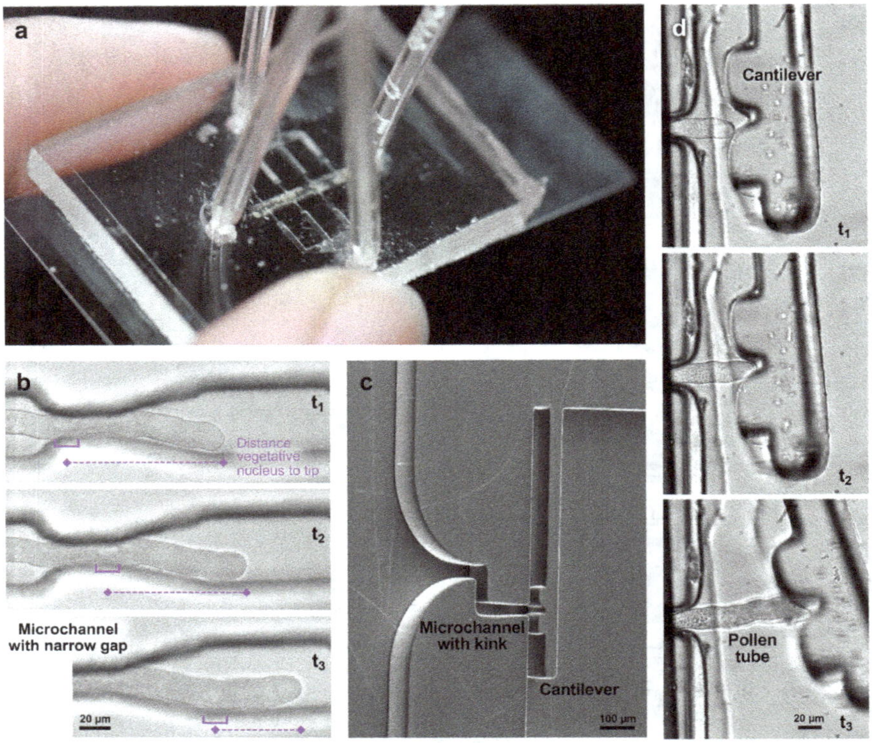

Figure 6.5 Applications of Lab-on-chip devices for the assessment of invasive growth behavior. (a) General design of the TipChip with PDMS layer containing microfluidic network adherent to cover slip. Tubes serve as inlets and outlets for liquids and for injection of pollen suspension. (b) *Camellia* pollen tube traversing a narrow gap. At t_1, the vegetative nucleus (purple bracket) has fallen behind its default distance from the growing tip because it got trapped in the narrow passage created by the gap. At t_2, the vegetative nucleus has almost made its way through the gap thanks to its elastic deformability. At t_3, the vegetative nucleus has reached its default distance from the pollen tube tip (60 µm) by accelerating its forward movement following gap passage. Image series provided by Amir Sanati Nezhad. Related data in Sanati Nezhad *et al.*[125] (c) Scanning electron micrograph of PDMS cantilever used to measure the growth force of pollen tubes in Ghanbari *et al.*[72] (d) Brightfield micrographs of *Camellia* pollen tube growing against cantilever shown in c). Images in d) reproduced from ref. 72 with permission from World Scientific Publishing Co., Inc., Copyright 2018.

through the gap (Figure 6.4b), suggesting the existence of a feedback mechanism that calibrates the invasive force through modulation of cell wall mechanical properties.[125] It was also shown that the vegetative nucleus and sperm cells were able to move forward through the tube while significantly constricted by the gap, demonstrating substantial elastic deformability (Figure 6.5b). Similar observations were made for root hairs and moss protonemata.[116]

The strain gauge principle was also used to measure the invasive force exerted by the hyphal apices.[25,126] In both pollen tubes and fungal hyphae, a technical difficulty consisted in stabilizing the longitudinal cell sufficiently to enable measurement of reliable data for the elongation force. Strain gauges and cantilevers have been combined with kinked microchannels (Figure 6.5c) or agarose to stabilize the base of the tubular cell. This is a critical element of the experimental design, as both pushback and buckling must be prevented to enable reliable and reproducible quantitative measurements. Because of their ability to strongly adhere to surfaces, the problem of stabilizing the base of the cell was less of a challenge in the case of the infection pegs formed from fungal appressoria. This allowed the use of a waveguide whose deformation by the emerging peg could be monitored optically and used to calculate the force[78,127] (Figure 6.4d).

Other attempts to measure the invasive force of tip-growing cells used optical tweezers. Fungal hyphae of *N. crassa* were faced with obstacles in the form of polystyrene beads trapped in a beam of laser light (Figure 6.4e). The force required to push a bead from its trapped position is directly proportional to the bead's size, suggesting that more force is required to dislocate a larger bead.[126] Regular hyphae displaced all the beads, because the strength of the optical trap is limited only to a few piconewton range,[25,126] but the force of a conidial germ tube was determined to be within the dynamic range of this assay, revealing that its invasive force is much lower than that of leading fungal hypha tips.[128]

Since the invasive force of tip-growing walled cells is the turgor pressure, establishing values for this parameter is an important component when characterizing the cell mechanical underpinnings of invasion. If the cell wall is completely pliable, the 'entire' internal pressure generated can be exerted to an outside substrate, but if the apical cell wall poses substantial resistance to deformation, the invasive force exerted by the tip-growing cell can be expected to be lower than its turgor.[25,126] The invasive force therefore does not equate to turgor. In order to measure the turgor pressure, several techniques have been employed, such as incipient plasmolysis or the pressure probe—an oil-filled microcapillary that is injected into the cell.[70,129,130] Since these methods are invasive in nature, they cannot be used repeatedly or over longer periods of time.[131] As a minimally invasive method, ball tonometry and other indentation techniques (Figure 6.4f) have been used to estimate turgor. In ball tonometry, a large spherical probe with a controlled load is applied to the cell and the contact area is measured to deduce turgor. Other indentation techniques are based on pressing a cylindrical probe into the cell and measuring the applied force and the indentation depth. Burri *et al.* (2019) coupled a modified ball tonometry approach with micro-indentation technique to calculate turgor pressure and measure cell wall elasticity. The authors employed a non-invasive microrobotic system based on cellular force microscopy (CFM) in combination with two force sensors with different geometries and force ranges for simultaneous biomechanical measurements on elongation pollen tubes.[131]

Cell wall mechanical properties are important parameters, and atomic force microscopy (AFM), an indentation technique with smaller indentation depth and higher spatial resolution than CFM, has been used to assess plant cells.[132–135] Depending on tip size and indentation depth, the deformation that is applied normal to the cell surface may be influenced by the turgor pressure and/or by the geometry of the tissue structure, and the extraction of absolute mechanical values from indentation measurements is not trivial,[136,137] but has delivered insights on the viscoelastic nature of the fungal hyphae cell wall.[25,138]

Exposing a tip-growing cell to a mechanical cue has the potential to trigger the cell to modulate its invasive force or other cellular parameter—a consideration that must be made when making force measurements. Evidence for this ability of invading cells to modulate their force stems from the observation that pollen tubes growing through increasingly narrow openings continue at a constant speed despite the increasing impedance.[125] This suggests that they may increase their invasive force during the process. This adaptive dynamic behavior was corroborated by the finding that once the obstacle is passed, the tube widens (Figure 6.4b) indicating that its cell wall had softened while it was pressing against the obstacle.

In addition to the absolute value of the invasive force, a comparative approach is therefore warranted. Pollen tubes exposed to an interface between two different concentrations of agarose were scored for their behavior and ability to penetrate into the stiffer medium, thus offering a test assay to assess the effect of pharmacological interference with specific cell features on invasive growth (Figure 6.4g). To put the measured values for the penetrative behavior of tip-growing cells in context, it will be valuable to measure the stiffness of the invaded matrix *in situ*.[24] In the case of pollen tubes, it will therefore be crucial to quantify the stiffness of the pistil transmitting tissue. Determining the stiffness of a uniform material is relatively straightforward and can be done by microindentation.[24] Results obtained with larger indenters can be extrapolated to calculate the invasion required by the microscopic invasive cell. This extrapolation is less obvious for matrices with complex micron-scale architecture, such as that generated by the cellularity of the transmitting tissue and the presence of the middle lamella. Here, indenter size will matter, and the interpretation of such measurements will require careful consideration of geometrical features.

6.6 Conclusion and Perspective

Quantifying physical and mechanical properties at cellular and subcellular levels is a challenge, as practical experimentation on these structures is difficult owing to the small size of the specimens. Despite these challenges, the parameters characterizing invasion and penetration events such as turgor-driven growth and the dynamic regulation of cell wall mechanical properties have been assessed successfully in a range of cell types. Recent developments in micromechanics technologies such as LoC and MEMS-based force sensors will enhance our ability to directly quantify the role of osmolyte concentration and turgor pressure and will also provide insight into the cellular regulation of biomechanical properties. Quantitative micromanipulation will increasingly be coupled with powerful image analysis software, for example those that enable the detection of the subpixel resolution in order to do quantitative analysis of cell behavior.[139]

One of the major challenges associated with studying cellular invasive growth is the observation of this behavior *in situ*, as the invasive growth events occur several layers deep within the invaded tissues. Two-photon excitation microscopy allows deeper penetration of plant tissues compared to conventional epi-fluorescence microscopy and confocal laser scanning microscopy and may thus be one of several possible avenues of research.[140] Novel high-resolution microscopy techniques such as lightsheet fluorescence microscopy are other options for deep tissue imaging, and the continuous improvement in spatial resolution is promising. Since lightsheet imaging requires only low doses of light in order to acquire an image in a single plane, this technique will be crucial to allow long-term time-lapse imaging.[141] The combination of micromechanics with powerful microscopy technology will open exciting avenues for single cell analysis.

Acknowledgements

Work in the Geitmann Lab is supported by a Discovery grant from the Natural Sciences and Engineering Research Council of Canada (NSERC) and by the Canada Research Chairs Program. We thank Louise Pelletier, Amir Sanati Nezhad and Mahmood Ghanbari for producing the micrographs shown in Figure 6.5.

References

1. M. Roper, C. Lee, P. C. Hickey and A. S. Gladfelter, *Curr. Opin. Microbiol.*, 2015, **26**, 116–122.
2. M. Zamil and A. Geitmann, *Phys. Biol.*, 2017, **14**, 015004.
3. A. Carminati, J. B. Passioura, M. Zarebanadkouki, M. A. Ahmed, P. R. Ryan, M. Watt and E. Delhaize, *New Phytol.*, 2017, **216**, 771–781.
4. C. Fife, J. McCarroll and M. Kavallaris, *Br. J. Pharmacol.*, 2014, **171**, 5507–5523.
5. M. Lasser, J. Tiber and L. A. Lowery, *Front. Cell. Neurosci.*, 2018, **12**, 165.
6. A. Sanati Nezhad and A. Geitmann, *J. Exp. Bot.*, 2013, **64**, 4709–4728.
7. Y. Chebli, A. J. Bidhendi, K. Kapoor and A. Geitmann, *Curr. Biol.*, 2021, **31**, R681–R695.
8. J. S. Han, *Proceedings of the Korean Society of Wood Science and Technology Annual Meeting*, 1998.
9. A. Snegireva, M. Ageeva, S. Amenitskii, T. Chernova, M. Ebskamp and T. Gorshkova, *Russ. J. Plant Physiol.*, 2010, **57**, 342–355.
10. K. Esau, *Plant Anatomy*, 1965.

11. T. Gorshkova, P. Mikshina, A. Petrova, T. Chernova, N. Mokshina and O. Gorshkov, *Plant Biomechanics*, Springer, 2018, pp. 141–163.
12. S. Lev-Yadun, *Russ. J. Plant Physiol.*, 2010, **57**, 305–315.
13. B. Clair, T. Alméras, H. Yamamoto, T. Okuyama and J. Sugiyama, *Biophys. J.*, 2006, **91**, 1128–1135.
14. E. J. Mellerowicz, P. Immerzeel and T. Hayashi, *Ann. Bot.*, 2008, **102**, 659–665.
15. A. F. Hill, *Economic Botany*, McGraw Hill, 1952, ch. 1.
16. H. Hayward, *The Structure of Economic Plants*, ed. L. Cramer, *Macmillan, New York*, 1938, pp. 550–579.
17. A. Fahn, *Secretory Tissues in Plants*, Academic Press London, 1979.
18. A. Geitmann, *Sex. Plant Reprod.*, 2010, **23**, 63–71.
19. Y. Chebli, J. Kroeger and A. Geitmann, *Mol. Plant*, 2013, **6**, 1037–1052.
20. A. Geitmann, *Mechanical Integration of Plant Cells and Plants*, Springer, 2011, pp. 117–132.
21. P. Fayant, O. Girlanda, Y. Chebli, C.-É. Aubin, I. Villemure and A. Geitmann, *Plant Cell*, 2010, **22**, 2579–2593.
22. Y. Chebli and A. Geitmann, *Funct. Plant Sci. Biotechnol.*, 2007, **1**, 232–245.
23. O. Gossot and A. Geitmann, *Planta*, 2007, **226**, 405–416.
24. R. Reimann, D. Kah, C. Mark, J. Dettmer, T. M. Reimann, R. C. Gerum, A. Geitmann, B. Fabry, P. Dietrich and B. Kost, *Plant Physiol.*, 2020, **183**, 558–569.
25. N. Money, *Biology of the Fungal Cell*, Springer, 2007, pp. 237–249.
26. R. R. Lew, *Biology of the Fungal Cell*, Springer, 2019, pp. 83–94.
27. L. Yafetto, D. J. Davis and N. P. Money, *Fungal Genet. Biol.*, 2009, **46**, 688–694.
28. C. G. Shaw III and G. A. Kile, *Armillaria Root Disease*, 1991.
29. N. Bolan, *Plant Soil*, 1991, **134**, 189–207.
30. A. Jungk, *J. Plant Nutr. Soil Sci.*, 2001, **164**, 121–129.
31. H. J. Dittmer, *Am. J. Bot.*, 1949, 152–155.
32. C. H. Farr, *Q. Rev. Biol.*, 1928, **3**, 343–376.
33. A. G. Bengough, K. Loades and B. M. McKenzie, *J. Exp. Bot.*, 2016, **67**, 1071–1078.
34. A. Creff, L. Brocard and G. Ingram, *Nat. Commun.*, 2015, **6**, 1–8.
35. M. Lucas, K. Kenobi, D. Von Wangenheim, U. Voβ, K. Swarup, I. De Smet, D. Van Damme, T. Lawrence, B. Péret and E. Moscardi, *Proc. Natl. Acad. Sci.*, 2013, **110**, 5229–5234.
36. L. X. Dupuy, M. Mimault, D. Patko, V. Ladmiral, B. Ameduri, M. P. MacDonald and M. Ptashnyk, *Curr. Opin. Genet. Dev.*, 2018, **51**, 18–25.
37. J. Lipiec, R. Horn, J. Pietrusiewicz and A. Siczek, *Soil Tillage Res.*, 2012, **121**, 74–81.
38. T. Colombi, N. Kirchgessner, A. Walter and T. Keller, *Plant Physiol.*, 2017, **174**, 2289–2301.
39. R. G. White and J. A. Kirkegaard, *Plant, Cell Environ.*, 2010, **33**, 133–148.
40. R. Stirzaker, J. Passioura and Y. Wilms, *Plant Soil*, 1996, **185**, 151–162.
41. S. Materechera, A. Alston, J. Kirby and A. Dexter, *Plant Soil*, 1992, **144**, 297–303.

42. J. G. Chimungu, K. W. Loades and J. P. Lynch, *J. Exp. Bot.*, 2015, **66**, 3151–3162.
43. L. Stolzy and K. Barley, *Soil Sci.*, 1968, **105**, 297–301.
44. E. Kolb, C. Hartmann and P. Genet, *Plant Soil*, 2012, **360**, 19–35.
45. F. Bizet, A. G. Bengough, I. Hummel, M.-B. Bogeat-Triboulot and L. X. Dupuy, *J. Exp. Bot.*, 2016, **67**, 5605–5614.
46. F. E. O'Callaghan, R. A. Braga, R. Neilson, S. A. MacFarlane and L. X. Dupuy, *Sci. Rep.*, 2018, **8**, 1–17.
47. J. Helliwell, C. J. Sturrock, S. Mairhofer, J. Craigon, R. Ashton, A. Miller, W. Whalley and S. J. Mooney, *Sci. Rep.*, 2017, **7**, 1–10.
48. C. J. Stubbs, D. D. Cook and K. J. Niklas, *J. Exp. Bot.*, 2019, **70**, 3439–3451.
49. A. Geitmann, M. Cresti and I. Heath, Cell biology of plant and fungal tip growth, Proceedings of the NATO Advanced Research Workshop, Siena, Italy, 19–23 June 2000, 2001.
50. A. J. Bidhendi and A. Geitmann, *Plant Physiol.*, 2018, **176**, 41–56.
51. J. H. Kroeger and A. Geitmann, *Mech. Res. Commun.*, 2012, **42**, 32–39.
52. J. H. Kroeger, F. B. Daher, M. Grant and A. Geitmann, *Biophys. J.*, 2009, **97**, 1822–1831.
53. S. Bartnicki-Garcia, D. D. Bartnicki, G. Gierz, R. A. López-Franco and C. E. Bracker, *Exp. Mycol.*, 1995, **19**, 153–159.
54. P. K. Hepler, L. Vidali and A. Y. Cheung, *Annu. Rev. Cell Dev. Biol.*, 2001, **17**, 159–187.
55. J. Bove, B. Vaillancourt, J. Kroeger, P. K. Hepler, P. W. Wiseman and A. Geitmann, *Plant Physiol.*, 2008, **147**, 1646–1658.
56. A. Geitmann and A. Nebenführ, *Mol. Biol. Cell*, 2015, **26**, 3373–3378.
57. A. Geitmann and J. K. Ortega, *Trends Plant Sci.*, 2009, **14**, 467–478.
58. J. H. Kroeger, R. Zerzour and A. Geitmann, *PLoS One*, 2011, **6**, e18549.
59. A. Geitmann and M. Steer, *The Pollen Tube*, Springer, 2006, pp. 177–200.
60. C. Y.-h. Chen, A. Y. Cheung and H.-M. Wu, *Plant Cell*, 2003, **15**, 237–249.
61. Y. Chebli, M. Kaneda, R. Zerzour and A. Geitmann, *Plant Physiol.*, 2012, **160**, 1940–1955.
62. P. W. De Groot, A. F. Ram and F. M. Klis, *Fungal Genet. Biol.*, 2005, **42**, 657–675.
63. E. Gomes, R. da Silva, J. de Cassia Pereira and G. Ladino-Orjuela, *Current Developments in Biotechnology and Bioengineering*, Elsevier, 2018, pp. 31–56.
64. M. Ageeva, B. Petrovska, H. Kieft, V. Sal'nikov, A. Snegireva, J. Van Dam, W. Van Veenendaal, A. Emons, T. Gorshkova and A. Van Lammeren, *Planta*, 2005, **222**, 565–574.
65. A. Siedlecka, S. Wiklund, M.-A. Péronne, F. Micheli, J. Leśniewska, I. Sethson, U. Edlund, L. Richard, B. Sundberg and E. J. Mellerowicz, *Plant Physiol.*, 2008, **146**, 554–565.
66. T. Gorshkova, N. Brutch, B. Chabbert, M. Deyholos, T. Hayashi, S. Lev-Yadun, E. J. Mellerowicz, C. Morvan, G. Neutelings and G. Pilate, *Crit. Rev. Plant Sci.*, 2012, **31**, 201–228.
67. D. J. Cosgrove, *Plant Physiol. Biochem.*, 2000, **38**, 109–124.

68. M. Hertzberg, H. Aspeborg, J. Schrader, A. Andersson, R. Erlandsson, K. Blomqvist, R. Bhalerao, M. Uhlén, T. T. Teeri and J. Lundeberg, *Proc. Natl. Acad. Sci.*, 2001, **98**, 14732–14737.
69. M. Israelsson, M. E. Eriksson, M. Hertzberg, H. Aspeborg, P. Nilsson and T. Moritz, *Plant Mol. Biol.*, 2003, **52**, 893–903.
70. R. Benkert, G. Obermeyer and F.-W. Bentrup, *Protoplasma*, 1997, **198**, 1–8.
71. J. T. Burri, H. Vogler, N. F. Läubli, C. Hu, U. Grossniklaus and B. J. Nelson, *New Phytol.*, 2018, **220**, 187–195.
72. M. Ghanbari, M. Packirisamy and A. Geitmann, *Technology*, 2018, **6**, 101–109.
73. R. Zerzour, J. Kroeger and A. Geitmann, *Dev. Biol.*, 2009, **334**, 437–446.
74. L. Beauzamy, N. Nakayama and A. Boudaoud, *Ann. Bot.*, 2014, **114**, 1517–1533.
75. A. E. Hill, B. Shachar-Hill, J. N. Skepper, J. Powell and Y. Shachar-Hill, *PLoS One*, 2012, **7**, e36585.
76. E. Parre and A. Geitmann, *Plant Physiol.*, 2005, **137**, 274–286.
77. M. Bastmeyer, H. B. Deising and C. Bechinger, *Annu. Rev. Biophys. Biomol. Struct.*, 2002, **31**, 321–341.
78. C. Bechinger, K.-F. Giebel, M. Schnell, P. Leiderer, H. B. Deising and M. Bastmeyer, *Science*, 1999, **285**, 1896–1899.
79. R. R. Lew and S. Nasserifar, *Microbiology*, 2009, **155**, 903–911.
80. J. Bronkhorst, M. Kasteel, S. van Veen, J. M. Clough, K. Kots, J. Buijs, J. van der Gucht, T. Ketelaar, F. Govers and J. Sprakel, *Nat. Microbiol.*, 2021, **6**, 1000–1006.
81. R. L. Harold, N. Money and F. Harold, *Protoplasma*, 1996, **191**, 105–114.
82. R. R. Lew, N. N. Levina, S. K. Walker and A. Garrill, *Fungal Genet. Biol.*, 2004, **41**, 1007–1015.
83. J. Gay and A. Greenwood, *Fungus Spore*, 1966, 95–108.
84. J. Li and I. B. Heath, *Exp. Mycol.*, 1994, **18**, 57–69.
85. A. Geitmann, *Fertilization in Higher Plants*, Springer, 1999, pp. 283–302.
86. B. C. Gibbon, D. R. Kovar and C. J. Staiger, *Plant Cell*, 1999, **11**, 2349–2363.
87. L. Vidali, S. T. McKenna and P. K. Hepler, *Mol. Biol. Cell*, 2001, **12**, 2534–2545.
88. S. K. Walker and A. Garrill, *Mycologist*, 2006, **20**, 26–31.
89. S. Suei and A. Garrill, *Protoplasma*, 2008, **232**, 165.
90. S. N. Grove and J. A. Sweigard, *Exp. Mycol.*, 1980, **4**, 239–250.
91. I. B. Heath, M. Bonham, A. Akram and G. D. Gupta, *Fungal Genet. Biol.*, 2003, **38**, 85–97.
92. C. Erbar, *Int. J. Plant Sci.*, 2003, **164**, S265–S277.
93. S. J. Hiscock, F. M. Dewey, J. O. Coleman and H. G. Dickinson, *Planta*, 1994, **193**, 377–384.
94. D. J. Cosgrove, P. Bedinger and D. M. Durachko, *Proc. Natl. Acad. Sci.*, 1997, **94**, 6559–6564.
95. M. Pezzotti, R. Feron and C. Mariani, *Plant Mol. Biol.*, 2002, **49**, 187–197.
96. F. Y. Bih, S. S. Wu, C. Ratnayake, L. L. Walling, E. A. Nothnagel and A. H. Huang, *J. Biol. Chem.*, 1999, **274**, 22884–22894.

97. R. Konar and R. G. Stanley, *Planta*, 1969, **84**, 304–310.
98. J. T. Greenberg, *Proc. Natl. Acad. Sci.*, 1996, **93**, 12094–12097.
99. M. Herrero and H. G. Dickinson, *J. Cell Sci.*, 1979, **36**, 1–18.
100. G. De Lorenzo, R. Castoria, D. Bellincampi and F. Cervone, *Plant Relationships, Springer*, 1997, pp. 61–83.
101. R. J. Linhardt, K. G. Rice, Z. M. Merchant, Y. S. Kim and D. L. Lohse, *J. Biol. Chem.*, 1986, **261**, 14448–14454.
102. R. Blanchette, A. Abad, R. Farrell and T. Leathers, *Appl. Environ. Microbiol.*, 1989, **55**, 1457–1465.
103. M. J. Flipphi, M. van Heuvel, P. van der Veen, J. Visser and L. H. de Graaff, *Curr. Genet.*, 1993, **24**, 525–532.
104. P. Apel, D. Panaccione, F. Holden and J. Walton, *Mol. Plant-Microbe Interact.*, 1993, **6**, 467–473.
105. H. Urbanek and J. Zalewska-Sobczak, *Biochem. Physiol. Pflanz.*, 1986, **181**, 321–329.
106. L. M. Rogers, M. A. Flaishman and P. E. Kolattukudy, *Plant Cell*, 1994, **6**, 935–945.
107. N. J. Tonukari, J. S. Scott-Craig and J. D. Walton, *Plant Cell*, 2000, **12**, 237–247.
108. K. Mendgen, M. Hahn and D. Holger, *Annu. Rev. Phytopathol.*, 1996, **34**, 367–386.
109. G. Daniel, *Secondary Xylem Biology*, Elsevier, 2016, pp. 131–167.
110. C. I. Ullrich, R. Aloni, M. E. Saeed, W. Ullrich and T. Efferth, *Phytomedicine*, 2019, **64**, 153081.
111. A. Matei, C. Ernst, M. Günl, B. Thiele, J. Altmüller, V. Walbot, B. Usadel and G. Doehlemann, *New Phytol.*, 2018, **217**, 1681–1695.
112. M. Horade, M. M. Kanaoka, M. Kuzuya, T. Higashiyama and N. Kaji, *R. Soc., Chem. Adv.*, 2013, **3**, 22301–22307.
113. C. G. Agudelo, A. Sanati Nezhad, M. Ghanbari, M. Naghavi, M. Packirisamy and A. Geitmann, *Plant J.*, 2013, **73**, 1057–1068.
114. A. Sanati Nezhad, *Lab Chip*, 2014, **14**, 3262–3274.
115. A. Geitmann, *Pollen Tip Growth*, Springer, 2017, pp. 87–103.
116. N. Yanagisawa, N. Sugimoto, H. Arata, T. Higashiyama and Y. Sato, *Sci. Rep.*, 2017, **7**, 1403.
117. N. Yanagisawa, E. Kozgunova, G. Grossmann, A. Geitmann and T. Higashiyama, *Plant Cell Physiol.*, 2021, **62**, 1239–1250.
118. C. Agudelo, A. Sanati Nezhad, M. Ghanbari, M. Packirisamy and A. Geitmann, *J. Micromech. Microeng.*, 2012, **22**, 115009.
119. C. G. Agudelo, M. Packirisamy and A. Geitmann, *Plant Cell Morphogenesis*, Springer, 2014, pp. 237–248.
120. H. Bertrand-Rakusová, Y. Chebli and A. Geitmann, *Pollen and Pollen Tube Biology*, Springer, 2020, pp. 211–221.
121. C. Agudelo, M. Packirisamy and A. Geitmann, *Sci. Rep.*, 2016, **6**, 19812.
122. A. Geitmann and R. Palanivelu, *Floriculture Ornam. Biotechnol.*, 2007, **1**, 77–89.

123. M. Ghanbari, A. Sanati Nezhad, C. G. Agudelo, M. Packirisamy and A. Geitmann, *J. Biosci. Bioeng.*, 2014, **117**, 504–511.
124. A. Sanati Nezhad, M. Ghanbari, C. G. Agudelo, M. Naghavi, M. Packirisamy, R. B. Bhat and A. Geitmann, *Biomedical Microdevices*, 2014, **16**, 23–33.
125. A. Sanati Nezhad, M. Naghavi, M. Packirisamy, R. Bhat and A. Geitmann, *Proc. Natl. Acad. Sci.*, 2013, **110**, 8093–8098.
126. G. D. Wright, J. Arlt, W. C. Poon and N. D. Read, *Optical Trapping and Optical Micromanipulation II*, 2005.
127. A. Geitmann, *Am. J. Bot.*, 2006, **93**, 1380–1390.
128. G. D. Wright, J. Arlt, W. C. Poon and N. D. Read, *Mycoscience*, 2007, **48**, 15–19.
129. P. B. Green, *Plant Physiol.*, 1968, **43**, 1169–1184.
130. A. D. Tomos and R. A. Leigh, *Annu. Rev. Plant Biol.*, 1999, **50**, 447–472.
131. J. T. Burri, H. Vogler, G. Munglani, N. F. Läubli, U. Grossniklaus and B. J. Nelson, *IEEE Rob. Autom. Lett.*, 2019, **4**, 641–646.
132. P. Milani, M. Gholamirad, J. Traas, A. Arnéodo, A. Boudaoud, F. Argoul and O. Hamant, *Plant J.*, 2011, **67**, 1116–1123.
133. A. N. Fernandes, X. Chen, C. A. Scotchford, J. Walker, D. M. Wells, C. J. Roberts and N. M. Everitt, *Phys. Rev. E*, 2012, **85**, 021916.
134. K. Radotić, C. Roduit, J. Simonović, P. Hornitschek, C. Fankhauser, D. Mutavdžić, G. Steinbach, G. Dietler and S. Kasas, *Biophys. J.*, 2012, **103**, 386–394.
135. M. Majda, P. Grones, I.-M. Sintorn, T. Vain, P. Milani, P. Krupinski, B. Zagórska-Marek, C. Viotti, H. Jönsson and E. J. Mellerowicz, *Dev. Cell*, 2017, **43**(290–304), e294.
136. A. Peaucelle, S. A. Braybrook, L. Le Guillou, E. Bron, C. Kuhlemeier and H. Höfte, *Curr. Biol.*, 2011, **21**, 1720–1726.
137. L. Beauzamy, J. Derr and A. Boudaoud, *Biophys. J.*, 2015, **108**, 2448–2456.
138. H. Ma, L. A. Snook, S. G. Kaminskyj and T. E. Dahms, *Microbiology*, 2005, **151**, 3679–3688.
139. D. S. Damineli, M. T. Portes and J. A. Feijó, *Pollen Tip Growth*, Springer, 2017, pp. 391–413.
140. A. Y. Cheung, S. Niroomand, Y. Zou and H.-M. Wu, *Proc. Natl. Acad. Sci.*, 2010, **107**, 16390–16395.
141. A. Candeo, F. G. Doccula, G. Valentini, A. Bassi and A. Costa, *Plant Cell Physiol.*, 2017, **58**, 1161–1172.
142. M. Held, O. Kašpar, C. Edwards and D. V. Nicolau, *Proc. Natl. Acad. Sci.*, 2019, **116**, 13543–13552.
143. L. Riglet, F. Rozier, C. Kodera, S. Bovio, J. Sechet, I. Fobis-Loisy and T. Gaude, *eLife*, 2020, **9**, e57282.
144. S. K. Walker and A. Garrill, *Mycologist*, 2006, **20**, 26–31.
145. K. Vaškovičová, V. Žárský, D. Rösel, M. Nikolič, R. Buccione, F. Cvrčková and J. Brábek, *Biol. Direct*, 2013, **8**, 1–21.
146. B. J. Sieberer, T. Ketelaar, J. J. Esseling and A. M. C. Emons, *New Phytol.*, 2005, **167**, 711–719.

CHAPTER 7

Hygroresponsive Movements of Plants and Soft Actuators

KEUNHWAN PARK,[a] JONGHYUN HA,[b] BEOMJUNE SHIN[c] AND HO-YOUNG KIM*[c]

[a] Department of Mechanical Engineering, Gachon University, Seongnam 13306, Korea; [b] Department of Mechanical Science and Engineering, University of Illinois at Urbana-Champaign, Urbana, Illinois 61801, USA; [c] Department of Mechanical Engineering, Seoul National University, Seoul 08826, Korea

*Email: hyk@snu.ac.kr

7.1 Introduction

Plants can generate motions, as evidenced by twirling circumnutation of growing tendrils and closing of Mimosa leaves and Venus flytraps. They are in motion all the time, albeit too slow or too fast to catch with the naked eye in many cases. They grow to receive more water, light and nutrients; they change their shapes in response to environmental stimuli, including light, humidity and heat; and they disperse their seeds to a great distance. Despite seemingly diverse ways to generate motions, these plant movements share a common feature in that they are essentially hydraulic. Namely, the motions are driven by the supply or deprival of water, which changes the volume of cells or tissues. Because plants do not resort to complex motor proteins of muscles, their relatively simple movements serve as a rich source of bio-inspired actuation technologies, which we aim to address in this chapter.

When live cells are involved in botanical motions, the differential turgor pressure builds internal stress and strain over the plant tissue, eventually

Soft Matter Series No. 15
Soft Matter in Plants: From Biophysics to Biomimetics
Edited by Kaare H. Jensen and Yoël Forterre

Published by the Royal Society of Chemistry, www.rsc.org

resulting in movements.[1] The turgor pressure changes by osmosis across permeable cell walls, resulting from active regulation or environmental conditions. For the active movements, as can be seen in stomata, for example, plants regulate the turgor pressure by controlling ion concentration in the cells.[2] When the ion concentration of guard cells in stomata is increased, the cells swell by absorbing water to deform in such a way that pores surrounded by the cells are opened, allowing gas exchange for photosynthesis. To block gas exchange, the ion concentration of guard cells is reduced so that the cells lose water through osmosis and close stomata. The turgor pressure can also change passively, as can be seen in the differential drying of the fruit of *Hura crepitans* due to environmental humidity change.[3] The passive movement of the fruit's catapult-like mechanism, triggered by drying and shrinkage, enables the seeds' explosive dispersal.

Plants can generate motions even with dead cells and tissues. Their motions are passively driven by the volume change of the cells and tissues which can swell by absorbing moisture from an external water source in the form of either liquid or vapor. That is, they explore the hygroresponsive nature of cellulosic materials constituting the plant cells instead of turgor pressure regulation. Figure 7.1a–c shows the seeds of some species with long appendages called awns, which are responsible for the seed locomotion by moving in response to environmental humidity change.[4–6] Simple swelling and shrinkage of the

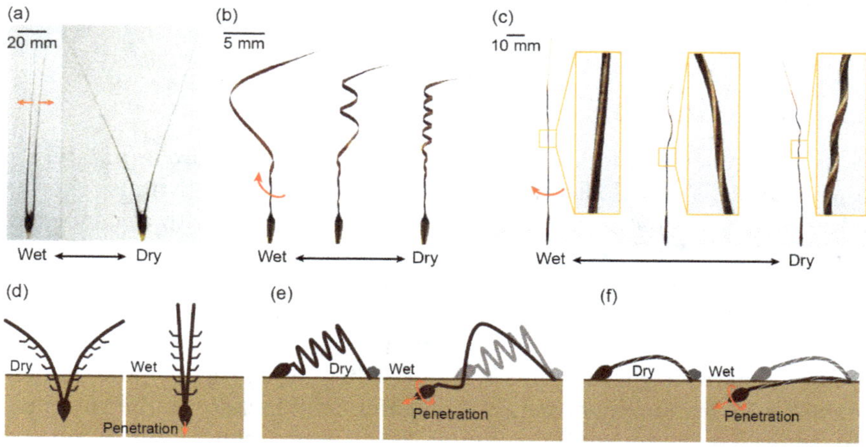

Figure 7.1 Hygroresponsive movements of seed awns with a change in environmental humidity. (a) Bending of the seed awns of wild wheat, *Triticum turgidum*.[4] (b) Helical coiling of the seed awn of *Pelargonium appendiculatum*.[10] (c) Twisting of the seed awn of *Stipa pulchra*. (d) Self-burial of the wild wheat seed that uses barbules as a ratchet. (e) The seed of *Pelargonium appendiculatum* that screws into soil with the end of the awn anchored at the soil. (f) The seed of *Stipa pulchra* that drills itself into soil. Panel (a) and (d) adapted from ref. 4 with permission from The American Association for the Advancement of Science, Copyright 2007. Panel (b) adapted from ref. 10 with permission from Elsevier, Copyright 2020.

constituent materials can be harnessed to generate diverse movements of the awns, including bending (Figure 7.1a), helical coiling (Figure 7.1b), and twisting (Figure 7.1c). Those movements allow the seeds to bury themselves *via* unidirectional propulsion as anchored to soil by barbules (Figure 7.1d) or at the awn tip (Figure 7.1e–f). The hygroresponsive movements of plants can also be observed in the bending of pine cones,[7] curling of resurrection plants,[8] and chiral deformation of seed pods of *Bauhinia variegata.*[9]

From the perspective of mimicking the plant movements for practical applications, the passive schemes based on the hygroresponsive materials are far more accessible and thus attractive than the active schemes for the following reasons. First, the passive schemes do not require an external control unit or energy source, allowing for low-cost devices of a simple structure. Second, the relatively simple motion generation strategy implies that miniature stand-alone devices are possible that use environmental humidity change as the trigger and energy source for movements. Third, environmentally friendly devices are possible with the continual development of eco-friendly materials that respond to humidity in the same way as plant cell walls.

Here, we introduce the physical principles behind hygroresponsive deformation of plants while emphasizing how plant tissues can program their deformations with structural constraints. We then survey artificial hygroresponsive actuators reported to date with classification according to their materials and mode of deformation. We compare the performances of biological and artificial actuators, and discuss the applications of bio-inspired actuators powered by environmental humidity.

7.2 Physical Principles of Hygroresponsive Deformation of Plants

We start with the fundamental mechanism for volume change of hygro-expansive materials in response to moisture content in a tissue surrounded by a wet environment. We then introduce ingenious strategies of plants to induce diverse movements by sophisticated arrangements of nanostructure in cell walls. We also briefly present mathematical models to understand the plant tissue's deformation based on the theory of elasticity.

7.2.1 Hygroscopic Swelling in Plants

The botanical movements powered by the environmental humidity change, as exemplified in Figure 7.1, originate from the physicochemical properties of the materials constituting the cell walls. The plant cell walls are composed of such organic compounds as cellulose, hemicellulose and lignin.[11] Figure 7.2a illustrates those constituents, where cellulose microfibrils are loosely surrounded by hemicellulose, and the microfibrils are connected by lignin that bonds to cellulose *via* hydroxyl groups.[12] As water molecules penetrate into cell walls through diffusion, they form hydrogen bonds with

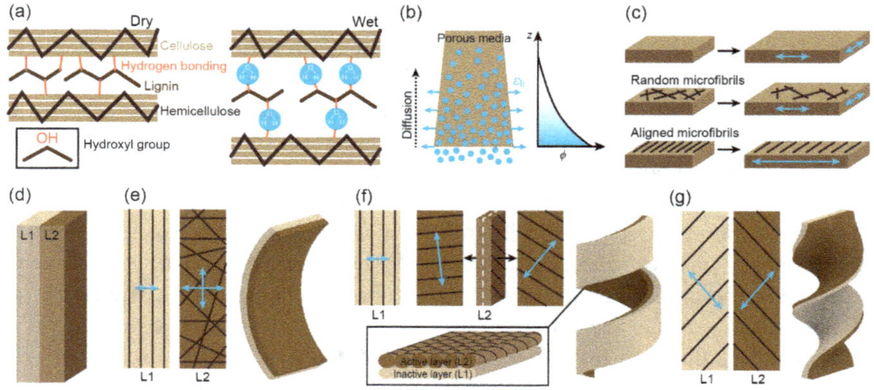

Figure 7.2 (a) The molecular mechanism of hygroscopic expansion of a plant cell wall. (b) Swelling of a porous plant tissue owing to diffusion of water molecules from the external water source at the bottom ($z=0$). (c) Hygroscopic expansion of isotropic (top and middle panels) and anisotropic (bottom panel) media. (d) A bilayer model used to analyze the mechanics of seed awn deformation. (e) Bending of the bilayer for isotropic contraction of L2 with drying. (f) Helical coiling for anisotropic contractions of the sublayers of L2 with tilted microfibril angles. Inset: Schematic of the bilayer structures of the awn. (g) Twisting of the bilayer with contractions of L1 and L2 having opposite microfibril angles. In (b–g), blue arrows indicate the strain directions.

hydroxyl groups, increasing the spacing between the microfibrils. Thus, cell walls can hygroscopically swell if water can participate in hydrogen bonding between lignin and microfibrils.

Now we consider a plant tissue composed of multiple cells, which is in contact with water at one end ($z=0$), as shown in Figure 7.2b. As dead plant tissues act as porous media that allow diffusion of water molecules,[13] the temporal evolution of the moisture concentration, ϕ, within the tissue is described by the diffusion equation: $d\phi/dt = D\nabla^2\phi$ with t being time. The diffusion coefficient, D, is determined by the size, distribution and network structure of pores.[14] As the amount of material expansion measured by the hygroexpansive strain, ε_h, is proportional to the change of the moisture concentration $\Delta\phi$, we write $\varepsilon_h = \alpha\Delta\phi$ with α being the hygroscopic expansion coefficient. If D and ε_h are known for a plant tissue with spatially uniform properties, we can obtain the instantaneous shape of the tissue (or the strain field) at a certain time t since the contact with water by simply solving the diffusion equation for ϕ (Figure 7.2b).

7.2.2 Deformation Modes Programmed by Structural Constraints

The cell walls expand as the gaps of cellulose microfibrils are increased by water molecules, as described in Figure 7.2a. This implies that the local hygroexpansion arises in a direction perpendicular to the microfibrils rather than isotropically. When the microfibrils in a single cell wall are aligned in one direction and those cells are also aligned, the expansion of the tissue comprising the cells can be guided anisotropically. Here we simplify the tissues with hygroresponsive cells as plates, and describe their hygroscopic deformations in Figure 7.2c–g. Figure 7.2c illustrates isotropic and anisotropic deformations of a hygroexpansive porous plate depending on the structural constraints. The upper and middle plates expand isotropically owing to the absence of constraints and the randomly embedded microfibrils, respectively. The lower plate swells only in a single direction perpendicular to the aligned microfibrils, thereby rendering the hygroscopic expansion coefficient α dependent on direction.

Plants need to go beyond the simple expansion behavior of such a porous plate to generate diverse movements. To this end, the motile parts of the plants consist of multiple layers as shown in Figure 7.2d, where the cell walls in layer L1 and those in L2 possess different microfibril arrangements. We note that the straightened configuration in Figure 7.2d corresponds to the wet state, which can exhibit diverse deformation modes in Figure 7.2e–g with drying.

Figure 7.2e explains the microstructure of the seed awn of wild wheat (Figure 7.1a). The long thin seed awn allows us to employ a plate model with the length much greater than the width. Microfibrils are aligned along the longitudinal direction in the cell walls of L1, thereby resulting in the transverse elongation and contraction. The cell walls in L2 have randomly arranged microfibrils, so that the expansion occurs isotropically. When the bilayer is dried owing to the decrease of environmental humidity, both layers contract. For the small width relative to the length, the transverse contraction of L1 is negligible and thus the contraction in L2 dominates, resulting in the bending of the dried awn concave toward the L2 side as shown in Figure 7.2e.

The helically coiling seed awn in Figure 7.1b also consists of two layers, but the cells in L2 are aligned as shown in the inset of Figure 7.2f. All of the microfibrils on the upper side of the cells in L2 appear aligned in one direction while those on the lower side appear aligned in another direction, as illustrated in Figure 7.2f. Hence, L2 can be conceptually decomposed into two sublayers with differently aligned microfibrils. L1 acts as an inactive layer for its negligible transverse contraction as previously explained. Because the microfibrils in L2 are tilted with respect to the longitudinal axis, the anisotropic contractions of the two sublayers of L2 cause the entire structure to helically coil when drying.[10]

The seed awn of *Stipa* in Figure 7.1c consists of two layers with tilted microfibrils, such that the microfibril angle in L1 (measured from the longitudinal axis) is opposite to that in L2, as illustrated in Figure 7.2g. The transverse contractions of the layers cancel each other, preventing the awn from coiling with a finite radius. Therefore, drying of the bilayer leads to twisting, or coiling with zero radius.[9]

We have seen that the movements of plant tissues consisting of many cells can be physically understood by modeling the tissue as a laminate structure with the constituent layers having peculiar arrangements of microfibrils. Thanks to the small thickness of the awns (measuring tens of μm) relative to their length and width, of centimeters and hundreds of micrometers, respectively, we use a theory of elasticity for laminated composite plates to mathematically analyze the deformation of the awns in response to environmental humidity. The theory can substantially simplify the problem because only the thickness among geometric parameters determines the elastic response. Here we briefly provide an overview of the theory.

A simple constitutive relation for elastic deformation is written as $\sigma = C\varepsilon$, where σ, C, and ε are the stress, the elastic modulus, and the total strain, respectively. The total strain at a distance z from the neutral plane is given by $\varepsilon = \varepsilon_0 + \kappa z$ with ε_0 and κ being the midplane strain and the curvature, respectively. The absence of any external loads leads us to write the equilibrium equations for the force, $\boldsymbol{F} = \int \boldsymbol{\sigma} \mathrm{d}z = 0$ and the moment, $\boldsymbol{M} = \int \boldsymbol{\sigma} z \mathrm{d}z = 0$. Combining the two equations results in the matrix equations of $\boldsymbol{F}$ and $\boldsymbol{M}$: $\boldsymbol{F} = \boldsymbol{A}\boldsymbol{\varepsilon}_0 + \boldsymbol{B}\boldsymbol{\kappa}$ and $\boldsymbol{M} = \boldsymbol{B}\boldsymbol{\varepsilon}_0 + \boldsymbol{D}\boldsymbol{\kappa}$, where $\boldsymbol{A}$, $\boldsymbol{B}$ and $\boldsymbol{D}$ are the matrices associated with the mechanical properties of the composite materials and the layer thickness.[10,15] We find the strain and curvature vector to be given by

$$\begin{Bmatrix} \varepsilon_0 \\ \boldsymbol{\kappa} \end{Bmatrix} = \begin{bmatrix} \boldsymbol{A} & \boldsymbol{B} \\ \boldsymbol{B} & \boldsymbol{D} \end{bmatrix}^{-1} \begin{Bmatrix} \boldsymbol{F} \\ \boldsymbol{M} \end{Bmatrix}. \tag{7.1}$$

Because the stress arises by hygroscopic swelling, we can also write $\boldsymbol{F}$ and $\boldsymbol{M}$ as $\boldsymbol{F} = \int \boldsymbol{\sigma}_\mathrm{h} \mathrm{d}z$ and $\boldsymbol{M} = \int \boldsymbol{\sigma}_\mathrm{h} z \mathrm{d}z$, where the hygroscopic stress is given by $\boldsymbol{\sigma}_\mathbf{h} = \bar{\boldsymbol{D}}\boldsymbol{\alpha}\Delta\phi$ with $\bar{\boldsymbol{D}}$ being the stiffness matrix dependent on the elastic modulus and Poisson's ratio. The alignment of microfibrils in each layer determines the hygroscopic expansion coefficient vector $\boldsymbol{\alpha}$, such that $\boldsymbol{\alpha}$ has a higher value in a direction perpendicular to the microfibril alignment. For a medium with randomly arranged microfibrils, α is a scalar without directionality. The total force and moment are obtained by integrating stresses over the entire cross-section of the composite laminates. Thus, the directional properties of hygroexpansion in each layer affect the directions of $\boldsymbol{F}$ and $\boldsymbol{M}$. The moisture concentration ϕ that changes with time for diffusion of water molecules also affects $\boldsymbol{F}$ and $\boldsymbol{M}$. We calculate $\boldsymbol{F}$ and $\boldsymbol{M}$ for the multilayers with time, which give the strain and curvature as expressed in eqn (7.1). They determine the temporal evolution of the shape of plant tissue with known microfibril arrangements and the elastic properties.

7.3 Hygroresponsive Soft Actuators

Soft actuators capable of flexible mechanical motions using soft materials are of great interest for their potential applications in soft robotics,[16,17] energy harvesting[18,19] and biomedical devices.[20,21] Here, we introduce soft actuators inspired by the hygroscopic movement mechanisms of plants, which harness water as energy sources. Just as plants utilize the hygroexpansive properties of cell walls and program their deformations using structural constraints, phytomimetic actuators should employ hygroexpansive materials and devise a way to guide their movements in the desired fashion. In the following, we classify the hygroscopic actuators depending on the type of materials, as listed in Table 7.1, and discuss the regulation mechanisms of their deformation, which are frequently linked to the materials processing scheme. Then we present the actuators' performance measured by energy density and response time, and discuss their practical applications.

One of the earliest hygroscopic actuators used paper as a hygroscopically active layer with a polymer film attached as an inactive layer.[22] Paper is composed of a network of cellulose fibers from plants including wood and fiber crops, which can imbibe water and swell. It was shown that the paper-polymer bilayer bent with swelling of wet paper, similarly to a bimetallic strip. Such actuators employing bio-based materials as an active medium are herein termed bio-hybrid actuators (Table 7.1). The examples include those using cellulose sponges,[23] suspension of *Bacillus* spores,[24] *E. coli* cells,[25] and

Table 7.1 Survey of hygroresponsive actuators.

Type	Mode of deformation	Active layer	Inactive layer	Reference
Bio-hybrid actuators	Bending	Paper	Polymer sheet	22
		Paper	—	27
		Beech wood	Spruce wood	26
		Paper	Wax	28
		Bacillus spores	Latex sheet	24
	Coiling	Sponge	Thread	23
Purely artificial actuators	Bending	Polyethylenimine/ poly(acrylic acid)	—	29
		Poly(acrylic acid)/ poly(allylamine hydrochloride)	NOA63	30
		Liquid crystal	—	31
		Pentaerythritol ethoxylate–polypyrrole	—	18
		Poly(allylamine hydrochloride)/ poly(acrylic acid)	Polytetrafluoroethylene	32
		Poly(*N*-isopropylacrylamide)-*co*-acrylic acid	Polymer film	33
		Polydopamine-modified reduced graphene oxide	NOA63	34
		Cellulose stearoyl ester	—	35
		Chitosan/graphene oxide	—	36
		Graphene oxide (wavy)	Graphene oxide (smooth)	37
		Polyvinyl alcohol	—	38
	Bending, coiling	Graphene oxide (wavy)	Graphene oxide (smooth)	39
		Liquid crystal polymer network	—	40
		Polyethylene oxide	Polyimide	10, 14

strips of beech[26] as active materials. While the other bio-hybrid actuators only exhibited simple bending with the expansion of the active layer, the cellulose sponge wound by a thread[23] was shown to helically coil for the spiral constraint to the swelling of the sponge.

A variety of artificial hygroexpansive materials has been exploited and developed as active media in purely artificial actuators. Such actuators listed in Table 7.1 utilize polymer materials, graphene oxide, or liquid crystal, which involve hydrogen bonding sites in which water molecules can participate. Innovations of these actuators are being sought: (1) to improve the response speed; (2) to increase the degree of deformation; and (3) to achieve diverse deformation modes, as surveyed in the following.

(1) The response speed of hygroresponsive actuators is mainly determined by water diffusion rate in the active layer. For increasing the diffusion rate, micropores were introduced in the drying process of a hygroscopic graphene oxide suspension[37] or by depositing nanofibers through electrospinning[14] (Figure 7.3a and b). The following relation was suggested to estimate the effective diffusivity of water vapor, D_e, in a porous medium of porosity η, $(D_e - D_s)/(D_e + 2D_s) \approx \eta(D_a - D_s)/(D_a + 2D_s)$, with D_s and D_a being the water vapor diffusivity in the air and the solid.[41] We can see that D_e can be greatly enhanced with the

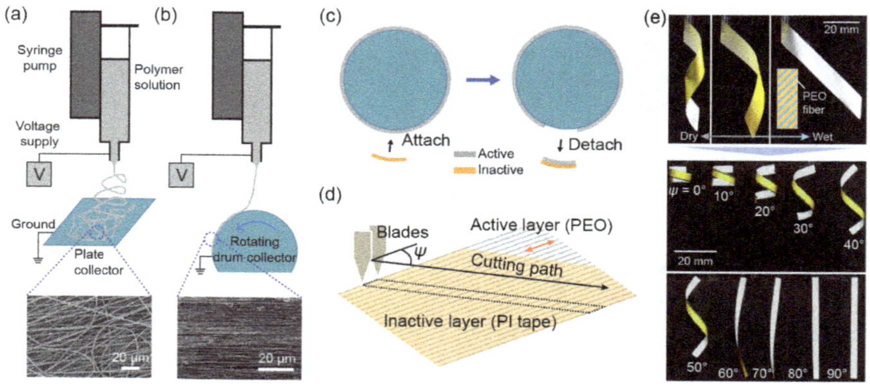

Figure 7.3 (a) Conventional electrospinning process to obtain a chaotic pile of polyethylene oxide (PEO) nanofibers. (b) Directional electrospinning process to deposit aligned PEO nanofibers on a rotating drum collector. (c) Fabrication of a bilayer actuator by attaching an inactive polyimide (PI) layer to the active PEO layer. (d) Cutting of the bilayer at an angle relative to the fiber alignment to obtain helical coiling. (e) Experimental images of the bilayer actuators that have coiled by drying under the relative humidity of 20%. The helix shape (radius and pitch) depends on the angle between the fiber alignment and the longitudinal axis of the actuator. Panel (b) and (c) adapted from ref. 14 with permission from The American Association for the Advancement of Science, Copyright 2018. Panel (d) and (e) adapted from ref. 10 with permission from Elsevier, Copyright 2020.

increase of η because $D_a \sim 10^{-5}\ m^2\ s^{-1}$ is of several orders higher than D_s of polymers on the order of $10^{-12}\ m^2\ s^{-1}$.

(2) The bending curvature of bilayer actuators is a function of the hygroexpansion coefficient, the thickness of each layer, and the ratio of elastic moduli, as dictated by a classical theory of Timoshenko.[42] Shin *et al.*[14] increased the hygroexpansion coefficient of the active layer by aligning nanofibers in the longitudinal direction through the process of directional electrospinning (Figure 7.3b). Patterning wrinkles on the active layer can enhance the degree of bending, as demonstrated by Qiu *et al.*[37]

(3) For more complicated deformations than mere bending, attempts were made to embed structural constraints in the active layer, just as observed in the seed awns of *Pelargonium* and *Stipa*. Alignments of liquid crystal polymer network,[40] electrospun polyethylene oxide (PEO) nanofibers,[10] and corrugations of graphene oxide deposits, in various angles with respect to the longitudinal axis of active layer, were shown to induce helical coiling of various pitches and radii. Figure 7.3e shows the bilayer actuators that coil helically depending on the tilt angle of aligned nanofibers with respect to the longitudinal axis.

An actuator's performance is frequently measured by how much mechanical energy it can produce per unit volume and time. Hence, we indicate the energy density and response time of the actuators whose pertinent information is available, in Figure 7.4. Such choice of the variables enables us to draw lines of constant slopes that correspond to lines of equal power density. We find that many of purely biological and bio-hybrid actuators based on *Bacillus* spores,[24] pine cones,[22] and wood[26] are relatively strong but slow by exhibiting a large energy density but a long response time. Artificial actuators with a short response time were reported, which adopted thin hygroexpansive layers.[14,18] A porous active layer produced by deposition of electrospun nanofibers achieved a response time of approximately 1 s for an actuator 25 mm in length and 68 μm in thickness.[14] Even for the same material, the energy density could be raised by an order of magnitude (blue triangle *versus* blue square) by aligning nanofibers in one direction through directional electrospinning, as depicted in Figure 7.3b.

The ubiquity of water in our environments (as humidity, sweat, fog, rain, *etc.*) makes it highly promising to utilize hygroresponsive actuators in practical applications. Attempts are being made to use the actuators for such areas as: (1) robotic locomotion; (2) energy harvesting; and (3) smart materials for architecture and garments.

(1) The continual variation of environmental humidity causes the actuators to alternate the deformation directions, *e.g.* between bending (coiling) and unbending (uncoiling), which should be rectified to achieve directional motion for locomotive applications. A bending actuator was shown to locomote in one direction under varying humidity when

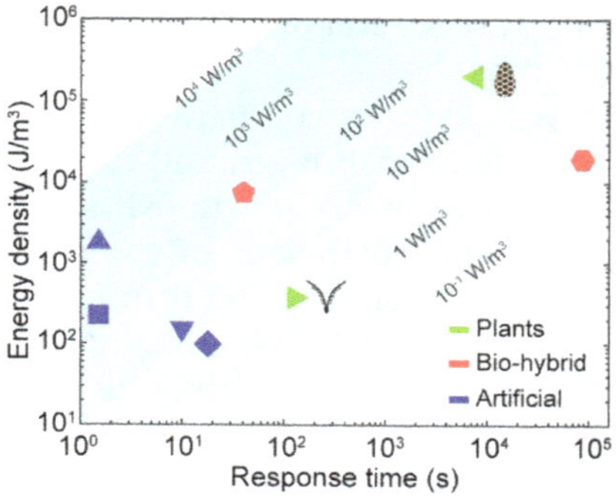

Symbol	Hygroscopic materials	Remarks	Ref.
◀	Pine cone	-	22
▶	Seed of wild wheat	-	4
⬢	Beech wood	-	26
⬟	*Bacillus* spore	-	24
▲	Polyethylene oxide	Aligned fibers	14
■	Polyethylene oxide	Random fibers	14
◆	Polyethylene oxide	Spin-coated	14
▼	Poly(acrylic acid) /Poly(allylamine hydrochloride)	Spin-coated	18

Figure 7.4 Energy density and response time of hygroresponsive soft actuators.

placed on a patterned substrate which allows only one-directional crawling.[30] Figure 7.5a shows a bending actuator with leg-like appendages at the ends, which exhibited a variable friction coefficient depending on the posture. The actuation system could wriggle forward at a velocity of 6 mm s^{-1} or 0.24 BL s^{-1}, with BL denoting the body length, under the humidity variation between 20% and 80% at 0.5 Hz.[14] Figure 7.5b shows a miniature car consisting of multiple bending actuators arranged on a wheel, which could rotate thanks to the humidity difference between the region surrounded by the wet paper and the outer region exposed to dry air.[43]

(2) Hygroresponsive actuation can be exploited to convert the energy of environmental humidity to electricity. Ma *et al.*[18] attached a piezoelectric film to a hygroscopically active layer, as shown in Figure 7.5c, and placed the actuator-generator system on a wet substrate. The gradient of relative humidity from the wet substrate to the dry environmental air drove the actuator's continuous flipping, so that electricity could be generated. Figure 7.5d shows an energy harvesting system placed on a water bath, in which a mechanical structure oscillated by an array of hygroresponsive actuator is connected to an electromagnetic generator.[43]

(3) The ability of hygroscopic actuators to change their shape in response to environmental humidity change can be harnessed to develop smart actuation systems that interact with surrounding natural conditions. Figure 7.5e shows a wall consisting of hygroscopic wood plates, which opens in dry weather but closes when wet.[44] The garment shown in Figure 7.5f has ventilation flaps made of hygroscopic actuators, which can open when wet with human sweat but close when dry.[25]

7.4 Conclusions

We have introduced the physical principle of hygroexpansive movements of plants and bio-inspired actuators powered by environmental humidity. The

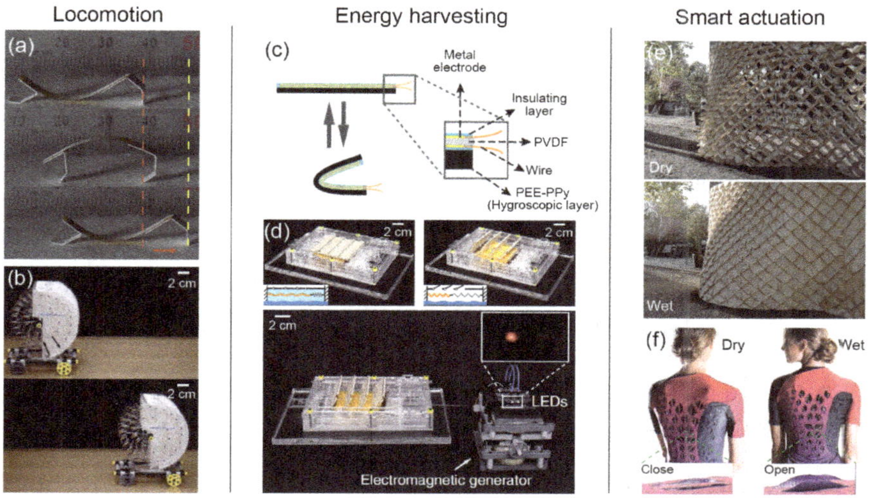

Figure 7.5 (a) A locomotive actuation system, named hygrobot, which propels itself by rectifying repeated bending and unbending with leg-like appendages under humidity variation. (b) A miniature car that is powered by rotation of a wheel consisting of multiple hygroresponsive actuators under different humidity conditions established by the white wet paper. (c) A cross-section of a composite layer consisting of the upper piezoelectric film with electrodes and the lower hygroscopically active layer. (d) An electromagnetic generator connected to a mechanical structure that oscillates with hygroexpansion of the actuator array. (e) A humidity-sensitive wall consisting of hygroscopic plates. (f) A garment with ventilation flaps that change shape in response to the surrounding humidity conditions. Panel (a) adapted from ref. 14 with permission from The American Association for the Advancement of Science, Copyright 2018. Panel (b) and (d) adapted from ref. 43, https://doi.org/10.1038/ncomms8346, under the terms of the CC BY 4.0 license https://creativecommons.org/licenses/by/4.0/. Panel (c) adapted from ref. 18 with permission from The American Association for the Advancement of Science, Copyright 2013. Panel (e) adapted from ref. 44 with permission from Elsevier, Copyright 2015. Panel (f) photo credit: Tangible Media Group, MIT Media Lab, adapted from ref. 25 with permission from The American Association for the Advancement of Science, Copyright © 2017, The Authors.

power density of artificial actuators is now comparable to their biological counterparts owing to advanced material synthesis and processing technology, and mathematical modeling for design optimization. Still, challenges lie ahead, as listed partially in the following, before hygroscopic actuators are widely appreciated as competitive elements for future sensors, actuators, and soft robots. First, the strength of the materials should be enhanced further to compete with other actuators, including artificial muscles and shape-memory polymers. Second, durability of the moisture-sensitive actuators should be improved under repeated cycles of humidity variation and excessive supply of water. Third, materials that change shape differently in response to different fluids can find diverse applications, including hazard detection, mechanochemical encryption, and versatile shape-morphing systems.

Acknowledgements

This work was supported by National Research Foundation of Korea (Grant Nos. 2018-052541 and 2021R1A4A3027074).

References

1. J. M. Skotheim and L. Mahadevan, Physical limits and design principles for plant and fungal movements, *Science*, 2005, **308**, 1308.
2. C. Willmer and M. Fricker, *Stomata*, Chapman & Hall, London, 1996.
3. M. Swaine and T. Beer, Explosive seed dispersal in *Hura crepitans* L.(Euphorbiaceae), *New Phytol.*, 1977, **78**, 695–708.
4. R. Elbaum, L. Zaltzman, I. Burgert and P. Fratzl, The role of wheat awns in the seed dispersal unit, *Science*, 2007, **316**, 884.
5. W. Jung, W. Kim and H.-Y. Kim, Self-burial mechanics of hygroscopically responsive awns, *Integr. Comp. Biol.*, 2014, **54**, 1034.
6. J. W. Bartolome, *Stipa pulchra*, a survivor from the pristine prairie, *Fremontia*, 1981, **9**, 3.
7. C. Dawson, J. F. Vincent and A. M. Rocca, How pine cones open, *Nature*, 1997, **390**, 668.
8. A. Rafsanjani, V. Brulé, T. L. Western and D. Pasini, Hydro-responsive curling of the resurrection plant *Selaginella lepidophylla*, *Sci. Rep*, 2015, **5**, 8064.
9. S. Armon, E. Efrati, R. Kupferman and E. Sharon, Geometry and mechanics in the opening of chiral seed pods, *Science*, 2011, **333**, 1726.
10. J. Ha, S. M. Choi, B. Shin, M. Lee, W. Jung and H.-Y. Kim, Hygro-responsive coiling of seed awns and soft actuators, *Extreme Mech. Lett.*, 2020, **38**, 100746.
11. M. McNeil, A. G. Darvill, S. C. Fry and P. Albersheim, Structure and function of the primary cell walls of plants, *Ann. Rev. Biochem.*, 1984, **53**, 625–663.
12. J. Pérez, J. Munoz-Dorado, T. Rubia and J. Martinez, Biodegradation and biological treatments of cellulose, hemicellulose and lignin: an overview, *Int. Microbiol.*, 2002, **5**, 53.
13. R. Elbaum and Y. Abraham, Insights into the microstructures of hygroscopic movement in plant seed dispersal, *Plant Sci.*, 2014, **223**, 124–133.
14. B. Shin, J. Ha, M. Lee, K. Park, G. H. Park, T. H. Choi, K.-J. Cho and H.-Y. Kim, Hygrobot: a self-locomotive ratcheted actuator powered by environmental humidity, *Sci. Rob.*, 2018, **3**, eaar2629.
15. R. M. Christensen, *Mechanics of Composite Materials*, Wiley, New York, 1979.
16. L. Hines, K. Peternsen, G. Z. Lum and M. Sitti, Soft actuators for small-scale robotics, *Adv. Mater.*, 2017, **29**, 1603483.

17. F. Schmitt, O. Piccin, L. Barbé and B. Bayle, Soft robots manufacturing: A review, *Front. Rob. Ai*, 2018, **5**, 84.
18. M. Ma, L. Guo, D. G. Anderson and R. Langer, Bio-inspired polymer composite actuator and generator driven by water gradients, *Science*, 2013, **339**, 186–189.
19. S. Bauer, S. B. Gogonea, I. Graz, M. Kaltenbrunner, C. Keplinger and R. Schwödiauer, 25th Anniversary article: A soft future: from robots and sensor skin to energy harvesters, *Adv. Mater.*, 2014, **26**, 149–162.
20. M. Cianchetti, C. Laschi, A. Menciassi and P. Dario, Biomedical applications of soft robotics, *Nat. Rev. Mater.*, 2018, **3**, 143–153.
21. G.-Z. Yang, J. Bellingham, P. E. Dupont, P. Fischer, L. Floridi, R. Full, N. Jacobstein, V. Kumar, M. McNutt, R. Merrifield, B. J. Nelson, B. Scassellati, M. Taddeo, R. Taylor, M. Veloso, Z. L. Wang and R. Wood, The grand challenges of Science Robotics, *Sci. Rob.*, 2018, **3**, eaar7650.
22. E. Reyssat and L. Mahadevan, Hygromorphs: from pine cones to biomimetic bilayers, *J. R. Soc. Interface*, 2009, **6**, 951–957.
23. Y. Abraham, C. Tamburu, E. Klein, J. W. Dunlop, P. Fratzl, U. Raviv and R. Elbaum, Tilted cellulose arrangement as a novel mechanism for hygroscopic coiling in the stork's bill awn, *J. R. Soc. Interface*, 2012, **9**, 640–647.
24. X. Chen, L. Mahadevan, A. Driks and O. Sahin, *Bacillus* spores as building blocks for stimuli-responsive materials and nanogenerators, *Nat. Nanotechnol.*, 2014, **9**, 137–141.
25. W. Wang, L. Yao, C.-Y. Cheng, T. Zhang, H. Atsumi, L. Wang, G. Wang, O. Anilionyte, H. Steiner, J. Ou, K. Zhou, C. Wawrousek, K. Petrecca, A. M. Belcher, R. Karnik, X. Zhao, D. I. C. Wang and H. Ishii, Harnessing the hygroscopic and biofluorescent behaviors of genetically tractable microbial cells to design biohybrid wearables, *Sci. Adv.*, 2017, **3**, e1601984.
26. M. Rüggeberg and I. Burgert, Bio-inspired wooden actuators for large scale applications, *PLoS One*, 2015, **10**, e0120718.
27. J. Y. Chung, H. King and L. Mahadevan, Evaporative microclimate driven hygrometers and hygromotors, *EPL*, 2014, **107**, 64002.
28. J. Ryu, M. Mohammadifar, M. Tahernia, H. Chun, Y. Gao and S. Choi, Paper robotics: self-folding, gripping, and locomotion, *Adv. Mater. Technol.*, 2020, **5**, 1901054.
29. L. Shen, J. Fu, K. Fu, C. Picart and J. Ji, Humidity responsive asymmetric free-standing multilayered film, *Langmuir*, 2010, **26**, 16634–16637.
30. Y. Ma, Y. Zhang, B. Wu, W. Sun, Z. Li and J. Sun, Polyelectrolyte multilayer films for building energetic walking devices, *Angew. Chem.*, 2011, **123**, 6378–6381.
31. D. J. Broer, C. M. W. Bastiaansen, M. G. Debije and A. P. H. J. Schenning, Functional organic materials based on polymerized liquid-crystal monomers: supramolecular hydrogen-bonded system, *Angew. Chem., Int. Ed.*, 2012, **51**, 7102–7109.
32. S.-W. Lee, J. H. Prosser, P. K. Purohit and D. Lee, Bioinspired hygromorphic actuator exhibiting controlled locomotion, *ACS Macro Lett.*, 2013, **2**, 960–965.

33. M. R. Islam and M. J. Serpe, Poly(*N-isopropylacrylamide*) microgel-based thin film actuators for humidity sensing, *RSC Adv.*, 2014, **4**, 31937.
34. M. Ji, N. Jiang, J. Chang and J. Sun, Near-infrared light-driven, highly efficient bilayer actuators based on polydopamine-modified reduced graphene oxide, *Adv. Funct. Mater.*, 2014, **24**, 5412–5419.
35. K. Zhang, A. Geissler, M. Standhardt, S. Mehlhase, M. Gallei, L. Chen and C. M. Thiele, Moisture-responsive films of cellulose stearoyl esters showing reversible shape transitions, *Sci. Rep.*, 2015, **5**, 11011.
36. Y. Zhang, H. Jiang, F. Li, Y. Xia, Y. Lei, X. Jin, G. Zhang and H. Li, Graphene oxide based moisture-responsive biomimetic film actuators with nacre-like layered structures, *J. Mater. Chem. A*, 2017, **5**, 14604.
37. Y. Qiu, M. Wang, W. Zhang, Y. Liu, Y. V. Li and K. Pan, An asymmetric graphene oxide film for developing moisture actuators, *Nanoscale*, 2018, **10**, 14060.
38. S. Wang, S. Yan, L. Zhang, H. Zhao, T. Yang, F. Li and H. Li, Bioinspired Poly(vinyl alcohol) film actuator powered by water evaporation under ambient conditions, *Macromol. Mater. Eng.*, 2020, **305**, 2000145.
39. M. Wang, Q. Li, J. Shi, X. Cao, L. Min, X. Li, L. Zhu, Y. Lv, Z. Qin, X. Chen and K. Pan, Bio-inspired high sensitivity of moisture-mechanical GO films with period-gradient structures, *ACS Appl. Mater. Interfaces*, 2020, **12**, 33104–33112.
40. L. T. de Haan, J. M. Verjans, D. J. Broer, C. W. Bastiaansen and A. P. Schenning, Humidity-responsive liquid crystalline polymer actuators with an asymmetry in the molecular trigger that bend, fold, and curl, *J. Am. Chem. Soc.*, 2014, **136**, 10585.
41. J. C. M. Garnett, Optical properties of thin solid films, *Philos. Trans. R. Soc. London*, 1904, **203**, 385.
42. S. Timoshenko, Analysis of bi-metal thermostats, *J. Opt. Soc. Am.*, 1925, **11**, 233.
43. X. Chen, D. Goodnight, Z. Gao, A. H. Cavusoglu, N. Sabharwal, M. DeLay, A. Driks and O. Sahin, Scaling up nanoscale water-driven energy conversion into evaporation-driven engines and generators, *Nat. Commun.*, 2015, **6**, 7346.
44. S. Reichert, A. Menges and D. Correa, Meteorosensitive architecture: Bio mimetic building skins based on materially embedded and hygroscopically enabled responsiveness, *Comput. Aided Des.*, 2015, **60**, 50–69.

Subject Index

www.ingramcontent.com/pod-product-compliance
Lightning Source LLC
LaVergne TN
LVHW011851200926
848484LV00002B/437